"Both vulnerability and resilience are 'slippery' topics that, to be useful, need extensive theorizing and careful analysis. This book takes a rigorous and comprehensive approach to their definition and elaboration, thereby making a very valuable contribution to the literature in this field."

*Edmund Penning-Rowsell,*
*Flood Hazard Research Centre, Middlesex University*

"This is an essential volume in which leading scholars from three continents grapple with vulnerability to natural hazards in a thorough, no-nonsense, fact-based manner. The Enlightenment tradition lives on despite both populist and post-modern scorn for science. Quantitative and qualitative assessment methods are clearly explained; whilst recent case examples, key messages, and innovative diagrams will please a wide audience."

*Ben Wisner,*
*University College London*

"Although being widely used in both risk research and management, the concepts of vulnerability and, particularly, resilience are the subject of ongoing debate with respect to their definition as well as their operationalization. In this intense discourse, few publications have aimed at a systematic view. Against this backdrop, the present book offers a comprehensive and multifaceted approach, and provides an important and timely contribution to the discussion on the relation between the concepts of vulnerability and resilience. Particularly the aspects of scale and time dependence will provide food for thought on their future role in science and practice."

*Jakob Rhyner,*
*United Nations University, Bonn*

# VULNERABILITY AND RESILIENCE TO NATURAL HAZARDS

In recent years, there has been growing recognition that disaster risk cannot be reduced by focusing solely on physical hazards without considering factors that influence socio-economic impact. Vulnerability (the susceptibility to the damaging impacts of hazards) and resilience (the ability to recover) have become popular concepts.

This book provides a comprehensive overview of the concepts of vulnerability and resilience and their application to natural hazards research. With contributions from both physical and social scientists, it provides an interdisciplinary discussion of the different types of vulnerability and resilience, explores the links between them, and concludes with the remaining challenges and future directions of the field. Examining global case studies from the US coast to Austria, this is a valuable reference for researchers and graduate students working in natural hazard and risk reduction from both the natural and social sciences.

SVEN FUCHS is a senior researcher at the Institute of Mountain Risk Engineering at the University of Natural Resources and Life Sciences in Vienna. His research interests include high-mountain geomorphology, mountain hazard risk management, the study of coupled human–environment systems, and vulnerability assessment for natural hazards. His research has taken him to the mountain regions of Europe, Southeast Asia, Central Africa, and the Russian Federation.

THOMAS THALER is a research fellow at the Institute of Mountain Risk Engineering at the University of Natural Resources and Life Sciences in Vienna. He is interested in policy issues related to flood risk management in different European countries. His research focuses on risk governance and natural hazards in Europe, with particular emphasis on questions relating to the design and effectiveness of governance systems.

# VULNERABILITY AND RESILIENCE TO NATURAL HAZARDS

*Edited by*

SVEN FUCHS
*University of Natural Resources and Life Sciences, Vienna*

THOMAS THALER
*University of Natural Resources and Life Sciences, Vienna*

CAMBRIDGE
UNIVERSITY PRESS

**CAMBRIDGE**
**UNIVERSITY PRESS**

University Printing House, Cambridge CB2 8BS, United Kingdom

One Liberty Plaza, 20th Floor, New York, NY 10006, USA

477 Williamstown Road, Port Melbourne, VIC 3207, Australia

314–321, 3rd Floor, Plot 3, Splendor Forum, Jasola District Centre, New Delhi – 110025, India

79 Anson Road, #06-04/06, Singapore 079906

Cambridge University Press is part of the University of Cambridge.

It furthers the University's mission by disseminating knowledge in the pursuit of education, learning, and research at the highest international levels of excellence.

www.cambridge.org
Information on this title: www.cambridge.org/9781107154896
DOI: 10.1017/9781316651148





First published 2018

Printed in the United Kingdom by TJ International Ltd. Padstow Cornwall

*A catalogue record for this publication is available from the British Library.*

*Library of Congress Cataloging-in-Publication Data*

Names: Fuchs, Sven, editor. | Thaler, Thomas, editor.
Title: Vulnerability and resilience to natural hazards / edited by Sven Fuchs, Thomas Thaler.
Description: Cambridge ; New York : Cambridge University Press, 2017. | Includes bibliographical references and index.
Identifiers: LCCN 2017040752 | ISBN 9781107154896 (hardback)
Subjects: LCSH: Hazard mitigation—Social aspects. | Hazard mitigation—Economic aspects. | Risk—Social aspects. | Disasters—Social aspects.
Classification: LCC HM1101 .V85 2017 | DDC 363.34/6—dc23 LC record available at https://lccn.loc.gov/2017040752

ISBN 978-1-107-15489-6 Hardback

# Contents

*Color plates are to be found between pages 144 and 145.*

# Figures

# Tables

# Contributors

**David Adamson**
Centre for Global Food and Resources, University of Adelaide

**Domenico Asprone**
Department of Structures for Engineering and Architecture, University of Naples Federico II

**Kayode Atoba**
Department of Landscape Architecture and Urban Planning, Texas A & M University

**Anna Bozza**
Department of Structures for Engineering and Architecture, University of Naples Federico II

**Samuel D. Brody**
Department of Marine Sciences, Texas A & M University

**Christopher Burton**
Department of Geosciences, Auburn University, Alabama

**Susan L. Cutter**
Department of Geography, University of South Carolina

**Christopher T. Emrich**
College of Health and Public Affairs, University of Central Florida

**Alexander Fekete**
Cologne University of Applied Sciences

**Tim Frazier**
Georgetown University, Washington D. C.

**Sven Fuchs**
Institute of Mountain Risk Engineering, University of Natural Resources and Life Sciences, Vienna

**Gérard Hutter**
Leibniz Institute of Ecological Urban and Regional Development, Dresden

**Brenden Jongman**
Institute for Environmental Studies, Free University of Amsterdam; Global Facility for Disaster Reduction and Recovery, The World Bank Group

**Adam Loch**
Centre for Global Food and Resources, University of Adelaide

**Daniel F. Lorenz**
Institute of Social and Cultural Anthropology, Free University of Berlin

**Gaetano Manfredi**
Department of Structures for Engineering and Architecture, University of Naples Federico II

**Burrell Montz**
East Carolina University

**Maria Papathoma-Köhle**
Institute of Mountain Risk Engineering, University of Natural Resources and Life Sciences, Vienna

**Carlos Dionisio Pérez-Blanco**
Eni Enrico Mattei Foundation; Euro-Mediterranean Center for Climate Change

**Samuel Rufat**
Department of Geography, The University of Cergy-Pontoise

**Laura Siebeneck**
University of North Texas

**Eric Tate**
Department of Geographical and Sustainability Sciences, University of Iowa

**Thomas Thaler**
Institute of Mountain Risk Engineering, University of Natural Resources and Life Sciences, Vienna

**Graham A. Tobin**
School of Geosciences, University of South Florida

# 1

# Introduction

SVEN FUCHS AND THOMAS THALER

Given the conditions of global environmental change such as outlined in the fifth assessment report of the United National Intergovernmental Panel on Climate Change (Stocker et al., 2013), impacts from natural hazards on natural and human systems are manifest worldwide (Field et al., 2014). There is a rising amount of losses due to these hazards reported even if considerable efforts have been undertaken globally and on the national level of affected countries in disaster risk reduction. Global trends show an increasing number of natural hazards, and thus increasing losses, from so-called disasters. This shift in events seems undoubtedly true on a global level if statistics from leading reinsurers are studied (e.g., Munich Re, 2017; Swiss Re, 2016, to name the most prominent), or if available worldwide data sources are consulted (e.g., the EM-DAT or NATHAN databases). The losses are a result of both the dynamics of frequency and magnitude of natural hazards and the dynamic exposure of the society or elements at risk such as buildings or infrastructure lines. Exposure to natural hazards is linked to vulnerability and resilience, concepts that (again) became popular in recent years both from a scientific point of view (Birkmann et al., 2013) as well as in policy development and disaster risk reduction (Birkmann et al., 2014; Zimmermann and Keiler, 2015).

Losses have been reported on a local, regional and national scale for decades. The level of losses is influenced by the level and type of economic activity existing in an area, as well as the institutional framework of the society, including previous decisions about specific adjustments to natural hazards. The simultaneous debate on vulnerability and resilience is not a new one. Burton et al. (1978) referred to continuously rising flood property losses during the 1970s and 1980s in the USA, after a decline in the decade 1956–65, and concluded that the wide development of floodplain management measures was offset by the continued vulnerability of older buildings, roads, and bridges. Moreover, the further invasion of fringes of demarcated 1-in-100-year floodplains and small tributary floodplains in expanding urban areas was made responsible for decreasing resilience. Russell (1970) linked the increased number of reports on disasters to an amplified awareness within the population due to (at that time) an increased availability of modern mass media such as television, leading to an increased concern about vulnerability and resilience. Moreover, he assumed that advancing technology, in combination with the Cold War, created some spur in possible threats due to natural hazards (more recently also argued by Pearce, 2003), since the effects on human society might somehow be similarly destructive. Third, and presumably most important, Russell (1970) assumed by taking an economic viewpoint, that a growing pressure to build new and more residential and nonresidential infrastructure in hazard-prone zones from both inside and outside government for improvement in the criteria on which public expenditure decisions are made, might trigger the growing vulnerability to natural hazards. Atkisson et al. (1984) concluded from their studies on natural hazards in the USA that the expected annual losses due to natural hazards in 1970 exceed losses from traffic accidents in the same year, and is approximately half the amount of all property taxes collected by state and local government in 1970. Moreover, they assumed a rise from the expected damage to buildings due to landslides and riverine flooding in the USA from US$2.5 billion in 1970 to US$16 billion in 2000 (in 2017 values) – a figure not far from reality today. Schuster and Fleming (1986) estimated the annual direct and indirect losses to public and private property arising from landslide processes at US$2.1 billion (in 2017 values), a value that had also been reported based on an unpublished UNESCO survey of landslide effects by Arnould (1976). Overall, these events (large as well as small ones) influence the individual behavior (Tapsell et al., 2002; Voss and Wagner, 2010).

However, apart from these meaningful numbers, and independent from the debates on vulnerability and resilience, the underlying trends should be carefully considered, as indicated by Downton and Pielke (2005) and Downton et al. (2005). To give an example, even if flood damages continued to increase despite extensive flood management efforts since 1900, particularly when measured in constant currency units, the trend is not as obvious once normalized. If the US flood data are presented in terms of damage per unit wealth, a slight and statistically insignificant downward trend is observed (Loucks and Stedinger, 2007), which suggests that floods might have a lessening or neutral impact on the overall personal wealth of citizens in the USA over the past decades. Some more recent studies have focused on regions with similar economic vulnerability to the USA, such as Canada

(Etkin et al., 2003) and Europe (Barredo, 2007; Becker and Grünewald, 2003; Mudelsee et al., 2003). Mudelsee et al. (2003) concluded from a hydrometeorological point of view by an analysis of flood magnitudes that there is no evidence from the observations for recent upward trends in the occurrence of large flood events in Central Europe. Similarly, Barredo (2009) concluded by using available information from EM-DAT and NATHAN, that no clear positive trend in flood losses in Europe exists if they are normalized by eliminating the socioeconomic influence of growing exposure in areas affected. It is still under debate, however, to what extent the recent increase in losses results from dynamic process behavior and thus increased magnitude and frequency, and to what extent these developments are due to increased use of areas prone to natural hazards for human settlement, economic activities, and infrastructure corridors, thus dynamics in vulnerability and resilience of affected communities. Moreover, it has repeatedly been debated that institutions and policies, even if not directly attributed to natural hazard risk management, influence resilience and vulnerability (Fuchs et al., 2017).

Resilience and vulnerability represent two related yet different approaches to understanding the response of systems and actors to natural hazards. Ideas and concepts of vulnerability and resilience are used by various scholars from different scientific disciplines – as well as by practitioners and institutions – and hence are used in multiple disciplinary models underpinning either a technical or a social origin of the concept, and resulting in a range of paradigms for either a qualitative or quantitative assessment, both scale-dependent. Vulnerability, broadly defined, is the potential for loss, and includes elements of exposure (people, places, infrastructure at risk from a hazard), sensitivity (the degree to which the people, places, or infrastructure are harmed), and coping (the skills, resources, and opportunities of people and places to survive, absorb the impacts, and manage the adverse outcomes). Resilience refers to the capacities of people, places, and infrastructure to not only cope with hazards, but also the longer-term adjustment and learning processes to adapt to future threats. Concept of capacities includes physical (e.g., shelter, food storages, etc.) and nonphysical factors (e.g., knowledge, motivational, network, economic, institutional, and procedural capacities) that permit individuals, buildings and/or society to cope with events. These simplified definitions capture the essence of vulnerability and resilience (Anderson and Woodrow, 1989; Cutter and Corendea, 2013). Despite the growing amount of studies recently published (e.g., Birkmann et al., 2013; Menoni et al., 2012), current approaches are still driven by a divide between natural and social sciences, even if some attempts have been made within to bridge this gap (e.g., Fuchs (2009) with respect to vulnerability, and Kuhlicke (2013) pointing to resilience). Whereas social scientists tend to view vulnerability and resilience as representing the set of socioeconomic factors that determine people's ability to cope with stress or changes (e.g., Cutter et al., 2003; Fekete and Hufschmidt, 2014), natural scientists and engineers often view both terms focusing on the likelihood of occurrence of specific hazards, and associated impacts on the built environment (e.g., Papathoma-Köhle et al., 2017). Representatives from each discipline define both vulnerability and resilience in a way that fits their individual disciplinary purposes (Fekete et al., 2014; Füssel, 2007). However, within academic discourse there is interconnection and feedback between vulnerability and coping capacity as an essential asset to understand natural hazard management (Gotham and Campanella, 2011; Turner II, 2010). For example, elderly people might be more physically frail, but through their experiences from past events they might better know what to do, ergo, show a lower vulnerability based on high individual coping capacity (Davis and Bellers, 1995; Dibben and Chester, 1999). Overall, scholars distinguish among four possible combinations between vulnerability and resilience. Communities classified under the first combination (high vulnerability and low resilience) are those that compound high social and physical vulnerability and low coping capacity to response to events, where natural hazards events cause high damages towards the population and constructions, such as the Indian Ocean earthquake and the related tsunami in 2004. The second combination (high vulnerability and high resilience) refers to local communities that demonstrate a high degree of coping capacity to respond to natural hazard events, but where the events have a tremendous impact, such as the Tohoku earthquake and the subsequent tsunami event in 2011. The third combination (low vulnerability and low resilience) includes those communities with a high degree of resilience-building policies, but which fail to develop a local network system and adequate demographic structure. Typically, this refers to natural hazard events affecting aging communities, such as Florida in the USA (Manuel et al., 2015; Tobin, 1999), or to nuclear disasters. The fourth combination (low vulnerability and high resilience) applies to situations that show a high social understanding and coping capacity of the society, which can fast absorb and recover from a natural hazard event (high resilience). The community is not inherently vulnerable towards natural hazard events.

Efforts to reduce the exposure to hazards and to create disaster-resilient communities require intersections among the different disciplines involved (Fuchs et al., 2011), since human activity cannot be seen independent from the environmental settings and vice versa. Simultaneously, approaches suitable within the development context may not fit to the climate change context (Wymann von Dach et al., 2017). Acknowledging different roots of disciplinary paradigms, methods determining structural, economic, institutional or social vulnerability and resilience should be inter-woven in order to enhance our understanding of vulnerability and resilience, and to adapt to ongoing global change processes. Therefore, there is a need to expand our vision on hazard and risk management integrating adaptation and mitigation approaches into the broader context of related governance arrangements. As such, it is increasingly recognized that disaster risk and threats to human security cannot be reduced by focusing solely on the hazards. The Sendai Framework for Disaster Risk Reduction 2015–30, which was formulated at the Third UN World Conference on Disaster Reduction in Sendai in 2015, underlines that the starting point for reducing disaster risk and for promoting a culture of disaster resilience is the knowledge of the hazards and the physical, social, economic, and environmental vulnerabilities to disasters that most societies bear (UNISDR, 2015a).

Following the axiom that natural hazard risk is a function of hazard and consequences (e.g., Varnes, 1984), the ability to determine vulnerability and resilience either quantitatively or qualitatively is an essential prerequisite for reducing these consequences, and therefore natural hazard risk. The assessment of vulnerability and resilience requires an ability to both identify and understand the susceptibility of elements at risk and – in a broader sense – of the society to these hazards. Studies related to vulnerability and resilience of human and natural systems to natural hazards, and of the ability of these systems to adapt to changes in the functional chain of hazards, are a relatively recent field of research that brings together experts from a wide range of disciplines, including natural science, social science, disaster management, policy development, and economics, to name only a few. Scholars from these fields bring their own conceptual models to study vulnerability and resilience, models that often address similar problems and processes using different languages (Brooks, 2003). As a result, nuanced differences in the definition of both terms between the hazard-risk research communities and the human-environment research communities remain (Cutter and Finch, 2008; Renn, 2008a). While the hazard-risk research communities place emphasis on mitigation and emergency management (e.g., Bankoff et al., 2004; Pelling, 2003a; Wisner et al., 2004), the human-environment research communities focus on large-scale global environmental processes, in particular with respect to climate change and its global to local impacts (e.g., Renn, 2008b; Turner II et al., 2003). Since studies on vulnerability and resilience are rooted in a multiplicity of disciplines, there is neither a common definition nor a common theory developed of how to measure them. However, most conceptualizations have in common that the vulnerability or resilience of a studied system cannot be observed directly, but rather has to be deduced from distinct environmental settings that are scale-dependent. If it is surmised that a system is vulnerable, projections of events the future might bring must be combined with a theory of how these events will make the system under study more resilient. Despite differences between the two research communities, both acknowledge that the composition of vulnerability and resilience is driven by exposure, sensitivity, resistance, and response, and requires measurements of both environmental and social systems.

In contrast to the integrative approaches recently published in journal papers (Birkmann et al., 2013; Menoni et al., 2012) and edited books and monographs (Birkmann et al., 2014; Gaillard, 2015; Rivera and Kapucu, 2015), this book builds on a comprehensive but separated presentation of both vulnerability and resilience. We would like to present the different concepts of vulnerability and resilience as a textbook, starting with the review of origins, concepts and perspective on resilience and vulnerability as well as possibility of measuring and methods by including examples from worldwide case studies. The readers will be guided through the topic and will gain insights into the concepts behind vulnerability and resilience, and how both approaches may be used within natural hazard management. As such, the book contains chapters for physical, social, economic, and institutional settings responsible for vulnerability and resilience, and is organized in a strictly logical way so that the reader will get an overview on the different schools of thought involved, and a summarizing and concluding chapter is provided where the possible integration of both concepts is discussed. In particular, for every chapter, the challenge of vulnerability and resilience research to integrate three different aspects is included, (1) the components such as exposure, sensitivity, or adaptive capacity; (2) the different methods used within different disciplines, and (3) the target dimension of assessment. We therefore explicitly asked the different contributors to discuss and outline (1) their understanding of vulnerability and resilience in order to present a variety of current assessment approaches, and (2) to formulate and deal with future challenges that may be addressed in order to get in-depth knowledge on the topic taking a broader viewpoint and opening up perspectives beyond the ordinary. Each chapter is supplemented by examples from individual case studies and concludes with key messages and paradigms, as well as the challenges associated with the individual topics presented, such as the dynamics of scale and data availability.

Fekete and Montz introduce the topic of vulnerability in Chapter 2 of this book. They distinguish between the phenomenon of vulnerability and the concept of vulnerability, and discuss the variety of existing definitions and conceptualizations. They argue that the pluralism of definitions should be seen as an asset rather than a challenge due to the multiple dimensions currently summarized under the term. Furthermore, the chapter shows the difference between the concept of vulnerability being an existing condition, and the approach defining vulnerability dynamically as an ongoing process. This distinction highlights the need for a broader understanding of the term, and shows the importance of both of these approaches. The contribution explains how the human dimension of vulnerability consists of both adaptive and coping capacities, including susceptibilities, but also skills. The focus of the physical dimension of vulnerability is on the susceptibility of structures and infrastructure to damage. Nevertheless, the multiple existing definitions of the term suggest the different contexts in which vulnerability is conceptualized and applied within disaster risk reduction. The authors conclude in their introductory contribution that vulnerability is both a phenomenon and a concept that has been influential in stimulating research on the impacts of hazardous events, and they provide examples of how the heterogeneity of definitions may be used to broaden the scientific debate, and also the practical applications in disaster risk reduction. Internal and external factors define vulnerability, and therefore both quantitative and qualitative approaches are relevant in order to understand the entire dimension of vulnerability.

The subsequent chapters are centered on different conceptualizations of vulnerability to natural hazards, such as physical, social, economic, and institutional, and provide an in-depth overview on the respective history in research, on applications and challenges for future research.

The contribution of Fuchs et al. on physical vulnerability starts with the notion that exposure and vulnerability are linked, starting with the degree of susceptibility of elements at risk within the physical environment, and as such to the negative impacts of hazards on buildings and other structures. Moreover, the contribution highlights how physical vulnerability is conceptualized. The first element centers on the nature of the hazard described as both the characteristics of the hazard agent (e.g. spatial distribution, probability, magnitude, and intensity) and the degree of exposure that the hazard exerts on a particular structure or community. The second element entails characteristics of the physical or built environment such as societal assets, agricultural resources, individual and community assets, and the surrounding natural environment. Furthermore, the authors discuss the different dimensions of exposure to hazards as root causes for physical vulnerability, namely exposure of physical assets, of population, of businesses and economies, of social and political systems, and of natural ecosystems. Different methods of vulnerability assessment are presented, following a distinction of three main approaches which are vulnerability matrices, vulnerability curves, and vulnerability indicators. Chapter 3 concludes with a case study in order to show how such concepts can be applied in research and practice.

In Chapter 4 on social vulnerability, Burton et al. give an overview on the framework and conceptual foundation of the term, and provide examples for assessment and evaluation. The authors show how – despite the considerably long history of research – the assessment of social vulnerability as a tool for disaster risk reduction continues to gain importance, in particular because of the worldwide increase in hazard exposure. They argue that an improved understanding of how social vulnerability indicators and indices conform to disaster losses could benefit both model development decisions and how social vulnerability indicators are interpreted during decision-making and public policy development processes. Starting with principal theoretical models of social vulnerability and the prominent construction approaches for defining social vulnerability models, the chapter highlights that results of different model types converge when compared spatially. It is shown, however, that social vulnerability indicators can only explain the social vulnerability of a population to an extent and cannot explain the whole picture. Moving forward, the authors argue the primary challenge of assessment, which is to derive a set of explanatory metrics (whether qualitative or quantitative) that is parsimonious, yet explains the phenomenon as clearly as possible. Illustrated by a case study on Hurricane Sandy, future research needs for improving social vulnerability assessments are identified.

Assessing economic vulnerability is fundamental for the scientific community as well as for practitioners and companies in terms of defining optimal risk management strategies and analyzing individual and societal risk levels. Chapter 5 on economic vulnerability by Thaler and Jongman addresses these issues comprehensively, starting with an overview on different damage categories and methods for their assessment. The authors show the similar importance of assessing direct, indirect, and nonmarket losses, and provide different methods for their measurement. They argue that the need to assess economic vulnerability has increased in our societies, mainly to compute the potential effectiveness of different adaptation options in reducing losses as well as to understand the potential pressures on current financing systems that are in place to compensate losses from natural hazard events. As most of the contributors do, they also discuss the role of scale in assessment as well as the challenges related to data availability in particular in less-developed countries. The chapter concludes with an example of economic vulnerability assessment for flood hazards in Germany and the UK, and shows how different methods may be used to gain insights in disaster losses.

In Chapter 6, Papathoma-Köhle and Thaler provide an overview on the institutional dimensions of vulnerability, defined as the impact of natural hazards combined with the weaknesses embedded in institutions and reducing the capacity of societies to resist, withstand, cope, or recover from natural hazards. These weaknesses are shaped by the sociopolitical context and may be rooted in drawbacks of the legal framework and its implementation, deficiencies in communication, and lack of participation of the relevant stakeholders and the public, as well as shortcomings of the available risk transfer mechanisms and poor land use planning. The authors argue that most methods for vulnerability assessment are targeted at quantifying the physical or the social dimension so far, whereas research focusing on how institutions influence the vulnerability of the society to natural hazards is still scarce. Political stability, transparency and impartiality are paramount features of resilient societies and may significantly reduce institutional vulnerability. Focusing on the post-disaster reconstruction and recovery phase, institutional vulnerability is often related to risk transfer mechanisms that are often translated into availability of private or public insurance schemes. A case study on the legislative framework in Austria is used to unveil some of the aspects discussed in this chapter. The authors conclude that multilayered rather than centralized institutions may contribute to equilibrium between knowledge, action, and social-ecological setting resulting in enhanced societal reliance; which in turn influences the accountability of authorities by ensuring a fair share of benefits and risks and contribute to an improvement of societal coping capacity.

The subsequent chapters are reporting different approaches of resilience to natural hazards, such as physical, social, economic, and institutional, and provide an in-depth overview on the respective research history, on scientific and practical applications and challenges for future research.

Chapter 7, written by Emrich and Tobin, provides an overall introduction to the concept of resilience in natural hazard research. The authors introduce the concept by a review of definitions of resilience, a discussion on the evolution of resilience frameworks, an examination of temporal and spatial scales as they relate to resilience, an overview of approaches and thematic areas under the umbrella of resilience, appraisal of the shortcomings of a resilience focus in hazards research, and an assessment of future research trends. They argue that while the relationship between resilience and vulnerability is contested, it is often too simplisticly seen as the opposite of vulnerability, that is low resilience equates with high vulnerability and high resilience corresponds with low vulnerability. While under certain circumstances this may be true, the emphasis on vulnerability in the earlier literature perhaps detracted somewhat from evolving concepts of resilience, which are presented successively in this chapter. As the lowest common denominator, resilience can be conceptualized by bounce-back and recovery concepts; adaptive capacity; coping and the ability to absorb change; and acceptance of levels of risk, all of which are fundamental in one or the other sense if we would like to achieve societies resilient against the impact of natural hazards. If seen from an interdisciplinary perspective, current gaps in resilience research include the focus on resilient cities, the importance of land-use regulations and their implementation, the acknowledgment of complex social, economic, and environmental factors in order to allow for participation, ownership and the leveling-out of societal inequities, and effective governance and institutions.

In Chapter 8, Bozza et al. offer an approach on how to understand and how to assess physical resilience to natural hazards. Resilience is conceptualized as the capacity of a system to withstand and recover from external stresses and achieve an equilibrium state (bouncing back) through mitigation and adaptive management of complex systems. Taking an engineering viewpoint, they develop a method to assess resilience against earthquakes for larger cities, which are represented by complex networks. In particular, the twenty-first century shows high challenges for urban environments to cope with natural hazard events. In order to characterize the resilience of an urban agglomeration, this chapter proposes an integrated framework based on the simulation of a seismic event and enabling for the assessment of expected damage and resilience. Based on a case study in Naples, Italy, the authors show two diverse recovery strategies that are modeled, with the former focusing on social dynamics and the latter on economies of scale related to a city service. The computation results presented highlight the direct correlation between the number of buildings located within an urban context and the systemic damage observed after a seismic event. As an outcome, urban agglomerations can be assessed and understood using engineering metrics to define their capabilities once threatened by natural hazards.

Hutter and Lorenz show background information on the concept of social resilience in Chapter 9. Their contribution conceptualizes social resilience as a matrix of different levels and capacities from a sociological point of view, and distinguishes between multiple social levels to consider different types of social entities (small social groups, organizations, and network of organizations). The authors present social resilience as a complex and dynamic set of capacities (coping, participative, adaptive, and transformative) of actors in modern society that is characterized by most influential social processes of perception, interpretation, and meaning-making as well as power structures. The contribution provides selected starting points for analyzing and assessing social resilience in the context of natural hazards, in particular with respect to multiple voices, conceptualizing social resilience, analyzing barriers to resilience and assessing social resilience. The authors conclude that the idea of social resilience encourages us to pay ample attention to the capacities and resources of individual persons, teams, organizations, and networks to proactively adapt to and recover from unexpected disturbances and unusual experiences.

Economic resilience refers to the ability of an economy to recover from damage and reach a new normal, as outlined by Pérez Blanco et al. in Chapter 10. With respect to a dynamic concept of resilience, the authors state that from a business perspective, static resilience makes the best possible use of available resources to reduce business interruption losses at a given moment in time; while dynamic resilience works after the event, relying on repair and reconstruction to speed up recovery and on the ability of businesses to reschedule, or recapture, foregone output. Similar to other types of resilience, economic resilience may be a (flexible and therefore dynamic) policy-induced ability that can be managed through public policy decisions that coordinate individual responses to shocks. However, in contrast to other resilience concepts, economic resilience is focused on time-specific property damage and continues until the economy has recovered, or reaches a new normal. The Yokohama Strategy and especially the 2005–15 Hyogo Framework for Action set a framework to enhance economic resilience through the use of instruments aiming at preventing or minimizing economic damage in the first stage, and economic losses in the second. Damage prevention instruments include infrastructure, regulations, information and awareness campaigns and economic instruments, while damage compensation instruments include insurance, tort law, state aid, and the action of NGOs, charities and private agents. The chapter introduces each of these instruments and gives an overview on how their performance can be assessed through economic analysis, taking the case study of the 2011 floods in Brisbane, Australia as an example.

The concept of institutional resilience is finally presented in Chapter 11 by Brody and Atoba, presenting the example of flood resiliency in the USA. The authors propose a framework for resilience reduction based on the four dimensions of resistance, avoidance, acceptance, and awareness. Focusing on a long-term perspective they argue that mitigation strategies within each of these four dimensions must work synergistically under an overall program to effectively reduce hazard risk and foster the development of resilient communities. Doing so, the possible portfolio of mitigation activities will depend on the specific characteristics associated with each institutional setting. Whatever strategies are implemented, it is critical that decision makers think beyond normal political cycles and consider the temporally changing socio-economic and physical conditions. The contribution also addresses the need of governmental intervention (organizational and financial) in order to increase institutional resilience. Such an approach should be accompanied by both systems-based structural interventions and land use planning techniques that seek to remove or avoid structures from areas most at risk. Finally, challenges faced by local jurisdictions, including the example of recovery-based approach to flood mitigation in the USA, high cost of maintaining and updating flood infrastructure, planning agency capacity, and inadequate incentives in applying flood mitigation strategies, are addressed.

In Chapter 12, Cutter is taking a critical viewpoint on the combination of vulnerability and resilience approaches in natural hazard management and disaster risk reduction. The chapter shows challenges in the scientific debate around both concepts, as well as on necessary adaptation and further development of the concepts with a particular focus on system dynamics. Divergent and convergent conceptualizations are discussed, and the overall challenge to move beyond the characteristics of existing levels of vulnerability and resilience in order to understand the underlying drivers of processes and pressures that increase vulnerability and affect capacities for enhancing resilience is outlined. The (re-)creation of vulnerability and the reduction in resilience depends on who, what, where, and for whom – all of which are spatially and temporally variant. Moreover, there is a strong need – not only from a policy perspective – to better focus on tipping points responsible for vulnerabilities and resilience, and to raise overall risk awareness and encourage a sense of shared responsibilities to manage natural hazard risk.

Chapter 13 provides a summary and outlook to future research needs. To summarize, this book demonstrates how vulnerability and resilience are conceptualized in natural hazard and risk management, and that sustainable risk reduction and adaptation strategies require a profound understanding of both of the concepts. The complex interplay between natural hazards and affected societies, together with a thorough understanding of dynamic environmental, social, economic, and institutional settings is imperative both to expand knowledge on vulnerability and resilience. Moreover, there is a strong need to adapt to the challenges of integration of diverse perspectives, such as different disciplinary views (physical/natural versus social sciences), as well as how different aspects of resilience (physical, social, economic, and institutional/organizational) interact with each other, and finally how vulnerability and resilience act together rather than viewing vulnerability as opposite of resilience.

Many research efforts recently focused on the outlined challenges, and EU FP7 projects such as MOVE (Methods for the improvement of vulnerability assessment in Europe) or ENSURE (Enhancing resilience of communities and territories facing natural and na-tech hazards) received funding. Apart from many resulting journal papers and policy briefs, these topics are not yet appropriately mirrored in-depth and comprehensively. The present book is intended to close this gap, relying on key concepts of both vulnerability and resilience research. The results presented subsequently are similarly targeted at (1) the research community in terms of a reference work, (2) practitioners working in the field of disaster risk reduction, and (3) the academic education at the graduate level. Therefore, the book will focus on the counterparts (vulnerability and resilience) in environmental hazards research to provide a comprehensive book for both physical and social scientists to guide readers through the different concepts, schools, and thoughts.

# 2

# Vulnerability

*An Introduction*

ALEXANDER FEKETE AND BURRELL MONTZ

## 2.1 Introduction

Vulnerability is both a concept and a phenomenon that has been very influential in stimulating research and in fostering understanding of the nature and distribution of impacts from hazardous events. Vulnerability (the phenomenon) exists but there are numerous mental constructs of it (the concept). Thus, as with any concept, the first phase of academic discourse deals with the term itself (its definition and meaning), complicated in this case by the differing applications of the word. Linguistics and semiotics teach us to look at the language origin in order to capture the essence of the meaning of "vulnerability". But there is also the colloquial use of the term that must be acknowledged, along with the perspectives attributed to its meaning in academic disciplines and public discourse. Since vulnerability encompasses a broad range of topics, including various types of hazards both natural and technological, and their repercussions and impacts in various social and environmental spheres, it should be no surprise that definitions of vulnerability vary as do contentions about what conditions and/or circumstances best characterize it.

While there exist substantial discussions relating to the plethora of definitions (Cannon, 2006; Füssel, 2007; Hinkel, 2011), it should be noted that similar paradigmatic terms such as sustainability, risk, and resilience face similar problems and critiques. It might just be symptomatic for such terms that resemble whole concepts of complex phenomena to stimulate discussion and multiple attributions to their meanings. As Birkmann (2006a: 56) notes, "vulnerability is multidimensional and often ill-defined," as exemplified by differences in perspectives and approaches if one is addressing "vulnerability from" or "vulnerability to" (Timmerman, 1981). While it is certainly difficult for scholars new to the term and its related concepts to understand what vulnerability means and, while it is understandable that standardization of such terms is requested by many, it might be that the pluralism of definitions of vulnerability could be one of its key assets, rather than a problem. Since vulnerability, in a natural hazards context, started out to broaden perspectives and integrate disciplines, its hallmark to trigger discussion about its meaning served to stimulate multilateral and interdisciplinary deliberation and collaboration. And while there is a lack of consensus among scholars regarding the definition, some authors (Adger, 2006) consider the diversity of epistemologies and methods as an advantage rather than a problem.

Before proceeding much further, it is important to distinguish between vulnerability and resilience. Timmerman (1981) pointed out they are by no means reciprocals, and scholarly debate continues, with Cutter et al. (2008) providing a useful framework for conceptualizing the dynamic relationship between vulnerability and resilience. For the purposes of this book, resilience is seen to be related to vulnerability (and vice versa), but the relationship is complex. Indeed, Cutter (2003: 6) notes that vulnerability science "integrates the constructs of risk (exposure), hazard, resilience, differential susceptibility, and recovery/mitigation." Thus, it is very difficult to separate these constructs and they are addressed here as they apply to our discussion of vulnerability.

This chapter first looks at the etymology of the term in order to get to the basics of what it means. The sections that follow deal with the different understandings of vulnerability in different contexts and schools of thought, and describe its basic components in relation to environmental hazards and social-environmental phenomena as well as the various methodologies applied to operationalize the concept.

## 2.2 Etymology of the Term "Vulnerability"

The root of the word vulnerability is the Latin word *vulnus*, which translates as "wound". Anything else to be written about it is additional interpretation. In the understanding of semiotics, there might be an object that can be observed, and words, meanings, and interpretations are assigned to it. A wound, as an object, is a phenomenon that can be visible or hidden. In one interpretation, it could be said that a wound is a somewhat infrequent phenomenon that, in general, triggers negative associations. Vulnerability, then, might be seen as a conditional possibility that can be attributed to the human body, but in principle, also to any other living subject or groups of objects. Some sources emphasize the usage of vulnerability in the context of people (Cannon et al., 2003). From a pragmatic perspective, in fact,

Table 2.1. *Associations of the term "vulnerable" with similar adjectives*

| Positive aspect | No direction | Negative aspect |
|---|---|---|
| | exposed | assailable |
| | open | compromising |
| | sensitive | conquerable |
| | undefendable, indefensible | dangerous |
| | | defenseless |
| | | endangered |
| | | insecure |
| | | penetrable |
| | | susceptible |
| | | threatened |
| | | under attack / fire |
| | | unguarded |
| | | unprotected |
| | | unsafe |
| | | weak |

Adapted from Visualtheasurus (2017).

vulnerability is already used for material objects such as bridges and other infrastructure, technical systems, and institutions, as we discuss later in this chapter. Hence the term has been applied to a range of entities, so the basis to argue against a transfer to other objects is diminishing.

The preceding suggests the importance of differentiating the meaning of the word vulnerability; is this a state or an ongoing process? If it is a state, is it before (an existing condition) or after (an outcome) the wound has been inflicted? To cite an example (Kelly and Adger, 2000), vulnerability is the state of a Roman soldier lying wounded on the ground thus being vulnerable to another attack. Yet, Cardona (2004: 46) argues that it is an ongoing process, "a condition, circumstance, or predisposition," which can be considered both a state and an ongoing process. This chapter, then, discusses vulnerability as it is conceptualized and defined from varying perspectives. An indication of the complexities involved can be seen in Table 2.1.

## 2.3 Historical Development of the Application of "Vulnerability" to Environmental Hazards

The foundations of what is presently captured under the flag of "vulnerability," it might be argued, go well before the first explicit mentions and academically accepted works on vulnerability. Risk forecasting and monitoring of human behavior in crises had been an accepted profession in the Roman

Empire, ancient Egypt, Mayan culture, and many other civilizations. Science was understood as natural science in the late Middle Ages and was guided by developments in observations of natural phenomena and mathematical and statistical methods. Probabilistic calculations used for insuring shipping and risks therefore focused on measurable and natural phenomena that influenced the way science is done and accepted to this day. At the same time, increasing recognition that the interaction between human and natural phenomena has different impacts in different places at different times, influenced the further development of risk assessment methodologies and set the stage for inclusion of vulnerability as a component of risk, though not without some tensions. For instance, during the UN International Decade for Natural Disaster Reduction (IDNDR) in the 1990s (United Nations General Assembly, 1989), it became apparent that existing hazard and risk assessments were falling short in explaining why risks and damages continued to mount. A paradigmatic turn shifted attention from the probability of the event and hazard side of things to the explanation of how the damage played out and could be deconstructed. Vulnerability filled this gap by differentiating between susceptibilities, capacities, and exposure, including the context and setting in which hazards affect humans and objects.

The concept of vulnerability has since been used in a number of disciplines, including anthropology, ecology, economics, geography, and sociology, and has been applied to topics such as climate change, development, food security, livelihoods, poverty, public health, and sustainability (Gain et al., 2012). The evolution of vulnerability from being a subcomponent and implicit risk factor to an encompassing concept is well documented in the literature (Birkmann, 2006a, 2013a).

It is impossible to trace the beginning of the use of the term "vulnerability" with respect to natural hazards research, though the various factors that contributed to vulnerability were analyzed in an early scholarly work on the Halifax Explosion in 1917 (Prince, 1920). In a natural hazards context, the work of Gilbert White (White, 1945) on increasing flood losses despite increasing investments into flood protection in the USA in the 1930s is often cited as a seminal work kick-starting the understanding of the human role in being architects of disaster themselves. The so-called "levee effect" of White (1945) recognizes the false sense of security created by protective measures such as dykes, which attract more settlements in risk-prone areas. Human adjustments to floods, or the lack thereof, as White noted, are one implicit notion of vulnerability. The acceptability of risks has seen a similar vein of development where researchers advocated the involvement of people's perceptions into technical risk calculations mostly based on probabilities.

Following White's work, a group of North American scientists in the 1970s were influential in promoting the notion of and focus on vulnerability, rather than on hazards only. The work of Ian Burton, Kenneth Hewitt, and Gilbert White (Burton et al., 1993; Hewitt and Burton, 1971; Hewitt, 1983; White, 1978; White et al., 2001) and many others is often cited for its consideration of vulnerability and a scholarly turn of perspective from the hazard side to the impact side of an event. The United Nations promoted the 1990s as the IDNDR, which also helped to converge global work on natural hazards. However, one-sided hazard-oriented technical solutions evolving from the IDNDR have been criticized as "not sufficient" to treat humanitarian crises (Blaikie et al., 1994; Wisner et al., 2004). These scholars emphasized the importance of finding the "root causes" of crises, in terms of societal conditions and drivers of political and economic development, which was influenced by their work in development contexts internationally, but is also a normative turn influenced by certain political perspectives. Some scholars have reflected on this "Western" political discourse and the differential political influences of vulnerability versus risk (Bankoff, 2001; Sarewitz et al., 2003). For instance, arguing for a more bottom-up approach and increasing local people's involvement may be seen to reflect certain political views, ranging from liberal economic to Marxist (Methmann and Oels, 2014). This process has also been a major driver of what has been subsequently termed "community resilience" (even by nonpolitically motivated scholars), and provides important contributions to the social and human side of disaster impacts and mitigation, recognizing the role of differential vulnerabilities (Wisner et al., 2012). Thus, as noted earlier, there are certain overlaps in definitions of resilience, vulnerability, and livelihood approaches that confuse practitioners and academics alike; however, these overlaps are difficult to avoid. As work by scholars, development organizations, and even relief authorities on vulnerability and vulnerabilities has evolved, increasing consideration has been given to understanding the components of vulnerability as they play out in various contexts. Some have suggested a double-sided concept of vulnerability and capacities (Anderson and Woodrow, 1989; Cannon, 1994); while Chambers, (1989) sees vulnerability as having two sides, an external side (risks, shocks and stress) and an internal side (defenselessness or lacking the means to cope). Similarly, Watts and Bohle (1993) and Bohle (2001) talk about risk exposure, coping capacity, and recovery potential as multidimensional and multilayered, providing choices and constraints that vary historically and socially. It is obvious that there have been different usages and applications of vulnerability components, but Adger (2006) finds that, in general, three main components of vulnerability commonly used are exposure, sensitivity, and capacities. Yet, there remains debate on this and on other issues surrounding the concept of vulnerability, including whether vulnerability is an outcome of a disaster perspective or the entry point as a process, and whether vulnerability can exist as a conceptual measure independent of a hazard or not.

Even as discussion continues relating to what comprises vulnerability and how it influences and is influenced by various physical, environmental, economic, social, and political factors, it is also increasingly linked to more global issues. In the field of social-ecological systems (SES), for example, vulnerability and resilience have been characterized as subcomponents of sustainability in a framework developed by a group of authors (Turner II et al., 2003). This framework, in which resilience is often termed "coping and adaptive capacities" has been quite influential and builds upon previous work in the fields of complexity and system theory. Many other interlinkages can be found with models developed in fields like biology, such as dose-response relationships or, sustainability politics and science, in the form of focusing on the "sustainability triangle", which translated into "vulnerability spheres" (social, environmental and economic vulnerability) as seen in the so-called "BBC framework" (Birkmann, 2006a) named for the authors Bogardi, Birkmann, and Cardona. The BBC model is the precursor to the MOVE framework (Birkmann et al., 2013) and the SREX framework (IPCC, 2012). Each of these models identifies elements that contribute to vulnerability as well as the relationships among them, and in different, but complementary ways, incorporate factors that can reduce vulnerability such as coping capacity, management, and adaptation.

Spurred by the Intergovernmental Panel on Climate Change (IPCC) processes amongst other factors, the fields of climate change adaptation (CCA) and disaster risk reduction (DRR) have become matched much more closely than before. On one hand, this generated differences between definitions and conceptualizations of vulnerability, but on the other hand, this then helped to align both understandings (Gain et al., 2012; Giupponi and Biscaro, 2015). Indeed, the special report on "Managing the Risks of Extreme Events and Disasters to Advance Climate Change Adaptation" (SREX), and the Fifth Assessment Report (AR5) of IPCC presented conceptual frameworks and interlinkages between contents and components of CCA and DRR and revealed the need to combine both, with the concept of vulnerability as the connecting concept.

With some significant exceptions, such as Adger (2006), in general in the 2000s, the role of resilience as an overall paradigm used in natural hazard contexts (UNISDR, 2005, 2015b) increased, while vulnerability has been

Fig. 2.1. Word cloud of terms found in definitions of vulnerability
*Source*: Adapted from Thywissen (2006) using the tool nVivo.

increasingly considered in the context of assessments and methods (Fekete et al., 2014). Some studies have therefore openly discussed and documented the different roles of resilience and vulnerabilities and the changing roles of these umbrella terms, including some studies discussing the recasting of vulnerability as resilience (Cutter et al., 2008), which some might ignore the complexities of both.

### *2.3.1 Definitions and Components*

Many have presented definitions of vulnerability, and several authors have presented compilations of a range of them (Cardona, 2005; Green, 2004; Schneiderbauer and Ehrlich, 2006; Thywissen, 2006; Weichselgartner, 2001a), so repetition here is unnecessary and would be incomplete. In order to sort out similarities and differences among various definitions, a word cloud was created from those presented by Thywissen (2006), illustrating the most common terms used (Figure 2.1). The overarching message to be gleaned from this figure is that there are a number of words that stand out as most common in the definitions, including human, capacity, people, reduction, exposure, community, resilience, and damage, most of which capture the human dimensions. The human dimension of vulnerability consists of both adaptive and coping capacities, including skills, but also susceptibilities. The physical dimension of vulnerability encompasses the susceptibility of structures and infrastructure to damage. What is also striking is the range of words used in the definitions, suggesting the different contexts in which vulnerability is conceptualized.

As Adger (2006) points out, there are several common elements to the definitions of vulnerability, no matter what words are used. These include "the stress to which a system is exposed, its sensitivity and its adaptive capacity" (Adger 2006: 269). It matters little whether the system is a human system or a physical one. The IPCC (Cardona et al., 2012) continues this theme with their definition as "the propensity or predisposition to be adversely affected [...] encompasses a variety of concepts and elements including sensitivity or susceptibility to harm and lack of capacity to cope and adapt." From these definitions and the word cloud above, it is clear that exposure and capacity are two critical elements defining vulnerability. Fekete et al. (2010) describe the nested, interwoven nature of vulnerable phenomena that contributes to their complexity, thus explaining the myriad definitions that place emphasis on different components. To complicate understanding even more, it is important to recognize the various factors that serve to define the state of vulnerability; those that lead to vulnerability, those that augment it, those that maintain it, and those that reduce it (Villagrán de León, 2006). To elucidate examples of each is beyond the scope of this chapter, but suffice it to say that these factors are multifaceted, interacting at varying scales, and embedded in political and socioeconomic systems, all of which are dynamic entities.

Despite these inherent complexities, we should consider Adger's three components of vulnerability. Exposure addresses the stress or stresses to which an individual, community, society, ecosystem, or structure, to name a few, is exposed, while sensitivity refers to the extent to which those exposed are affected and/or modified by the stress. Adaptive capacity, or the lack of such capacity, is quite variable depending on numerous characteristics of those exposed. The dynamic nature of each of the components differs both individually and as they interact over space and time, which compounds the difficulties in delineating the most salient elements in a given situation, and therefore, in addressing the "root causes" (Wisner et al., 2004). Much of this goes a long way in explaining the numerous and varying definitions, especially since it is an event that often reveals the underlying vulnerabilities, leading to the difference between "vulnerability as outcome" and "contextual vulnerability" that O'Brien et al. (2007) discuss.

### 2.3.2 Who Is Vulnerable? What Is Vulnerable?

An issue that arises when considering definitions of vulnerability is the subject of concern, recognizing, of course, that all are interrelated. Typically, when looking at who is vulnerable, the focus is on individuals, social groups, and institutions. Some definitions of vulnerability address system exposure and susceptibility – sometimes referring to societal systems, but other times referring to physical, engineered, and/or economic systems. So the question of what is vulnerable might encompass the built environment, infrastructure, health systems, and ecosystems, along with the people and communities connected to them, suggesting vulnerability of the coupled human-environment system (Turner II et al., 2003). If infrastructure is compromised, vulnerability is likely to increase irrespective of socioeconomic contexts, given the disruptive impacts of a loss of some service, yet the relative impact will likely vary widely. Engineers have long addressed the potential for failure when designing infrastructure such as water or transportation systems (Plate, 2002), with more recent emphasis on the networked nature of such engineered systems (Dalziell and McManus, 2004), recognizing the potential for what are called cascading failures (Little, 2002, 2004). We have also seen evolution in the field of critical infrastructure wherein the perspective has shifted from a hazard and threat focus that prevailed immediately after the September 11, 2001 attacks, to an all-hazard approach, which mainly promoted resilience as an overall paradigm, but where the vulnerability of "vital systems" is still a main focus (Collier and Lakoff, 2008). Vulnerability of infrastructure is hazard-dependent, such that susceptibility to damage will vary with the type of event (Ciurean et al., 2013; Fuchs, 2009), while vulnerability of people, as measured by such characteristics as coping capacity and extent of disenfranchisement, can exist irrespective of the type of hazard.

At first blush it may seem that it is rather straightforward to manage the vulnerability of engineered systems, the more complex the systems, the greater the possibility of unexpected failures and, in turn, the greater the impact on those who rely on those systems in one way or another. This is not a new revelation, as the "levee effect" identified by Gilbert White illustrates, and before as does the number of failures of "tightly coupled, complex systems" identified by Perrow (1999). At the same time, infrastructure vulnerability is, at least in theory, more easily managed than the vulnerability of social, economic, and institutional systems. To borrow from Wisner et al. (2004), infrastructure vulnerability represents a dynamic pressure resulting from insufficient investment, while vulnerability of the other systems exemplifies

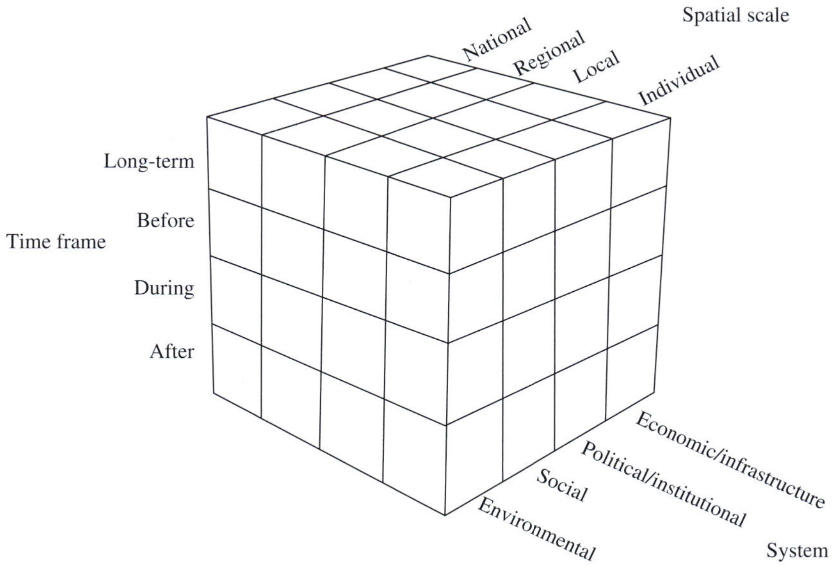


Fig. 2.2. A "Rubik's Cube" of vulnerability elements

root causes. The diagram in Figure 2.2, based on Rubik's Cube, provides a means of conceptualizing some of the various elements comprising the considerations of who and what are vulnerable, taking into account both spatial and temporal scales. While this oversimplifies the complexity of relationships and represents only broad categories, it serves to illustrate the range of potential interactions. Vulnerability is seen at various spatial scales from national (and even global), to the individual and at different time frames, shown in the diagram as ranging from the long-term, to the period immediately before an event through to the time after the event, when recovery is taking place. In reality, each of these is variable, depending upon both the spatial scale of the event and the sector of concern. As previously discussed, vulnerability varies among systems, and again, the nature of a system's vulnerability will fluctuate with varying spatial and temporal scales. What is not shown in this diagram are the linkages among the systems themselves, such that, a vulnerable economic system will have a knock-on effect on other systems. Nevertheless, the cube, as a whole, and the various combinations that are possible, show the interwoven nature of who and what are vulnerable, and serve to explain the rather extensive number of definitions of vulnerability. Each depends on the perspective of the author, and the various permutations shown in the cube present many such perspectives. Indeed, this is not the first vulnerability cube that has been suggested. The vulnerability cube presented by

Kienberger et al. (2013) provides a framework for positioning vulnerability assessments while the one presented here addresses various arrangements of elements of the systems involved and spatial and temporal scales.

### *2.3.3 Methods and Tools*

While there are many definitions of vulnerability, there are fewer ways to assess it, despite a plethora of indices that have been proposed to address overall vulnerability (Cutter et al., 2003) and vulnerability to various risks, including: climate change (Young et al., 2015), floods (Fekete, 2009), heat (Johnson et al., 2012), mountain hazards (Fuchs, 2009), and storm surges (Rygel et al., 2006), and vulnerability of particular locations such as coastal cities (Balica et al., 2012), low income nations (Dabla-Norris and Gündüz, 2014), and small island states. Other methods have addressed vulnerability from the perspective of livelihoods (Hahn et al., 2009), health (Reid et al., 2012), and particular resources such as water (Jun et al., 2011). If nothing else, this rather incomplete list suggests that if defining vulnerability is fraught with difficulty, measuring it is even more difficult. Of the numerous methods, some are quantitative, some are qualitative, and others a mixture of both. What is clear is that there is no agreed upon means to measure such a complex phenomenon comprehensively (nor perhaps should there be), yet many have tried. At the same time, there have been criticisms about some methods, particularly the use of indices (King, 2001), because of problems associated with spatial and context differences (de Sherbinin, 2014), the aggregation of data at various scales of analysis (Barnett et al., 2008), statistical artifacts (Lüchters and Menkhoff, 1996), uncertainties associated with decisions on data selection, transformation, and weighting (Tate, 2012), and a focus on specific variables that may mask or overlook underlying issues (Adger et al., 2006). Further, as Tapsell et al. (2010: 22) point out with respect to social vulnerability, "there is still no consensus on a) the primary factors that influence social vulnerability, b) the methodology to assess social vulnerability, or c) an equation that incorporates quantitative estimates of social vulnerability into either overall vulnerability assessment or risk." These concerns are not unique to social vulnerability analyses – they cross into all vulnerability assessments. Nevertheless, despite these and other challenges, the need to measure vulnerability is seen as an essential step in reducing risk (Birkmann, 2006a, 2013a) by documenting change, both spatially and temporally, thereby facilitating the setting of priorities and developing of strategies for investment. Thus, the discussion that follows is based on a recognition of the need for quantitative indicators while at the same time incorporating qualitative factors that contribute to defining vulnerability. Overviews of methodological approaches for each are presented here.

### *2.3.4 Considerations for Quantitative Assessments*

A range of examples of quantitative methods for measuring vulnerability (see Section 2.3.2) illustrates the importance such methods have had in understanding vulnerability and many of its components. Quantitative assessments of vulnerability require the selection of representative indicators, benchmarking/normalization of indicator values, and then weighting and aggregating them. The selection of indicators can be either deductive, based on a theoretical understanding of relationships, or inductive, based on empirical observations or statistical relationships among a large number of variables. Often the variables are chosen based on availability, such that they encompass the desired temporal or spatial range (data collection options are discussed below). To avoid this, involvement of stakeholders can be important for defining the area of concern and for the selection of most appropriate indicators.

Once the indicators have been chosen, it is likely necessary to normalize them prior to aggregation (Saisana and Tarantola, 2002), as the indicators in a data set often have different measurement units. Once completed, there are several weighting and aggregation algorithms available for data management. Commonly used weighting procedures are based on either statistical methods (Aleksandrova et al., 2016) or participatory methods (Schmidt-Thomé, 2006; Giupponi et al., 2015). These are important choices as the statistically based methods do not incorporate stakeholders' input and attitudes. So, as with data selection, participatory-based procedures may be effective, if more time-consuming. An example of a participatory based, non-additive aggregation procedure has been presented by Gain and Giupponi (2015). At the same time, there exist numerous other examples of aggregated indices using different aggregation methods (Saisana and Tarantola, 2002; Birkmann, 2006a, 2013a).

As discussed above, there are a number of challenges associated with quantitative vulnerability assessments. In addition to those relating to data choice and management, others are perhaps more systemic, stemming from, for example, preferences for specific methods and/or research designs, the embedded concept of vulnerability and its components, and the extent to which there is user involvement. Other challenges involve communication and the need for integration of disciplines and applications in new geographic and thematic areas (Fekete, 2012; Fekete et al., 2015). Yet such challenges are not in any way new, as deterministic, aleatoric, and epistemic uncertainties are well recognized, for instance, in the context of seismic probability hazard analysis (SSHAC–Senior Seismic Hazard Analysis Committee, 1997).

While quantitative data are essential to many vulnerability assessments, care must be taken in indicator selection, aggregation, construction, and sensitivity analyses (Saisana and Tarantola, 2002). Spatial autocorrelation effects, scaling effects, and the selection of the research unit (Fekete et al., 2010; Gibson et al., 2000) are critical considerations. Indeed, the results of analyses using spatial indicators at a regional level can contradict findings from local household assessments (Fekete, 2012; Steinführer and Kuhlicke, 2007), although comparability and upscaling may be possible to a limited extent (Fekete et al., 2010). Much can be learned from validations of vulnerability assessments and indicators (Fekete, 2009; Papathoma-Köhle et al., 2017), but they are lacking, as are overviews of existing studies, datasets, and longitudinal observations.

The most critical component of such vulnerability assessments is the data. Often, the data chosen are the data most readily available. Unfortunately, the "age" of the data as well as their spatial dimensions may not be most representative of the temporal and spatial patterns of vulnerability, but they are the best available. This creates a dilemma for both the researcher and the policymaker. While it is well recognized that vulnerability is "a dynamic phenomenon often in a continuous state of flux" (Adger, 2006: 274), the data available cannot represent its dynamic nature, which limits the use of "socioeconomic indicators to predict community vulnerability" (King, 2001: 147). Yet, this is not to suggest that such assessments should not be undertaken. Indeed, most provide useful indications of the state of vulnerability in a given place at a given time, thus having potential policy implications. Furthermore, it is only through continued work that we can improve our methods. Tate (2012) suggested that global sensitivity analyses can serve to internally validate deductive, hierarchical, and inductive methods used in the most common social vulnerability index designs. Similarly, de Sherbinin (2014) noted that spatial comparisons between countries and specific challenges associated with the use of indicators are increasingly conducted and critically reflected. Thus, much more needs to be done and we continue to learn from and build on previous work.

### *2.3.5 Consideration for Qualitative Assessments*

Despite the advantages of such quantitative methods of vulnerability assessment, among those concerned with community-based vulnerability research,

Table 2.2. *Vulnerabilities, capacities and exposure analysis (VCE) matrix*

| | Vulnerabilities | Capacities | Exposures |
|---|---|---|---|
| **Motivational/Attitudinal** | | | |
| Individual skills and characteristics | | | |
| **Social/Organizational** | | | |
| Networks, community cohesion and organization | | | |
| **Physical/Material** | | | |
| Resources, environmental conditions, infrastructure | | | |
| **External drivers** | | | |
| Economic, institutional, environmental, or political constraints | | | |

Adapted from Anderson and Woodrow (1989).

there is skepticism relating to what is seen as top-down research designs – skepticism that is justified, given that many indicators are applied at a scale that often leaves out the household level, including their needs as well as their coping capacities (Wisner et al., 2004). The need for this was addressed in the 1980s and 1990s when Anderson and Woodrow (1989) presented their Vulnerabilities and Capacities Analysis (VCA) that included physical/material, social/organizational, and motivational/attitudinal components (see modified version in Table 2.2), clearly an approach focused on qualitative information. This framework helped to foster subsequent work relating to the factors that differentiate vulnerabilities by recognizing that those with vulnerabilities also have capacities, however limited in some categories.

The importance of covering the context of data and of persons and values at stake has been recognized as imperative, presenting an area of needed advancement with respect to vulnerability indicators, which, in turn, integrates different disciplinary perspectives and mindsets (Kuhlicke et al., 2011). Achieving the integration of these qualitative factors is important to overcome skepticism and critiques of the measurability of vulnerability and the science-policy interface (Hinkel, 2011), and to improve the ability to address user needs within the science-policy nexus, as well as the capability and application of research and science to policy (Fekete et al., 2015). Those experts involved in vulnerability analyses should be well aware of the methods to gather the kind of information needed to incorporate the variables delineated in the VCA for different groups, whether it is the homeless in megacities (Wisner, 1998), marginalized populations in floodplains (Rufat et al., 2015), or people surviving on subsistence livelihoods (Wilson, 2014).

These include interviews, focus groups, and expert opinions. While all of these can be time-consuming to carry out and to analyze, the results will likely be much richer than those resulting from secondary data sources or even from surveys. At the same time, they may suffer from similar temporal and spatial constraints as do the quantitative data discussed above. One approach to perhaps overcome this is the use of scenarios. Yet, scenarios applied to assessing vulnerability and adaptive change in the context of socioeconomic development are lacking, despite their use in many scientific assessments for climate change adaptation and risk reduction (Birkmann et al., 2013). In fact, some researchers (Mechler and Bouwer, 2014) found that vulnerability scenarios have not been considered explicitly for assessing risk dynamically over longer time horizons, despite the opportunities such an approach offers.

It should be clear that there is a variety of methods that can be employed to obtain the kind of less quantitative information needed to more fully assess and, indeed measure vulnerability, whether that is vulnerability of a place, of a sector, or of a group of people or individuals. There are also challenges associated with obtaining this information (Birkmann et al., 2013) and with getting it into a form that is in some way comparable or least compatible with the quantitative analyses, so that these critical factors are not lost. This creates a bit of a Catch-22 wherein, as noted above, the ability to measure vulnerability is critical (Birkmann, 2006a), but how does one "measure" qualitative criteria? Lack of measurement does not equate to lack of significance, and in some cases (such as Wisner and colleagues' root causes), an unquantifiable characteristic may contribute more significantly to vulnerability (or to capacity) than do some that are readily quantifiable. This leads to a call for continuing research on, and development of, comprehensive methods, taking both the advantages and the caveats associated with each approach to assessing vulnerability into account.

## 2.4 Conclusion

Because of the complexity of the word vulnerability, in this chapter we have presented it through only some of its forms and understandings. Within the context of this book, this chapter has mainly tried to outline specific characteristics of vulnerability that have shaped parts of risk research, specifically, risk assessments. Such aspects include a focus on the evolution of hazards research to address the impacts of hazard events rather than a purely hazard-driven research focus. Moreover, this differentiation of the impacts and resulting processes are hallmarks of what is termed "vulnerability".

However, assessments and foci on human behavior, resources and habits, stability of materials, and impact cascades have existed long before vulnerability became a term in this context. For example, in traditional risk matrices addressing the damage extent, the concept of vulnerability had already existed while perhaps not being identified as such, by the term 'criticality' or even as part of estimating the probability. Vulnerability covers aspects that help explain why something (human being, community, or physical object or technical system) is adversely affected by a hazard and how this might play out. Vulnerability has been applied in ecosystem contexts (and especially in a systems theory context), and is deeply linked with a conceptual differentiation between internal and external factors, that is, between a system of interest and its "environment". Such a differentiation, however, can also be problematic. For instance, humans and buildings are both integral parts of what is termed "environment". Even risk analysts who focus on analyzing physical objects per se, find themselves establishing the contexts of such assessments as they intersect or overlap with human interests. For example, the vulnerability of flora and fauna is often analyzed (in relation to DRR, at least) in the context of the services such environmental systems provide to humans. The evolution of work on vulnerability has nevertheless sharpened the recognition of the role of the human side of risk, and environmental hazards have particularly shaped this discourse.

Not questioning those achievements of vulnerability, we nevertheless want to take the opportunity to try to think a bit beyond it and cast a look into the near future. What has research on vulnerability already achieved in a way that progress might continue, or shift to other fields?

An assessment of some strategic documents of the United Nations and the European Union, alongside documents from the USA and UK, has produced indications that vulnerability and resilience have been influential terms and concepts. But the developments vary; while resilience has increasingly become a trend in usage of a more encompassing "umbrella term", sometimes even replacing "risk" for describing a holistic concept, vulnerability is found to be stagnating somewhat and being used much more in the context of risk assessments rather than as a policy or strategic term (Fekete and Hufschmidt, 2014).

Vulnerability and assessments of it have been key elements in strategies of UNISDR (UNISDR, 2015a, b) and are strongly promoted within an overall DRR paradigm. Reduction is just one of the strategic options in risk management or governance, next to risk acceptance, avoidance, and transfer, among others. In a broader context of civil protection and national security research, DRR is strongly linked to what is termed "preparedness" and "prevention" in the "disaster cycle". This denotes the ex-ante side of risk, before a hazard or threat turns into an impact. Much effort and political, as well as scientific, activity have been invested in the past two decades at least to promote such an approach, complementing the previously dominant view of hazard-based risk probability assessments, as well as disaster management focusing on reaction immediately after an impact, thus conducting "crisis management" rather than long-term strategic preparation and planning. Planning and preparation undoubtedly deserve this emphasis provided by DRR and other paradigms, as well as disaster management, mitigation, and recovery. However, it might also be wise to ask, from a research perspective: has vulnerability research had the impact it should or could have had? Is more to be expected on the "learning curve" or on the "action and impact" side of measures to be derived from vulnerability and capacity assessments? We can, certainly in context of this final subchapter, only indicate the repercussions of asking such a demanding question. Our motivation for asking such questions stems not only from our work with practitioners questioning complicated concepts such as vulnerability but also the results of evaluation studies of humanitarian aid and the activities of development organizations, specifically. Furthermore, some semi-political platforms such as the IPCC or UNISDR have given vulnerability some global attention, but it must also be questioned if also some saturation in awareness has been achieved. We have made progress in understanding and assessing vulnerability and in bringing its importance to the political stage. Despite these efforts, and in light of the extensive research aimed at understanding vulnerability as both a concept and a phenomenon, we still need to ask: which gaps remain? What has to be done on national agendas and with international platforms in order to improve risk and security conditions, that is, to reduce vulnerability?

A future challenge therefore might lie in "understanding risk" (c.f. the Sendai Framework for Disaster Risk Reduction of the United Nations, Priority 1, UNISDR, 2015a; see also Zimmermann and Keiler, 2015), but in a two-fold sense: first gathering even more information about hazards and vulnerabilities, and second, fostering knowledge transfer between actors, disciplines, and even countries. Vulnerability scholars place high priority on knowledge sharing and transfer, but this is often stymied by the differing definitions and contexts in which vulnerability is being considered. However, in a project on improving knowledge management between actors in civil protection, an "Atlas of Vulnerability and Resilience" has been developed (Fekete and Hufschmidt, 2016) that explains different approaches on assessments but also different conceptualizations of both vulnerability and resilience. While vulnerability has been criticized for its lack of definitions and varying applications in different disciplines and contexts, this atlas turns this notion around and projects this diversity as being one of the strengths of vulnerability (and resilience). Vulnerability allows for learning through its heterogeneity; by being "exposed" to different applications in psychology, engineering, flood, or epidemic studies' novel approaches for improving existing vulnerability assessments, their range of criteria or theoretical background emerge. A future application of such meta-studies and collections of the body of knowledge on vulnerability assessments lies in having sufficient data and studies at hand for meta-evaluations that emerge more and more in disaster risk science (Rufat et al., 2015). It is through these kinds of efforts that we can answer the questions we pose in this section.

Except for continuing developments in vulnerability research and application such as needs for evaluation, sensitivity analysis of its tools, and knowledge transfer, what other developments are possible? There remain some of the old unsolved challenges, such as the quest for a unifying theoretical foundation behind vulnerability (King and MacGregor, 2000). Currently it appears more focus is on fitting vulnerability into resilience terminology and concepts, rather than advancing the conceptualization of vulnerability. Both are needed. Yet, what else could be demanded from vulnerability studies? Identifying benchmarks of vulnerability for better comparability, possibly derived from critical thresholds and tipping points, is one important avenue that can lead us to helping humans and structures become "invulnerable". Therefore, key points to understanding vulnerability are:

- Vulnerability is both a concept and a phenomenon that has been very influential in stimulating research and in fostering understanding of the nature and distribution of impacts from hazardous events.
- Finding a common definition of vulnerability seems to be problematic, but the heterogeneity of definitions used may also serve to broaden the debate and interdisciplinary perspectives.
- Vulnerability is deeply linked with a conceptual differentiation between internal and external factors, that is, between a system of interest and its "environment".
- Both qualitative and quantitative approaches are relevant in assessing and handling vulnerability.
- Much work remains to be done in understanding the elements of vulnerability and in working to reduce vulnerability.

# 3

# Physical Vulnerability

SVEN FUCHS, TIM FRAZIER, AND LAURA SIEBENECK

## 3.1 Introduction

It has been argued that vulnerability is a socially constructed phenomenon by which humans, through their presence and their introduction of societal assets into the hazard zone, create risk. As such, the interactions between the hazard and the physical and built environment, along with the social characteristics of a location (as described in Chapter 4), inform the overall vulnerability within a community (Cutter et al., 2000). This chapter focuses on the physical dimension of hazard vulnerability. First, physical vulnerability and the various means by which this term is defined and conceptualized are presented. Second, this chapter describes indicators commonly used to measure physical vulnerability. Next, we provide an overview of methods most utilized in the assessment of physical vulnerability. A brief case study examining the measurement of physical vulnerability in Sarasota, Florida, is then provided as a demonstration of how the analyses presented in this chapter could be applied to inform mitigation. The chapter then concludes by summarizing future research needs and practical applications.

## 3.2 Defining Physical Vulnerability

Physical vulnerability, sometimes called biophysical vulnerability (Adger et al., 2005; Cutter et al., 2000), refers to the degree of susceptibility within the physical environment and as such to the negative impacts of hazards. Similar to the long-standing debates surrounding the use of the term vulnerability by scholars across multiple disciplines (Birkmann et al., 2013; Cutter, 1996; Cutter et al., 2000; Fuchs, 2009; Hufschmidt, 2011; Wisner and Luce, 1993), the term physical vulnerability has also been defined, conceptualized, and measured differently across the existing literature (Akbas et al., 2009;

Brooks, 2003; Davidson and Shah, 1997; Wang et al., 2013), with a lack of complete consensus on the use of this term (Füssel, 2007). In terms of physical vulnerability, the most widely used definition was introduced by UNDRO in 1980, in which physical vulnerability is defined as "the degree of loss to a given element, or set of elements, within the area affected by a hazard. It is expressed on a scale of 0 (no loss) to 1 (total loss)." This definition suggests that vulnerability equals the degree of loss of an element at risk resulting from the occurrence of a natural process (ex-post), and – once the relation between hazard impact and rate of destruction is known – allows for an ex-ante evaluation of possible future loss. Consequently, this approach requires empirical data in order to create a relation between the impacts of a hazard process on the element at risk. However, according to UNISDR (2009) vulnerability can be conceptualized in a more universal and qualitative way as the "characteristics and circumstances of a community, system or asset that makes it susceptible to the damaging effects of a hazard." This definition considers vulnerability as a preexisting condition (ex-ante) suggesting that it directly relates to the characteristics of the elements at risk, giving less emphasis to the process magnitude or intensity (Fuchs et al., 2011).

Despite differences in how the term is defined across the various research disciplines, two key elements emerge in the way physical vulnerability is conceptualized. The first element centers on the nature of the hazard, which is often described as both the characteristics of the hazard agent (e.g., spatial distribution, probability, magnitude, and intensity) and the degree of exposure that the hazard exerts on a particular structure or community. As shown in Table 3.1, the definitions offered by Brooks (2003) and Fuchs et al. (2007) mention the role of the hazard or event itself as being a key component in defining physical vulnerability, as the characteristics of the hazard influence the extent of susceptibility of the surrounding environment or elements at risk. The second key element entails characteristics of the physical or built environment (Birkmann et al., 2013; Clark et al., 1998). The physical environment comprises various dimensions such as societal assets (Ebert et al., 2009) including lifelines, human life and well-being, agricultural resources, individual and community assets, and the surrounding natural environment (Kapucu and Ozerdem, 2011). Several definitions provided in Table 3.1 place emphasis on the potential loss and expected damage a physical asset or structure might sustain due to stresses incurred by the hazard agent (Akbas et al., 2009; Birkmann et al., 2013; UNDRO, 1980). The next section expands on indicators pertaining to the two key elements of physical vulnerability: hazard exposure and the physical and built environment.

Table 3.1. *Selected definitions of physical vulnerability*

| Definition | Source |
|---|---|
| "The term 'biophysical' [vulnerability] ... suggests both a physical component associated with the nature of the hazard and its first-order physical impacts, and a biological or social component associated with the properties of the affected system that act to amplify or reduce the damage resulting from these first-order impacts." | Brooks (2003: 4) |
| "the degree of loss to a given element at risk, particularly to built structures, resulting from the occurrence of a hazard of a given magnitude." | Akbas et al. (2009: 1) |
| "potential for damage to physical assets including built-up areas, infrastructure and open spaces." | Birkmann et al. (2013: 200) |
| "the degree of exposure and fragility of the exposed elements to the action of the phenomena." | Cardona et al. (2004: 3) |
| "Social and bio-physical vulnerability are broader in scope and refer to social groups and landscapes that have the potential for loss from environmental hazards events." | Cutter et al. (2000: 715) |
| "degree of loss to a given element, or set of elements, within the area affected by a hazard. It is expressed on a scale of 0 (no loss) to 1 (total loss)." | UNDRO (1984) as cited in Papathoma-Köhle et al. (2011) |
| "Thus, vulnerability – often referred to as 'technical' or 'physical' vulnerability – is defined as the expected degree of loss for an element at risk as a consequence of a certain event." | Fell (1994), Varnes (1984), as synthesized in Fuchs et al. (2007: 496) |
| Vulnerability "is defined as the expected degree of loss for an element at risk, occurring due to the impact of a defined hazardous event. These events are themselves conditioned by a certain intensity, frequency and duration, all of which affect vulnerability." | Fuchs (2009: 338) |

## 3.3 Hazard Exposure

Knowledge about a location's degree of exposure to a hazard is essential in the assessment of physical vulnerability. In the hazards literature, exposure refers to a location's or system's degree of risk from a hazard (Adger, 2006; Clark et al., 1998). Within these locations or systems, exposure is often measured by the number of people, buildings, assets, infrastructure, and economic variables contained within or in proximity to the hazard zone (Birkmann et al., 2013; Fuchs, 2009; Hossain et al., 2015; Weichselgartner, 2001b). Assessments of a community's extent of physical vulnerability often examine the sensitivity and fragility of the buildings, infrastructure, and population given the stress and physical demands the hazard exerts on the surrounding environment.

Hazard exposure includes both spatial and temporal qualities (Birkmann et al., 2013; Fuchs et al., 2013). Measures of physical exposure to the environment often look at the location of a population or assets within a hazardous zone. For example, studies that examine the vulnerability of a physical structure to flood risk usually consider the structure's location within the 1-in-100-year or 1-in-500-year flood zone (Clark et al., 1998; Fuchs et al., 2017). In their measurement of hazard exposure along the Texas Gulf Coast to the effects of sea level rise, Brody et al. (2008) included other geographic dimensions such as proximity to the hazard (coastline), elevation, location within an area at risk to inundation as a result of sea level rise, and location with respect to base flood elevation levels for the 100-year floodplain. Similarly, studies that examine exposure to geologic hazards consider people and structure's location relative to a mapped landslide body (Guillard-Gonçalves et al., 2015), proximity to a fault rupture zone during an earthquake, and location within an earthquake liquefaction potential zone (Davidson and Shah, 1997). In cases where there is no clearly delineated boundary for the hazard, such as in the case of tornado risk when an entire county or region could be at equal probability for experiencing a tornado, assessments accounting for hazard exposure will often consider the historical dimensions of the hazard (Cutter and Emrich, 2006; Siebeneck, 2016). As noted by Peduzzi et al. (2009), the degree of physical exposure to a hazard is the product of the hazard frequency and the total population at risk. Given the hazard's historical frequency, the temporal aspects of a hazard can be determined by calculating recurrence intervals and probabilities of the hazard's occurrences.

Measurements of hazard exposure also account for the magnitude and intensity of the hazard. The amount of stress a hazard can exert on the surrounding built environment matters when examining the likelihood of an individual surviving a disaster or whether a building or particular infrastructure is able to withstand the forces exerted on it (Adger, 2006; Cardona, 2004). The means by which the magnitude or intensity of hazard exposure is incorporated into vulnerability models is largely dependent upon the nature of the hazard. For example, models of seismic vulnerability often account for the magnitude and intensity of the event using the Richter Scale and Modified Mercalli Index (Davidson and Shah, 1997). Landslide exposure often includes metrics such as landslide depth and damage potential ratings (Guillard-Gonçalves et al., 2015) and landslide volume (Winter et al., 2014).

Similarly, flood intensity can be measured by accounting for floodwater's velocity and hydrodynamic actions (Kelman and Spence, 2004), or in case of fluvial sediment transport taking the height of sediment deposition as a proxy (Totschnig and Fuchs, 2013).

As shown in the introductory paragraphs, the key component in understanding physical vulnerability to hazards is through the examination of the people, structures, and assets within a community exposed to hazards that comprise the physical and built environment. When examining the impacts disasters have on the physical environment, scholars often focus on five features: physical infrastructure, population exposure, economic exposure, social–political systems, and natural ecosystems. The following section describes these features and identifies the related indicators scholars use in the measurement of physical vulnerability.

### *3.3.1 Exposure of Physical Infrastructure*

Measurements of physical infrastructure often include indicators that reflect a community's type and quality of building infrastructure, transportation infrastructure, and utility/resource infrastructure (Davidson and Shah, 1997; Roberts et al., 2009). The ability of physical infrastructure to withstand varying degrees of exposure to a hazard is partially dependent upon the physical characteristics of these structures (Mazzorana et al., 2014). Because of this, the research literature offers numerous indicators that inform the ability of physical infrastructure to survive the impacts of a hazard or disaster. Davidson and Shah (1997) propose that some of the more salient variables in measuring the degree of vulnerability that buildings have to the effects of hazards include age of the structure, the purpose of the structure, how well the structure was constructed, and the location of the structure relative to a potential hazard. The latter is often considered when vulnerability for transportation infrastructure is addressed. When examining the location of buildings and structures, Papathoma-Köhle et al. (2011) argue that accounting for the area surrounding buildings is essential, as characteristics of the surrounding environment that may enhance or reduce the impacts of the hazard. For example, the Papathoma Tsunami Vulnerability Assessment model includes factors such as a structure's proximity to a slope or river and locations of nearby objects that can be moved by either water or debris flows, as both of these factors can increase the magnitude of damage caused by a landslide or flood event. However, other factors such as whether the structure is protected by vegetation (such as forest cover which could slow debris flow and landslide movements), and mitigation measures undertaken in the surrounding area, such as a retaining wall or drainage system (Holub et al., 2012), can reduce the physical vulnerability and impacts to an individual structure.

Indicators identified in other models and frameworks included other variables such as a building's location with respect to other buildings (Fuchs and McAlpin, 2005; Kappes et al., 2012), the cost of the building, the number of people the building serves (Douglas, 2007), and the building type (e.g., residential homes, cottages, etc., Totschnig et al., 2011). Similarly, knowledge of the materials used in the construction of the building (Mavrouli and Corominas, 2010), the extent to which the building has been maintained (Holub and Fuchs, 2008), and the number of floors present (Papathoma-Köhle et al., 2015) are also important factors considered in vulnerability assessments. The geometric shape of the structure itself as well as the strength of the building's foundation should be included in the examination of susceptibility to hazard impacts (Bertrand et al., 2010). Some scholars also suggest that the contents of the building should be accounted for in the measurement of vulnerability to physical infrastructures (Kappes et al., 2012), as assets and materials within a building may be damaged or lost as a result of the impacts a hazard has on the structure.

### *3.3.2 Exposure of Population*

The second dimension of the physical environment accounted for in vulnerability assessments includes population exposure. This type of assessment generally examines the number of people located within a hazard zone that have the potential to be exposed to the effects of a hazard (Birkmann et al., 2013; Fuchs et al., 2015; Jongman et al., 2014). Spatial characteristics of the population, such as the geographic distribution of individuals and population densities with respect to the proximity of the hazard are often accounted for in these analyses. For example, in a study examining vulnerability to climate change and rising sea levels along the Texas Gulf Coast, population exposure was captured by measurements that described an individual household's location within the floodplain, the elevation of their home, and their distance to the hazard (Brody et al., 2008a). Similarly, Fuchs et al. (2015) used information of the nationwide population register in Austria to model flood hazard exposure. Another study examining evacuation vulnerability during wildfire events modeled population vulnerability as a function of the numbers of households per exit in a community (Cova et al., 2013). As the ratio of households per exit increased, the time needed to safely evacuate the community also increased, thus increasing the population's evacuation vulnerability to wildfire.

The vulnerability of a population to a hazard changes as people move from place to place throughout the day. As such, Davidson and Shah (1997) and Keiler et al. (2005) note the importance of considering the temporal dimensions of population, such as day-to-day movements of people and seasonal migrations. A recent study by Kobayashi et al. (2011) examines the movements of people throughout the day in Salt Lake City, Utah, and argue that analyses of these movements (e.g., commuters to and from work) are essential, as the location of a population throughout the day is not static and their presence within hazardous areas varies. Similar results were obtained by Keiler et al. (2005) for European mountain regions during wintertime, where not only tourists travel for holiday purpose in areas endangered by various hazard types, but also on a diurnal basis they move between, for example, the hotels and the skiing areas. As result, Kobayashi et al. (2011) and Fuchs et al. (2013) suggest that emergency managers need to account for temporal changes in population exposure to hazards in their decision-making process in the mitigation and preparedness process as well as during times of crisis.

### *3.3.3 Economic Exposure*

Similar to buildings and populations, businesses and economies can be exposed to the negative effects of disaster. Clarke (1998) argues that there is a value to examining economic exposure to hazards, as hazards can impact the production, availability, and movement of goods and services across multiple economic sectors (Davidson and Shah, 1997; Weichselgartner, 2001b). Disasters often interrupt economic activities and in many cases result in individuals and communities experiencing adverse short-term and long-term economic impacts (Hossain et al., 2015). Direct economic losses often stem from damaged business (Karagiorgos et al., 2016) as well as from the interruptions of business-related activities, including effects on the entire production chain, whereas indirect losses may result from the number of employee casualties during a disaster (Davidson and Lambert, 2001), or from the recovery time needed to reconstruct the production facilities.

Several recent studies focus on the exposure of the economic sector to disaster. For example, Nicholls et al. (2008) use the gross domestic product (GDP) of port cities to estimate potential losses these cities could sustain as a result of storm surge and high wind events. They argue that negative impacts from these hazards could not only have adverse consequences for the GDP of cities in which the ports are located, but could also significantly impact global trade activities. Briguglio et al. (2009) echo this sentiment in their study of economic vulnerability to disaster, and suggest that disasters can create significant fluctuations in a country's overall GDP, especially in smaller or less-developed countries, and adversely affect the cost of imported goods and services such as energy, agricultural products, and industrial-related materials. Similarly in an examination of economic exposure as a result of drought hazards, Peduzzi et al. (2009) included variables such as GDP per capita and measurements of exposure of crops and the agricultural sector to the effects of drought. In all three of these studies, the authors include indicators reflective of the local economy and the hazards present in their community. Considering these two components allows for more in-depth examination into the hazards that pose the greatest short-term and long-term threats to the economy and the potential impacts these events have on individuals and businesses in a community.

### *3.3.4 Exposure of Social and Political Systems*

Davidson and Shah (1997) note that it is important to consider the impacts hazards have on political institutions and their ability to carry out governmental activities. While studies of physical vulnerability as they pertain to this category are not as numerous as studies examining population and building exposure and vulnerability, these institutions provide numerous services to the public and the loss of these services can hinder the community's ability to function and ultimately recover after a disaster. Physical structures, such as government offices, the mayor's office, city hall, and local courthouses house and provide important services to the public (McEntire and Myers, 2004), and depending on their degree of exposure to the hazard, are susceptible to disaster impacts. Other structures that can be considered within this category include schools, medical facilities, police and fire stations, and homeless shelters.

### *3.3.5 Exposure of Natural Ecosystems*

The exposure of the natural environment to the impacts of hazards can result in losses of various natural resources and landscapes. In a study examining the physical risk of coastal communities to hurricanes, Boruff et al. (2015) discuss that the wave action associated with hurricane storm surges can accelerate shoreline erosion rates, thus resulting in the loss of beaches along the coast. Similarly, storm surges and sea level rise resulting from climate change can negatively impact barrier island wetlands (Calhoon et al., 2006). A study of barrier islands along the US Gulf Coast from Florida to Louisiana found that the storm surge from Hurricane Katrina resulted in significant losses to barrier islands in this region, thus increasing the exposure and vulnerability of coastal communities to future storm surges (Fritz et al., 2007). Riparian zones have also been identified as a natural ecosystem at risk to both natural and man-made hazards. Exposure of riparian zones to wildfires can have devastating impacts on riparian zones, as the burning of vegetation and introduction of hydrophobic soil to the area can accelerate erosion in riparian zones (Petit and Naiman, 2007) and reduce the ability of the natural environment to control flooding. Additionally, riparian zones are at risk to man-made activities such as the construction of dams and reservoirs as well as other human modifications to the landscape (Goodwin et al., 1997). Loss of these ecosystems can reduce a location's ability to absorb floodwater, therefore increasing the vulnerability of the surrounding physical environment to various flood-related hazards.

## 3.4 Methods of Vulnerability Assessment and Analysis

Numerous methods for analysis, assessment, and quantification of vulnerability exist (Fuchs et al., 2011; Papathoma-Köhle, 2016); however, this chapter will focus on three of the more widely utilized approaches for assessing and measuring it: (1) vulnerability matrices, (2) vulnerability indicators, and (3) vulnerability curves. Table 3.2 provides an overview of the three categories and their main benefits and challenges.

### *3.4.1 Vulnerability Matrices*

Vulnerability matrices were among the first approaches applied in the assessment of physical vulnerability to natural hazards. The qualitative results and the use of expert judgment to evaluate empirical data made them popular among scientists and practitioners. This methodology contributed to the understanding of the interactions between the process and the elements at risk, and in some cases, they have been used to determine thresholds for damage classes. These matrices provide a sound overview of the consequences of a hazard. The intensity of the hazard may be expressed in different ways (impact pressure, flow depth, or velocity) usually as a range of values. Compared to vulnerability curves (see Section 3.4.2), the use of matrices bears a number of advantages since they make the relationship between process and consequence clear and easy to understand by

Table 3.2. *Overview of methods for the assessment of physical vulnerability*

| Method | Benefits | Challenges |
|---|---|---|
| Vulnerability matrices | Qualitative method, no need for ex-ante data or detailed information. | Results may not be translated into monetary loss. Assessment of damage under specific intensities or process characteristics is objective. |
| Vulnerability curves | The method is quantitative, connecting the hazard intensity to an expected degree of loss, which is a relation between the economic value of the element at risk and the loss height. | Important characteristics of the natural process (e.g., velocity, duration, direction, etc.) as well as of the element at risk (number of floors, construction material) are ignored. Highly demanding in ex-post information. |
| Vulnerability indicators | Characteristics of the element at risk are taken into consideration. | The intensity of the process is not considered, demanding in data (detail, amount, quality). |

Adapted from Papathoma-Köhle et al. (2017).

nonexperts. A precise evaluation of the hazard intensity and also data on economic values or loss height is not required for the application of matrices. However, the method is highly subjective since the assessment of damage level as high, medium, or insignificant may differ among experts (Papathoma-Köhle, 2016). For this reason, comparisons between different matrices and transferability of results are limited.

Throughout the past several decades, vulnerability matrices have been applied in a variety of different contexts (Özyurt and Ergin, 2010). One of the earlier applications of this technique was carried out in studies conducted by Leone et al. (1995, 1996) in order to assess landslide vulnerability. To do this, a matrix was proposed using damage intensity and resilience to hazard impact. The damage dimension was qualitatively expressed based on the potential of structural failure of buildings and the extent and volume of repairs that would be necessary to restore materials, such as road networks. While the output of this assessment resulted in a numerical value that described the degree of physical vulnerability, one criticism of this technique was that the resulting values were unrelated to process intensities (Bründl et al., 2008). Similarly, a recent study utilized a vulnerability matrix in the examination of climate impacts on coastal areas in the North Adriatic Sea (Torresan et al., 2012). Though vulnerability was assessed using a combination of multiple approaches in this study, the vulnerability matrix employed

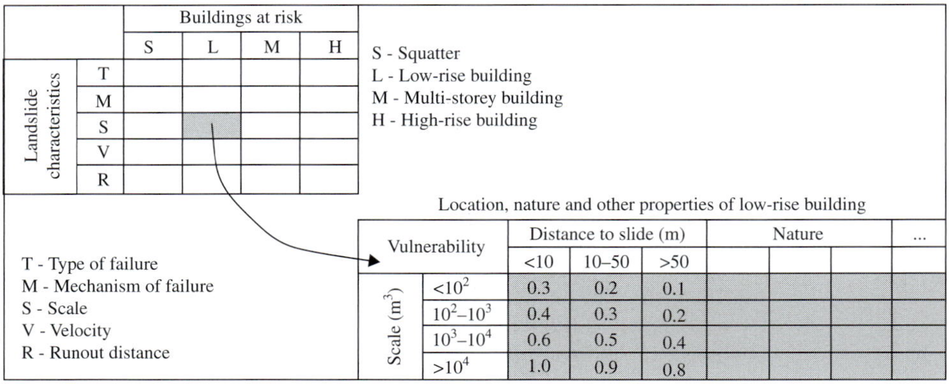


Fig. 3.1. A vulnerability matrix for elements at risk of landslides
*Source*: Adapted from Dai et al. (2002: 75).

in the research included categorical and quantitative measurements of various vulnerabilities associated with hydrodynamic impacts of sea level rise. In terms of the categorical dimensions, the authors relied upon their own judgments, those of experts in the local area, and previous literature to guide the section of qualitative categories, which were rated on a scale of 1–5. The authors note that although this method may not be as precise as quantitative measurements, it does allow for integration of local expert knowledge about the hazard and environment to be reflected in the assessment, as well as allows a means for making judgments related to uncertainty often present in these measurements (Torresan et al., 2012). A similar technique employing historical records to assess vulnerability was used to examine the vulnerability of buildings to debris flows in Australia (Fell and Hartford, 1997). This study utilized a qualitative assessment in which history of the hazard and local expert judgment were used to determine hazard intensity, which was described using three categories: low, medium, and high. A general procedure of assessing physical vulnerability using matrices is shown in Figure 3.1.

### *3.4.2 Vulnerability Curves*

Vulnerability curves are used for assessing physical vulnerability in a quantitative way. Vulnerability curves, sometimes named stage-damage functions, translate structural damage into monetary loss in relation to hazard intensity, which is either modeled or evaluated ex-post after an event. They link the hazard intensity (x-axis) to the damage response of an element at risk (y-axis), and require a substantial amount of empirical data to be robust. In case of hazards that impact a high number of buildings and infrastructure (e.g., earthquakes, floods), the amount of empirical data is often sufficient to establish this link using statistical fitting or building a statistical model (Apel et al., 2009). For other hazard types such as mountain hazards, however, there is only a very limited amount of data available that results in a high data spread and consequently in considerable uncertainty (Totschnig et al., 2011). Moreover, if vulnerability curves are used to quantify the expected loss, they are not always expressed in relationship to the same hazard characteristics. With respect to debris flow hazards, for example, vulnerability is computed using the hazard intensity in terms of deposit height as a proxy for the hazard magnitude (Fuchs et al., 2007). Other properties of the hazard such as flow velocity or impact pressure are not accounted for due to missing empirical information, since for many hazard types an area-wide or even local monitoring is not available. However the literature provides a variety of examples of vulnerability curves as they apply to riverine flooding (Papathoma-Köhle et al., 2011). The majority of these studies are based on the relationship between expected damage and inundation depth (Thieken et al., 2008). As Papathoma-Köhle et al. (2016) note, the high number of flood studies utilizing vulnerability curves may be partially explained by the fact that floods occur more frequently and result in more building damage and losses when compared to other hazards (Douglas, 2007). An example for a vulnerability curve based on the FLEMOps model is given in Figure 3.2 together with stage-damage functions from other flood hazard models (Apel et al., 2009).

### *3.4.3 Vulnerability Indicators*

As shown in Figure 3.3, the lack of substantial empirical data often leads to physical vulnerability assessment using indicators that supplement the presented methods of vulnerability assessment through careful identification, weighting, and aggregation of representative indicators (Papathoma-Köhle et al., 2011). These indicators include (explicitly or implicitly) goals that should be reached in reducing vulnerability (or increasing resilience) to natural hazards. According to Birkmann (2006b: 57), such indicators can be defined as variables which are "an operational representation of a characteristic or quality of a system able to provide information regarding the susceptibility, coping capacity and resilience of a system to an impact of … a hazard." Consequently, an indicator's quality is based on the degree to which it accurately represents the envisioned goal. Therefore, limitations exist in the calculation of a single value estimate of social vulnerability, as outside contexts and frameworks such as Social Vulnerability Index (Cutter et al.,

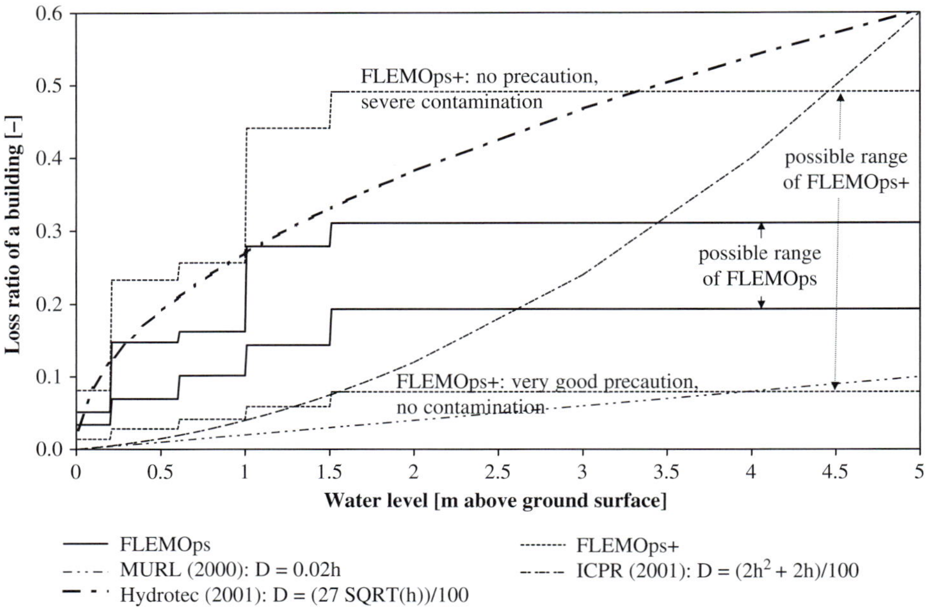


Fig. 3.2. Vulnerability curves (stage-damage functions) used to evaluate the susceptibility of buildings exposed to river flooding
*Source*: Adapted from Apel et al. (2009: 87).

2003) or the Spatially Explicit Resilience and Vulnerability or SERV model (Frazier et al., 2014), both commonly used in the USA, are often necessary in order to properly interpret the resulting models (Birkmann, 2006b).

## 3.5 Case Study: Measuring Physical Vulnerability for Coastal Areas

In spite of our best efforts to model and understand vulnerability and resilience, as outlined in this chapter, there are aspects of these areas of study that remain incomplete. There are spatial and temporal limitations that should be addressed if we are to remedy some of the deficiencies in traditional vulnerability and resilience analysis. We must understand the current gaps in our vulnerability and resilience science and the need for more detailed assessments to overcome many of the challenges associated with competing needs that often serve to limit the implementation of mitigation or adaptation actions post-assessments. The relationship between humans and their environments is not static; rather it is constantly being altered as humans respond to and shape their environments. This results in a unique context of vulnerability for each and every place in which humans reside. Place vulnerability can be closely tied to the resilience of place. Vulnerability is a condition of

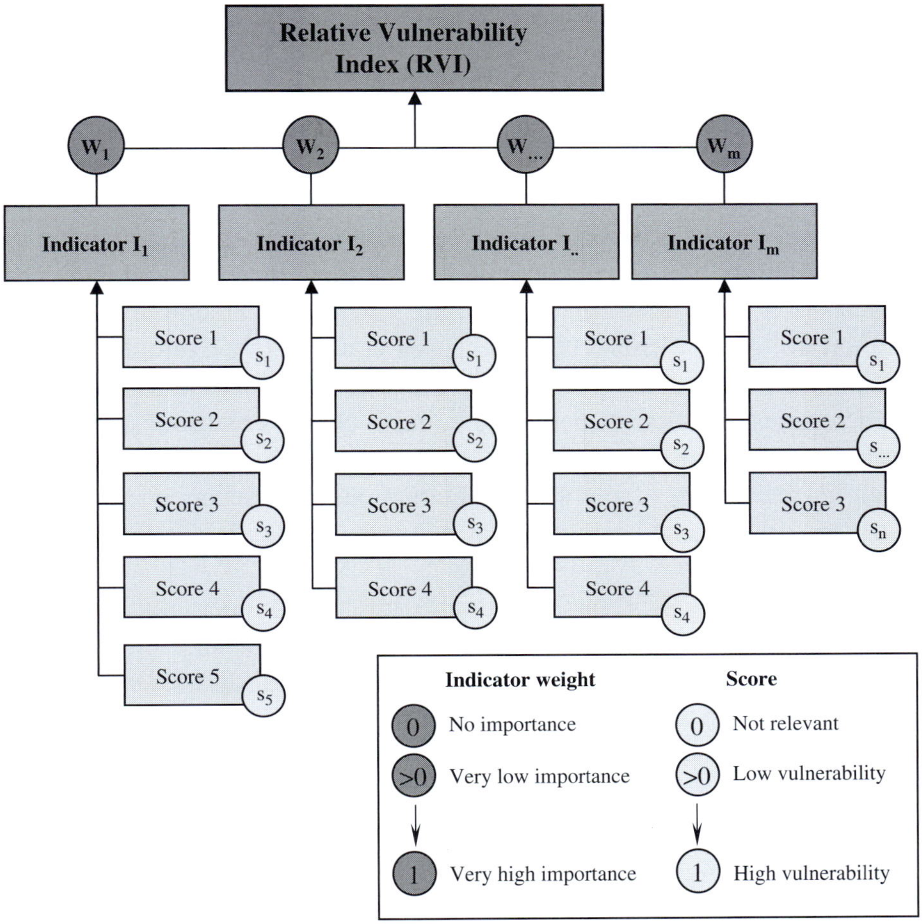


Fig. 3.3. Vulnerability indicators used for the assessment of multiple hazards
*Source*: Adapted from Kappes et al. (2012: 582).

combined exposure and sensitivity, as well as stifled adaptive capacity. Resilience plays a role in decreasing vulnerability by contributing to adaptive capacity.

Due to data limitations, existing research often focuses on comparing vulnerability across geographic locations, most prominently, at the county level. This is due to the fact that census data, which is widely available, is aggregated at the county level, therefore making it one of the most convenient and utilized scales in these analyses. The selection of scale and indicators can be problematic when carrying out vulnerability analyses. As detailed in Frazier et al. (2013: 96), one prevalent issue when conducting geographical research is "the modifiable areal unit problem (MAUP), which states that the results of a geographic study may vary depending on the scale at which the data is aggregated" (Fotheringham et al., 2002). It is possible this issue also exists in vulnerability indicators. Therefore, "it is important to look at locally-derived factors, rather than relying solely on nationally collected data" (Frazier et al. 2013: 96). Disaster vulnerability of a place is a complex construct that cannot be easily defined using such generalizable indicators. While broadly comparable indicators can be important for determining disaster vulnerability of a place, factors reflecting multiple scales and local characteristics of a place must be carefully integrated into vulnerability modeling (Cutter et al., 2008; Füssel, 2007; Siebeneck et al., 2015). Examples of place-based factors that reflect the unique characteristics of a location include hazard regimes, distinctive social and cultural characteristics of populations, location-specific biophysical factors, the surrounding natural environment, as well as the nature of the spatial interactions and dependencies of these factors to other neighboring places (Cutter et al., 2008; Füssel, 2007; Godschalk, 2003).

Furthermore, much of the previous research on vulnerability indicators has neglected to use differential weighting for the importance of each factor (see Cutter et al., 2010). Each factor relating to natural disaster vulnerability has a different capacity with which to contribute to or detract from resilience. Some resilience indicators may have a large effect on vulnerability, indicating that investment into these areas might have a positive effect on resilience enhancement. Others may not have as large of an impact, so investment in these areas may not be as effective in improving resilience. Vulnerability indicators also exhibit varying degrees of importance during each of the disaster phases, predisaster, emergency, recovery, and reconstruction (Haas et al., 1977). While certain indicators may be more important overall, some indicators may also have different levels of significance depending on the disaster recovery phase at hand. Indicators may be crucial during the first one or two phases, while their importance wanes during the later phases, and vice versa. Little has been done in previous disaster resilience research to incorporate these differential weighting and temporal components to disaster vulnerability indicators.

Finally, even when resilience and vulnerability has been reasonably assessed it is often difficult for mitigation and adaptation actions to be enacted in a meaningful way for vulnerability reduction. Some of this difficulty stems from competing interest (Frazier et al., 2010a), a lack of certainty in the vulnerability and resilience models (Frazier et al., 2014), and/or nonspatial or temporal specific assessments at spatial scales specific enough to drive mitigation and adaptation actions. Frazier et al. (2013, see Figures 3.4 and 3.5) determined

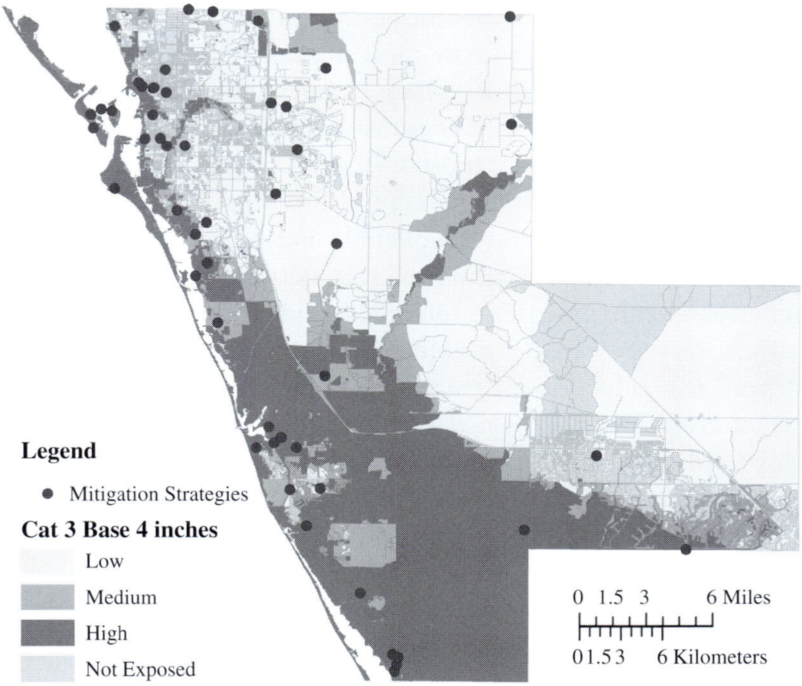


Fig. 3.4. Physical exposure from Category 3 storm surge with inland precipitation (A black and white version of this figure will appear in some formats. For the color version, please refer to the plate section.)

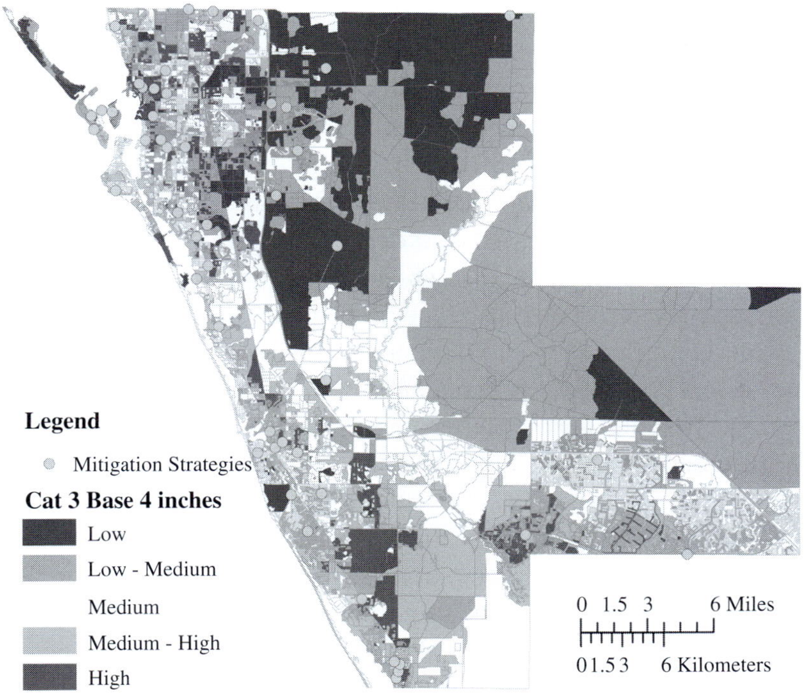


Fig. 3.5. Physical vulnerability to Category 3 storm surge with inland precipitation (A black and white version of this figure will appear in some formats. For the color version, please refer to the plate section.)

when examining local mitigation strategies for Sarasota County, Florida, that the county was consistently enacting mitigation actions in areas where physical exposure was high but not necessarily where vulnerability was the highest within the community. The fact that Sarasota County is a coastal county, and as such, much of the physical exposure is from coastal flooding, means that examined mitigation actions in the county are serving to mitigate in large part the private property of those most able to recover or areas with overall low vulnerability scores. In this case the lack of appropriate spatial scale vulnerability assessments translates into limited guidance in actionable mitigation that, as such, serves to create social and environmental justice issues within the context of risk reduction for marginalized populations.

In the study of Frazier et al. (2013), researchers reviewed Sarasota County's CEMP or their Comprehensive Emergency Management Plan (their hazard mitigation plan) and their Unified Local Mitigation Strategy plans to determine in-progress and planned mitigation actions for the county. Mitigation strategies from both plans were geocoded in GIS and compared to the county's physical exposure to sea level rise, hurricane storm surge, and flooding from inland precipitation (Figure 3.4). Physical exposure was determined through other studies (Frazier et al., 2010b; Thompson and Frazier, 2014). Researchers then ran the Spatially Explicit Resilience and Vulnerability or SERV model developed by Frazier et al. (2014) to determine countywide vulnerability at the census block spatial scale. The higher spatial resolution of the SERV model was necessary to evaluate the spatial relationship of the county's mitigation strategies to their physical and social vulnerability. After vulnerability was determined, researchers compared geocoded mitigation strategies with physical and social vulnerability at the census block level to determine if mitigation was planned or occurring at areas of high physical and/or high social vulnerability (Figures 3.4 and 3.5). The results of the study indicated that county mitigation strategies are predominately occurring and planned in areas of high physical exposure and not in areas where social or overall vulnerability was the highest. As such, the county is targeting mitigation to areas that already likely possess the highest resilience to sea level rise, hurricane storm surge, and flooding from inland precipitation. Without higher spatial resolution vulnerability modeling, better understanding of the consequences of such mitigation decisions would not be possible. Even given this heightened understanding, the county still may privilege mitigation to high physical exposure areas but could do so with a more informed perspective as to the consequences to its most marginalized populations for disaster risk reduction, disaster impacts, and disaster recovery.

Previous research indicates that individual indicators of vulnerability play specific and quantifiable roles in a community's ability to adequately recover from a disaster event (Cutter et al., 2010). However, there has been little research concerning the implications of specific indicators of resilience on the speed and effectiveness of recovery, both in general and in a temporal and spatial context. Improved understanding of the spatial and temporal scales of disaster vulnerability indicators would enable improved estimation and modeling of baseline vulnerability levels. It would provide the opportunity for communities to learn which indicators should be given high priority and where investment would be most beneficial in strengthening their specific community preparedness. This understanding could offer a community the means to cope without the need for recovery or improve its ability to recover and adapt. Better spatial and temporal assessments with an effort to include differential weighting of indicators across the disaster preparedness and recovery cycles as such are needed to impact risk reduction at the local community level where most mitigation and adaptation decisions are made (Frazier et al., 2013).

### 3.6 Discussion

It has been shown that vulnerability can be assessed using indicators, curves, and matrices. As discussed earlier, there is a need for the development and further refinement of methods that better capture the complexities related to varying geographic and temporal scales of physical vulnerability. Because hazards and disasters vary in terms of areal risk and impacts, modeling needs to ensure the location dimension is accounted for appropriately and can accommodate analysis needs at the varying local and regional scales. Secondly, further research on the temporal dimensions of physical vulnerability is necessary. Different hazards vary in terms of their speed of onset. Hazards such as flash floods, earthquakes, and tornados can occur with very little lead time, whereas hazards related to drought and climate change may take months or even years to come to realization. Understanding differences in how the onset of these hazards impact and stress physical infrastructure is needed.

Additionally, studies should examine how changes in land use, in particular growing urbanization and increasing population in vulnerability areas, influence physical vulnerability. In spite of the utility or regional and local vulnerability models, what is needed are more models that are multi-scalar that take into account how vulnerability is variable across the landscape. Frazier et al. (2010, 2014) identified this need by exploring inter- and intracommunity geospatially dependent vulnerability. The notion that vulnerability indicators operate only at a limited spatial scale assumes no spatial autocorrelation of these indicators and ignores the impact of both near and far externalities on overall vulnerability and resilience.

Determining whether the values resulting from a vulnerability assessment are high or low is challenging, and as described in this chapter, many approaches for defining, identifying, measuring, comparing, and interpreting relative vulnerability exist. When compared to vulnerability matrices and indicator approaches, vulnerability curves repeatedly have shown vulnerability in absolute values, even if the value for the loss height is given in terms of a degree of loss. Nevertheless, vulnerability is based on a concept that accounts for the associations and complex relationships inherent between hazard and exposed elements (Cardona, 2004), which has been expressed by Alexander (2005) in a way that one could not "be vulnerable if one is not threatened, and one cannot be threatened if one is not exposed and vulnerable" (Cardona, 2004: 38). Hazard and vulnerability are highly correlated, and are only conceptually defined as independent so that the risk equation can be better solved. With respect to physical vulnerability, it is consequently possible to reduce exposure (by, for example, building a dam to protect a floodplain) or to reduce the susceptibility of elements at risk (by, for example, strengthening the building envelope) and therefore reducing physical vulnerability (Holub et al., 2012).

Prospective analyses in particular can offer insights into the future extent of physical vulnerability, and in turn can inform future mitigation decisions related to strengthening building codes, land use, local ordinances, and imminent structural engineering projects. Improvements in the accuracy of vulnerability models can serve as a basis for informed and better decision-making related to reducing exposure impacts to natural and man-made hazards. However, for this to occur, scholars need to do a better job sharing and communicating the results of physical vulnerability analyses with the practitioner community. Getting sound research in the hands of individuals with the ability to guide policy and implement practical solutions in a community needs to be prioritized.

Many of the methods highlighted in this chapter are complex and are not always feasible to carry out in the practitioner communities. Increased partnerships between universities and the emergency management community could lend to new collaborations focused on modeling physical vulnerability in a manner that would be most beneficial to local and regional entities.

### 3.7 Conclusion

Recent catastrophic events such as those experienced in New Zealand and Japan (earthquake and tsunami), USA (Hurricane Katrina and Hurricane Sandy), Thailand and Pakistan (floods), have focused the attention of research and practice to vulnerability. Despite decades of research on physical vulnerability, the question remains: why has there been so little progress in our ability to mitigate and adapt to natural hazards (Fuchs et al., 2011)? White et al. (2001) discuss at length this paradox in their article entitled "Knowing better and losing even more – the use of knowledge in hazards management." Besides many other reasons for this paradox, truly interdisciplinary research is required in order to accurately model, analyze, and address the multifaceted nature of vulnerability.

As demonstrated in the research literature, the indicators and methods used in the assessment of physical vulnerability widely vary and often place emphasis on addressing vulnerabilities to buildings, infrastructure, population, and lifelines (for a review see Davidson and Shah, 1997, and Papathoma-Köhle, 2016). Consequently, fewer studies examine other dimensions susceptible to the effect of natural hazards, such as agricultural vulnerability (Dutta et al., 2003; Fisher et al., 2002), the economic dimensions of vulnerability manifested in the short-term and long-term impacts to businesses, employment, tourism, etc. (Karagiorgos et al., 2016), and negative environmental impacts of disasters (Davidson and Shah, 1997). Furthermore, the scholarly literature is limited in terms of studies that consider the multidimensional nature of vulnerability (Papathoma-Köhle, 2011) and therefore constitutes an area where the integration of these various dimensions into future research is needed. One example of how to do this is offered by the SERV model, which incorporates both socioeconomic and biophysical vulnerability into one overall community vulnerability score (Frazier et al., 2014). Future studies also need to utilize methods that allow for more scalable models. As mentioned in this chapter, the lack of scalability in these analyses can create barriers when trying to recreate and apply these methods to different hazards and differing geographic scales. While the majority of protective action decisions that emergency managers make during times of disaster rely upon local vulnerability assessments and knowledge (Fuchs, 2009), regional assessments are valuable resources for central and regional governments when making decisions about resource and funding allocations on a larger scale (Thaler et al., 2016). Lastly, due to the complexity inherent in identifying, measuring, and analyzing physical vulnerability, scholars and researchers need to improve collaborations with the practitioner community in order to inform best practices in reducing physical vulnerability. Improved methodology will help advance the knowledge and theory pertaining to physical vulnerability; however, collaboration and sharing information with the practitioner communities will help reduce the risks and threats people and places face as a result of natural hazards.

Key messages of this chapter can be summarized as follows:

- In order to assess physical vulnerability, vulnerability matrices, vulnerability indicators and vulnerability curves may be used. Spanning from qualitative expert judgment to empirical quantification, each method has inherent challenges and the success is, furthermore, dependent on the overall aim of risk analysis.
- Physical vulnerability has a temporal and spatial dimension. The most important variables for both are given by the exposure of elements at risk, people at risk, and economic activities, social and political systems, and ecosystems exposed to negative effects of natural hazards.
- Finally, studies are limited in terms of works that consider the multidimensional nature of vulnerability, with physical vulnerability as one pillar of various dimensions of the concept, and roots in science as well as in practical implementation.

# 4

# Social Vulnerability

CHRISTOPHER BURTON, SAMUEL RUFAT, AND ERIC TATE

## 4.1 Introduction

Extreme climatic and geophysical events exert enormous negative consequences on people around the world. Faced with ever-increasing societal impacts from natural hazards, a wealth of research and analysis has focused on understanding causal processes and impacts. Historically, these efforts centered on characterizing hazard dynamics and using physical structures such as levees to mitigate damage. The result has been a proliferation of technocratic approaches, while financial and human losses from natural hazards continue to escalate. Over time, this has given rise to an alternative explanation that mounting losses are related less to natural event dynamics, and more to the vulnerability of exposed human populations. There is a continuing need for disaster risk reduction strategies to shift emphasis from assessing hazard events toward reducing vulnerabilities within social systems (Briceño, 2015). Conceptualizing and measuring social vulnerability is an important stage along this path.

Broadly defined, vulnerability is the potential to suffer loss or harm. This can include structural vulnerability of buildings and lifelines, biophysical vulnerability (physical exposure) of people and places to natural events, and social vulnerability describing differential susceptibility based on social, economic, and political factors (Cutter, 2001; O'Keefe et al., 1976). As applied in social science research, the term vulnerability generally describes a state of people and populations rather than physical structures, economies, or regions of the earth (Wisner et al., 2004). Vulnerability can vary significantly across both social and geographic space (Bohle et al., 1994; Cutter, 1996; Liverman, 1990). Social space refers to who is vulnerable, and is defined by the political, economic, and institutional capabilities of people at a specific time and place (Bohle et al., 1994; Wisner et al., 2004). By contrast, geographic space describes the location and scale at which people and places are vulnerable (Cutter, 1996).

Our current understanding of vulnerability is guided by theories, methodologies, and measurement standards derived from different schools of thought. This chapter provides a summary of core concepts on vulnerability and its measurement from a social perspective (i.e., social vulnerability). The chapter has three main sections that delve into conceptual frameworks for hazard vulnerability (Section 4.2), approaches for measuring social vulnerability (Section 4.3), and a case study analysis of social vulnerability in Hurricane Sandy (Section 4.4).

## 4.2 Conceptual Frameworks

Many definitions of vulnerability to hazards appear in the literature of this field of study, with most including components of exposure, susceptibility, and adaptive capacity (Table 4.1). Exposure is the degree to which constructed environmental elements and the people who inhabit them intersect

Table 4.1. *Selected vulnerability definitions*

| Source | Definition | Tradition |
|---|---|---|
| Mileti, 1999 | Measure of the capacity to weather, resist, or recover from the impacts of a hazard in the long term as well as the short term. | Risk-hazard |
| Alexander, 2002 | Susceptibility of people and things to losses attributable to a given level of danger, a given probability that a hazard will manifest itself at a particular time or place, in a particular way, and with a particular magnitude. | Risk-hazard |
| Bohle et al., 1994 | Aggregate measure of human welfare that integrates environmental, social, economic, and political exposure to a range of harmful perturbations. | Political economy |
| Wisner et al., 2004 | Characteristics of a person or group and their situation that influence their capacity to anticipate, cope with, resist, and recover from the impact of a natural hazard. | Political ecology |
| Turner II et al., 2003 | Degree to which a system, subsystem, or system component is likely to experience harm due to exposure or hazard; either a perturbation or stress/stressor. | Social-ecological systems |
| Adger, 2006 | State of susceptibility to harm from exposure to stresses associated with environmental and social change, and from the absence of capacity to adapt. | Social-ecological systems |

the spatial extent of a hazard event. Susceptibility (also expressed as sensitivity or fragility) is the predisposition of exposed people and places to suffer adverse effects from a hazard. Adaptive capacity and coping capacity refer to the ability of people, communities, and systems to adjust to adverse hazard impacts.

Few definitions have been widely agreed upon in large part because the study of hazard vulnerability has proceeded under at least three distinct conceptual framings. These include risk-hazard, political ecology, and social-ecological systems frameworks (Eakin and Luers, 2006), although other typologies have been suggested (Adger, 2006; Cutter, 1996; Liverman, 1990; Turner II et al., 2003). Each is situated within a larger human-ecological modeling framework in which hazards and disasters stem from complex interactions between environment and society. However, the frameworks differ in their conceptualization of the dominant components, processes, and relationships that define human-environmental interactions. Conceptual framing is important because it drives the types of questions that are posed, the analytical methods used to answer them, and the solutions that are implemented. Over time, conceptual frameworks of vulnerability evolved from an initial focus on physical and managerial aspects to more explicit inclusion of social drivers and differential impacts. The following sections describe the evolution of prominent conceptual frameworks that have shaped social vulnerability analyses.

### *4.2.1 Risk-Hazard*

Vulnerability in risk-hazard is generally defined as the potential for loss or other adverse impacts, or the capacity to suffer harm. Expressed mathematically: risk = hazard × vulnerability. Risk is a function of biophysical, social, and economic factors, but in risk-hazard analyses, the hazard assumes the central role. Many studies have highlighted characterization of the spatial distribution of biophysical risk, with the most vulnerable populations considered to coincide with areas of biophysical exposure (Liverman, 1990). Due to the assumed dependence on exposure, risk-hazard approaches often favor solutions that seek to modify physical exposure through such actions as monitoring, forecasting, engineering, and land-use planning (Hewitt, 1983).

Within the risk-hazard field of study, geographers Gilbert White, Ian Burton, and Robert Kates pioneered the study of hazards and human adjustments made in response to them. Following a series of major floods along the Mississippi River, the Flood Control Act of 1936 (FCA 1936) was passed

(United States Congress, 1936). The statute granted authority to the US Army Corps of Engineers and other federal agencies to build and maintain large-scale flood control structures such as dams, dikes, floodwalls, and levees. This structural mitigation approach was indicative of the dominant mentalities of the day: the root causes of disasters are extreme physical events; and, societies can master flooding through technology. For Gilbert White, there was an alternative range of adjustments to flood hazards, including evacuation, building codes, zoning, public relief, and insurance (White, 1945). These were nonstructural approaches that recognized the role of both nature and society in creating hazards.

Social science research in the risk-hazard domain has delved into questions of why people inhabit hazardous areas, what factors influence the adoption of human adaptations to hazards, and how risk perception influences behavior (White, 1973). As such, the elements that define risk should expand from a dominance of external natural processes to also include societal dimensions of perception, policies, and management (Cardona, 2005). The basic idea is that people exacerbate hazards through behaviors such as developing in floodplains and translating risk perception into action. The focus on societal behavior led this perspective to also be referred to as the "behavioral paradigm." In the risk-hazard domain, risk and vulnerability assessments generally proceed through a sequential analysis that progresses from event characterization to exposure to fragility to impacts to potential adjustments (Eakin and Luers, 2006; Hewitt, 1983). Vulnerability reduction, in turn, is achieved through decision-making processes that consider natural event characteristics, risk perception, and cost.

The Human Adjustment to Natural Hazards model is indicative of the risk-hazard perspective (Figure 4.1). It defines a hazard as the intersection of systems of natural events and human activity, governed by a process of human adjustment to both. This interplay of systems produces hazard effects such as impacts to people and their belongings, to economic activity, and to social systems. The model is continuous and reflexive in that it assumes feedback effects of human adjustments to future states of natural events and human systems.

The natural event system is described by the magnitude, frequency, duration, areal extent, and spatial dispersion of geologic, hydrologic, meteorological, and biological events. By contrast the human use system is defined by the distribution of people, human activities, material elements, and decision-making ranging from household scales to national scales. A natural hazard occurs when these systems collide and create direct and indirect threats to people and the things they value. Impacts that reach a threshold of human

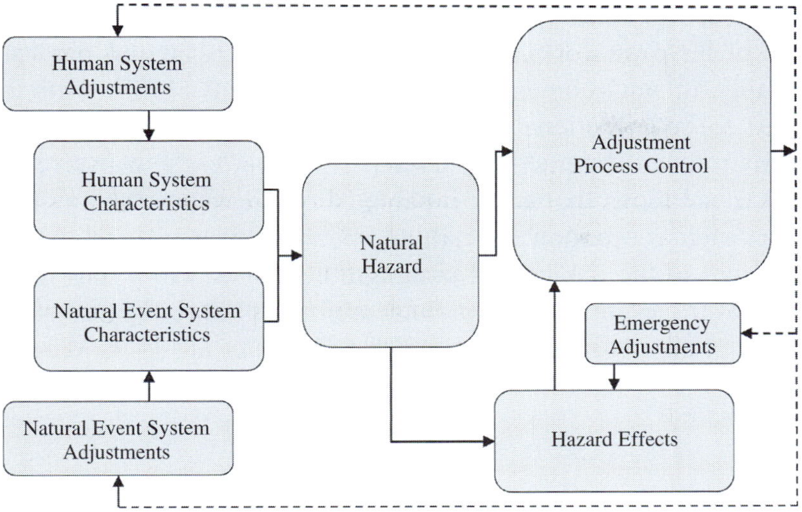


Fig. 4.1. Human adjustment to natural hazards
*Source*: Adapted from Kates (1971).

risk perception trigger an evaluation of potential adjustments based on environmental, economic, technological, and social constraints.

Emergency adjustments include post-disaster rescue and relief operations, while long-term adjustments (adaptation) more fundamentally alter natural and human systems. Examples of adjustments to natural systems include flood control structures, reservoir storage for drought, and forest thinning to reduce wildfire risk. Adjustments to the human use system include loss reduction efforts (e.g., building codes, early warning systems, evacuation), changes in land use (e.g., zoning, wetland restoration, property acquisition), and loss redistribution via insurance and disaster relief (Burton et al., 1993; White, 1973). The aggregate of adjustments implemented and avoided at different scales determines the severity of natural hazards.

Overall, the risk-hazard approach has tended to underemphasize the contribution of human use systems to hazards. Critics have argued that risk-hazard approaches poorly describe how people exacerbate or attenuate hazard effects, and fail to consider the role of political economy in shaping differential exposure and impacts (Turner II et al., 2003). In particular, it poorly addresses the societal context in which hazards take place. For example, the range of hazard impacts and available adjustments to a natural event may be influenced by factors exogenous to the system, such as public and private agendas and concurrent economic downturns. However, analysis at the system level may overlook differential hazard exposure and impacts among population subgroups within the system.

Hewitt issued perhaps the strongest condemnation of risk-hazard (including some of his own work), arguing the approach is far too predicated on the causality of environmental determinism (Hewitt, 1983), with hazards considered to be exceptional phenomena that occur outside the domain of human-environment interactions. This perspective leads to overreliance on technocratic solutions that are incapable of addressing important contributing factors such as economic, social, and political structure and values. Over time, critiques of the risk-hazard paradigm helped lead to a shift in perspective from natural events to human vulnerability as the root cause of disasters (Kasperson et al., 2001; Liverman, 1990; O'Keefe et al., 1976; Wisner et al., 2004).

### 4.2.2 Political Economy and Ecology

A focus on the social determinants of vulnerability helps explain why people with similar levels of exposure may experience very different levels of adverse impact. The social vulnerability perspective has its roots in the domains of political economy and political ecology. Political economy researchers generally focus on how political, economic, social, historical, and institutional factors (Table 4.2) produce differential exposure and susceptibility, with particular attention paid to cross-scalar interactions. Political ecologists extend the analysis by examining how these structural determinants generate, exacerbate, and attenuate environmental and other hazards (Collins, 2008). From both perspectives, disasters result from failures of political and economic systems, producing inequality, marginality, and differential constraints on individual action.

Central research questions related to social vulnerability focus on who is vulnerable, why they are vulnerable, how they are vulnerable, and how

Table 4.2. *Political economy determinants of social vulnerability*

| Determinant | Components |
|---|---|
| Political | Strength of democratic system, human rights, legitimacy of government action, corruption, citizen participation in decision making, linkages to local governments and civil organizations |
| Economic | Income, wealth, debt, credit access, economic reserves, trade policy |
| Social | Class, gender, ethnicity, age, religion, immigration status, literacy, education, health |
| Institutions | Rules, regulations, practices, programs, decision making procedures (e.g., property rights, water rights) |

institutions contribute to vulnerability (Eakin and Luers, 2006). The elevation of sociopolitical and economic factors is illustrated in the Causal Structure of Vulnerability conceptual model (Bohle et al., 1994), which posits vulnerability as a social space bounded by components of risk exposure, coping capacity, and recovery potential. These components are governed by a place and time-specific causal structure dependent on human ecology, expanded entitlements, and political economy. Trailblazing researchers in the political economy/ecology tradition include Piers Blaikie, Hans Bohle, Ken Hewitt, Amartya Sen, Michael Watts, and Ben Wisner.

Within the political economy domain, vulnerability is often conceptualized as a lack of entitlements in livelihoods. Entitlements refer to the set of commodities or resources – acquired through personal labor, production of owned resources, trade, or transfer – that a person can legally command (Sen, 1981). Ownership of these entitlements results from a combination of historical, institutional, and economic factors. Command over entitlements allows for the diversification of livelihoods in the face of a hazard or other stressor, increasing coping ability (Eriksen et al., 2005). Lack of entitlements occurs as a result of social factors resulting in vulnerability via unequal exposure to hazards (Watts and Bohle, 1993).

Context is another factor explicitly considered in political economy/ecology analyses. Societal issues such as policy agendas, resource management, land-use patterns, wealth distribution, and economic development may overlap with natural events, combining with place and time to make individual hazards unique (Mitchell et al., 1989; Tobin and Montz, 1997). There is also the potential for multiple stressors to occur simultaneously. For example, the occurrence of an extreme natural event concurrent with economic marginality can result in double exposure (O'Brien et al., 2004). Therefore, in order to truly understand hazards, stressors cannot be considered in isolation (Tobin and Montz, 1997).

While the Human Adjustment model evaluates vulnerability largely as a function of natural events and managerial decisions, a structural perspective on political economy posits vulnerability as an intrinsic property of human systems. Vulnerability is thus an underlying condition revealed by, rather than resulting from, stresses to the system, of which natural events are only one type. Other stresses may originate from unequal access to resources, economic structure, policy agendas, resource management, and violent conflict. For example, the occurrence of an extreme natural event concurrent with economic downturn can result in double exposure to hazards (O'Brien et al., 2004). Accordingly, vulnerability analysis should consider how macroeconomic, political, and institutional factors amplify and attenuate vulnerability

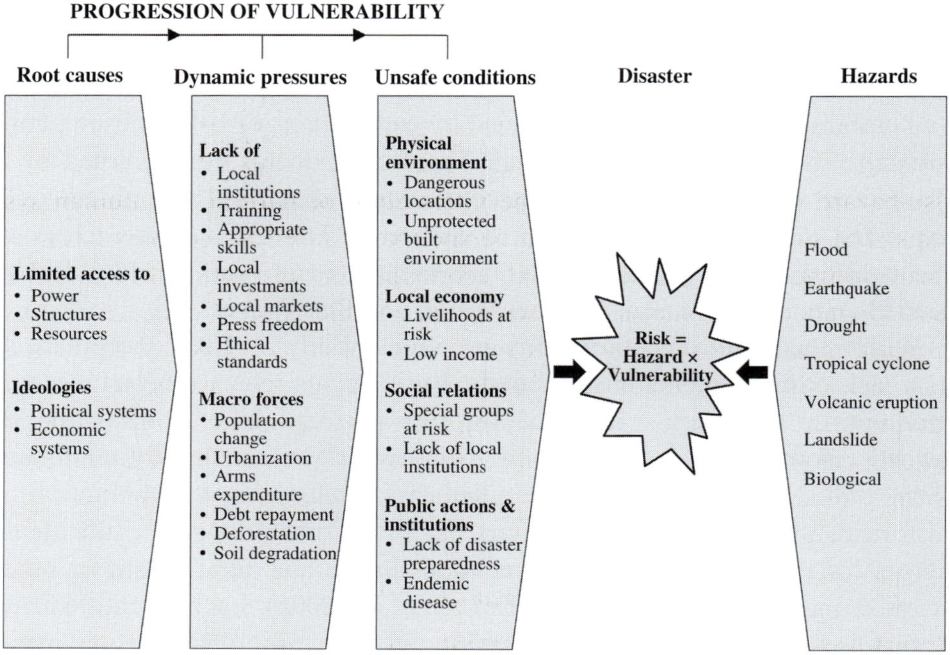


Fig. 4.2. Disaster pressure and release model
*Source*: Adapted from Blaikie et al. (1994).

to hazards. Emblematic of the political ecology viewpoint, the Disaster Pressure and Release model (PAR) highlights a sequence of societal conditions that determine disaster vulnerability (Blaikie et al., 1994). The components and relationships of the PAR are shown in Figure 4.2.

In the PAR, human vulnerability and physical hazards are opposing and additive pressures that combine to produce disasters. Social vulnerability arises as the result of a progression that proceeds from root causes through dynamic pressures to unsafe conditions. Root causes are societal scale and interrelated political, economic, and demographic structures that establish and sustain power relationships, and govern the allocation of resources. These driving forces are often spatially and temporally distant from the hazard event. Dynamic pressures are institutional, social, and environmental processes that channel generalized root causes into environmental degradation (Kasperson et al., 1995) and specific unsafe conditions. Unsafe conditions are the physical, economic, social, and managerial factors that determine differential exposure to hazards at the local scale. The progression of vulnerability combines with hazard severity to exert pressure on people through disaster impacts and constraints on coping capacity. Only by reducing vulnerability can the pressure of disaster risk be released.

With a focus on the social construction of vulnerability, the PAR is well suited to analyze chronic, slow onset, and spatially diffuse hazards (e.g., drought, climate change) that are more difficult to address with adjustments to the natural events system. Within political economy research, it is often equated with extreme natural events themselves. This is a departure from the risk-hazard conceptualization as the intersection of natural and human systems. The strength of the structural perspective of political economy lies in its focus on underlying political, social, economic, and institutional factors that lead to vulnerability, inclusion of scale, and consideration of concurrent stressors other than natural events that may compound vulnerability. However, it has been criticized for downplaying physical aspects of natural-human systems interactions and underemphasizing system feedbacks common in human adjustment models (Turner II et al., 2003). Other critics argue that the PAR is well suited for descriptive analysis, but less amenable to quantification.

### *4.2.3 Hazards of Place*

The Human Adjustment model and PAR share a focus on system-level interactions between people and environment as the source of vulnerability. But questions remain about the vulnerability of what, and the vulnerability of whom? Vulnerability is frequently described in the academic literature as a dynamic phenomenon that can vary significantly across time and space. Analysis at the system level can provide a good understanding of physical and social causal processes, but not necessarily how they vary spatially and manifest at local scales. To fully understand hazard effects and associated human adjustments the analysis should expand from a focus on single hazards to all hazards that affect a place (Hewitt and Burton, 1971).

The Hazards of Place model (Cutter, 1996) addresses this concern, combining vulnerability from biophysical and social systems to produce vulnerability specific to a particular place and time (Figure 4.3). In the model, risk (defined as hazard sources, probability, and consequences) and hazard mitigation (defined as planning, structural, and technological improvements) interact to determine the initial hazard potential for a place. This is roughly analogous to the ideas of natural system characteristics and human use system adjustment in the Human Adjustment model. Biophysical vulnerability is generated by the translation of hazard potential through physical characteristics such as elevation, location, and threat proximity, which comprise the geographic context. Likewise, social vulnerability is produced via the filtering of the hazard potential through social characteristics, such as socioeconomic conditions, risk perception, and coping capacity, which collectively

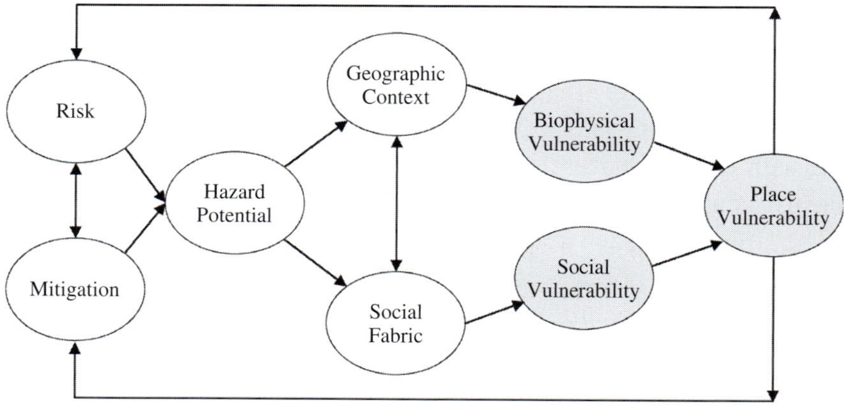


Fig. 4.3. Hazards of place model
*Source*: Adapted from Cutter (1996).

constitute the social fabric of a place. The social fabric is then distilled into place-specific characteristics of marginalized populations (Wisner, 1998) and used to measure social vulnerability. Biophysical and social vulnerability dynamically interact to create the vulnerability of a place, which feeds back to influence future states of risk and mitigation.

"Place" in the Hazards of Place model is not tied to a particular scale, but is intended for application at the community level. Such analysis allows for the comparison of hazard vulnerability of different locales, while revealing how factors contributing to vulnerability combine in different ways depending on geography and scale. The Hazards of Place model has typically been implemented in a Geographic Information System (GIS) by spatially integrating biophysical and social vulnerability. This produces results that are explicitly spatial, but doing so can omit vulnerability characteristics and interactions that are difficult to quantify (e.g., social capital) or have limited spatial variation (e.g., governance).

One of the first implementations of the Hazards of Place model was a case study of Georgetown County, South Carolina (Cutter et al., 2000), which mapped indicators of social and natural hazard characteristics. The investigators spatially aggregated frequencies of historical hazard events to create a composite map of biophysical vulnerability. Then they developed a composite index of social vulnerability at the sub-county scale using census demographic data. The social and biophysical maps were combined and the values summed to produce a composite map representing overall place vulnerability. This place-based approach using geospatial analysis has been since applied in numerous studies of integrated vulnerability (Chakraborty et al., 2005; Collins et al., 2009; Koks et al., 2015).

### 4.2.4 Other Vulnerability Frameworks

The evolution in conceptual frameworks has helped shape understanding the social dimensions of vulnerability to natural hazards. However, other frameworks have also been influential. The Social Amplification of Risk model describes how psychological, social, and cultural processes interact with hazard events in ways that amplify or attenuate public perception of risk (Kasperson et al., 1988). Risk perception can modulate the level of social and economic impacts from hazards, potentially limiting response even when capacity exists (Adger, 2006).

The social-ecological systems perspective posits natural and human systems as a highly dynamic, tightly coupled system in which people and the biophysical environment are constantly interacting and co-evolving (Turner II et al., 2003). Central research questions involve who and what are vulnerable to environmental change, how to identify thresholds that signal change, how changes are attenuated or amplified by human and environmental conditions, and the degree to which resilience is a useful concept for developing proactive strategies for vulnerability reduction.

The Bogardi/Birkmann/Cardona (BBC) conceptual framework adopts a sustainability structure, situating exposure, susceptibility, and coping capacity among environmental, economic, and social dimensions (Birkmann, 2013a). Through temporal feedbacks, the model highlights emergency management interventions for reducing impacts, and it is intended to be applied to specific hazards. The Methods for the Improvement of Vulnerability Assessment in Europe (MOVE) framework extends the BBC, adding cultural and institutional dimensions of vulnerability to the analysis (Birkmann et al., 2013). Under MOVE, interactions between natural hazards and society produce risk, which is amplified or ameliorated over the long term by risk governance and adaptation.

How vulnerability is defined has a major effect on how it is addressed. This is why conceptual frameworks matter. Frameworks describe the most influential causal factors and the relationships among them. Most vulnerability frameworks share the human ecology underpinnings of the Human Adjustment Model: hazard impacts are a result of human-environmental interactions. Over time, conceptual frameworks of vulnerability evolved toward greater detail and complexity, inclusion of cross-scalar effects and temporal feedbacks, and links to resilience, sustainability, and adaptation. However, even as conceptual understanding of vulnerability processes increases, challenges remain in operationalizing conceptual frameworks into vulnerability assessments.

## 4.3 Assessment and Measurement

In transitioning from conceptual framing to assessment, much more effort and resources have been invested in understanding physical processes underlying hazards vulnerability than the array of social determinants. Focusing on social vulnerability, measurement is often a precursor to the development of strategies to reduce hazard risk. Measurement first requires the ability to identify characteristics that determine a population's risk, as well as an understanding of what enhances or reduces the ability of people to respond to and recover from environmental threats. Translating this understanding of the social fabric of communities into quantitative metrics of social vulnerability is the embodiment of a central aspect of the Hazards of Place model.

Social vulnerability is highly multidimensional and constituted by both tangible and intangible characteristics. This multidimensionality makes it difficult or impossible to represent the concept with a single equation or universal set of metrics across scales and hazards (Birkmann, 2014). As a result, there are few consensus guidelines for measuring social vulnerability. There is agreement, however, that quantitative measurement of the concept involves a number of stages including the selection or compilation of relevant data, data standardization, and aggregation of data to derive a final value (Tate, 2012). Due to difficulty with direct measurement, proxy measures are selected to convey the overall capacity of populations to prepare for, respond to, and recover from damaging events. There is a rich tradition of research focused on the factors that increase or decrease the social vulnerability of populations.

This section provides insights into techniques used to understand, measure, and communicate the social vulnerability of populations to hazards and disasters. It is not intended to be an exhaustive review of the literature related to measuring social vulnerability. Rather, we selected methodologies most relevant to the development of composite indicators associated with social vulnerability frameworks discussed in the preceding section. An indicator is a quantitative or qualitative measure derived from observed facts that can simplify and explain a complex reality (Freudenberg, 2003). A composite indicator (index) is a mathematical aggregation of individual indicators or thematic sets of indicators that represent different dimensions of the concept being measured.

### *4.3.1 Deductive Models*

A leading characteristic distinguishing composite indicators of social vulnerability is the structural design, which includes deductive, inductive, and hierarchical arrangements (Tate, 2012). Deductive approaches are top-down and

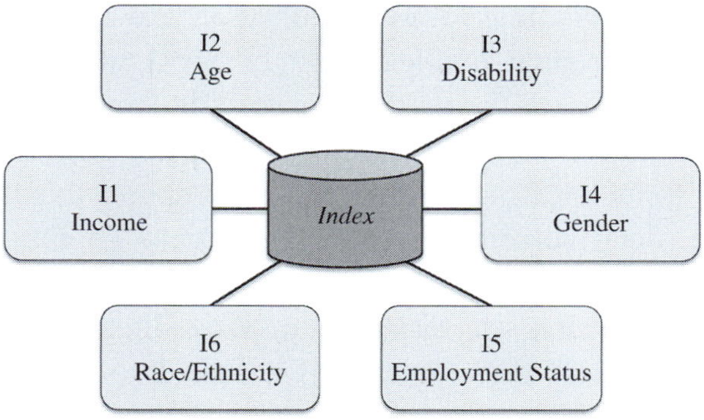


Fig. 4.4. Deductive approach for social vulnerability assessment
*Source*: Adapted from Tate (2012).

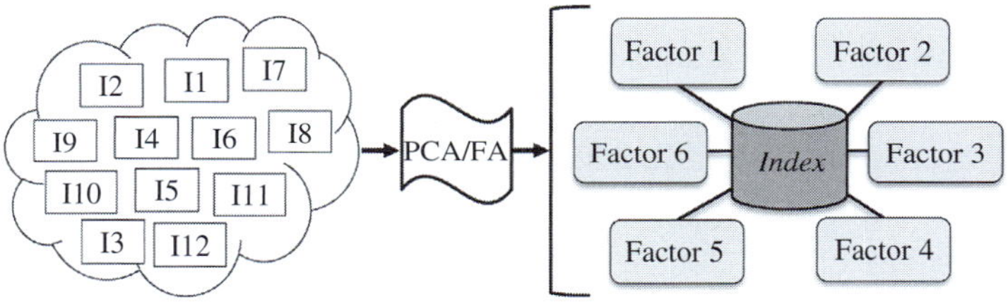


Fig. 4.5. Inductive approach for social vulnerability assessment
*Source*: Adapted from Tate (2012).

they heavily emphasize the theoretical frameworks upon which their variables are selected. The most important dimensions of social vulnerability are identified *a priori*, with typically ten or fewer variables that are standardized and aggregated into an index (Figure 4.4).

### *4.3.2 Inductive Models*

Inductive approaches begin with a larger indicator set (typically 20 or more) that is reduced to a smaller set of values that helps to explain the social vulnerability across a given area. This is accomplished using a Principal Components Analysis (PCA) and/or a Factor Analysis (FA) to decompose a large dataset into a smaller number of uncorrelated factors that explain the variance within the data. The resultant factors are aggregated to build an index (Figure 4.5).

Inductive approaches for modeling social vulnerability were popularized by the Social Vulnerability Index (SoVI, Cutter et al., 2003). The SoVI is largely based on social dimensions identified in the PAR and Hazards of Place models. The original configuration of SoVI utilized between 29 to 42 variables, depending on the analysis scale. The variables are input into a Factor Analysis that estimates the variability that the original input data has in common due to its intercorrelations, and decomposes this data into factors that explain the social vulnerability of a study area. To ensure that the sign of the factor loadings corresponded with their known influences on social vulnerability, factor scores are scaled by making slight adjustments to their directionality. A positive directionality is assigned to all factors known to increase social vulnerability (race and poverty, for example), and a negative directionality is assigned to factors known to decrease vulnerability (indicators such as wealth). This is accomplished by multiplying the loadings by −1, since replacing a component with its inverse subtracts from the overall index. All components are then summed and mapped as standard deviations from the mean allowing for a quantified and visual representation of how social vulnerability varies across space.

Since its inception, the inductive approach for social vulnerability assessment has been expanded to address questions regarding scale (Fekete et al., 2010; Schmidtlein et al., 2008), temporal change (Cutter and Finch, 2008), specific hazards (Rygel et al., 2006; Schmidtlein et al., 2011), and integration of social vulnerability with physical vulnerability (Burton and Cutter, 2008; Finch et al., 2010). Despite its usefulness in reducing data to factors that potentially explain the social vulnerability of an area, however, the inductive approaches rely on subjective decision making at critical stages which could lead to uncertainties within the model outputs (Tate, 2013).

### *4.3.3 Hierarchical Models*

Hierarchical models typically employ 10 to 20 indicators aggregated into pillars that share the same underlying dimension (e.g., special needs populations, economy, institutional factors). The pillars are then aggregated to create the index (Figure 4.6). Hierarchical models thus require a greater level of theoretical organization than purely deductive models. Hierarchical models may rely on participatory approaches and/or expert knowledge both for the selection of variables and for weighting them (Bankoff et al., 2004). Participatory approaches require more time and resources than the deductive or inductive methods, but the increased use of stakeholder-selected indicators and expert-derived weights may result in an index with greater transparency, acceptability, and incorporation of local context.

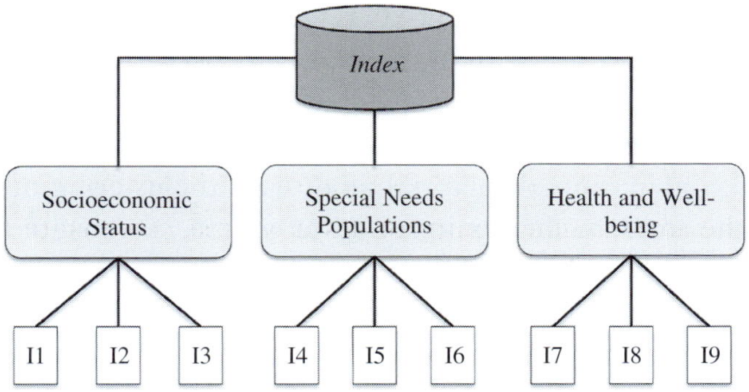


Fig. 4.6. Hierarchical approach for social vulnerability assessment
*Source*: Adapted from Tate (2012).

### *4.3.4 Social Vulnerability Profiling*

The models described thus far share a design goal of aggregating indicators to measure the spatial variation in the magnitude (e.g., high, medium, low) of social vulnerability. A different approach is to model the spatial variation of the dominant vulnerability drivers. The result is spatially varying typologies or profiles of social vulnerability (Tuccillo and Buttenfield, 2016). Emblematic of this approach is the Spectroscopy of Vulnerability (Rufat, 2013), which directs attention to interactions among indicators (rather than indicator aggregation) to identify social vulnerability profiles.

The aim of profiling is to identify (1) locally convergent characteristics that give rise to social vulnerability, (2) the intersectionality of characteristics that generate adverse impacts, (3) clusters of different vulnerability profiles, and (4) why some places might be more vulnerable than others. The outputs are well suited to inform the implementation of targeted mitigation policies. The profiling approach begins by extracting the most relevant vulnerability drivers for a given case study according to the interactions described by the PAR and Hazards of Place frameworks. It then reduces the selected variables to around 15 to 20 based on statistical correlations, seeking balance in the count of vulnerability dimensions (i.e., around five demographic, five socioeconomic, and five cultural or institutional variables).

Vulnerability profiles are generated through a Principal Component Analysis (PCA) on the input demographic data, followed by application of a hierarchical clustering (HAC) method. The factor coordinates of spatial units in the first components extracted from the PCA (i.e., components explaining around 75 percent of the combined variance) are used as a distance matrix for the classification (HAC). The HAC is a nonspatial hierarchical clustering algorithm (according to Ward's method) without the use of weights. The optimal partitioning is determined without *a priori* expectations. The larger thresholds in Ward's level indexes point to the ideal number of clusters. The combination of factorial analysis and clustering makes it possible to eliminate random fluctuations and to obtain more stable partitioning and compact groups of vulnerability profiles. The output are three to seven relative vulnerability profiles that are mapped alongside a visual representation of the interactions specific to each profile.

Vulnerability profiling requires interpretation of the interactions of each relative vulnerability profile. It also requires explaining why some places might be more vulnerable than others, making it a more holistic approach than the inductive and deductive methods. In that sense, the social vulnerability profiles provide scientists and practitioners a tool to associate natural hazard risk with social vulnerability factors as they diverge across space.

### *4.3.5 Qualitative Case Study Approaches*

There is considerable diversity of approaches for social vulnerability analysis (Fekete, 2012). The models described thus far include only top-down quantitative approaches. However, qualitative methods can also be employed to assess social vulnerability in case studies that are in-depth and bottom-up. They encompass a set of context-oriented methods that typically include structured interviews and focus groups. Pre- and post-disaster case studies attempt to uncover drivers of social vulnerability. These driving factors are often described in terms of the real-world processes from which vulnerable conditions are derived. Here research is aimed at better understanding of the complexity of the social system being evaluated. Because of the time-consuming nature of a structured interview, these types of qualitative studies typically occur at the community or household scale, making the findings highly context-dependent. Qualitative methods allow for:

- Collecting risk information from experts or community members
- Communicating risk and social vulnerability with local stakeholders
- Applying tools to facilitate local use of social vulnerability information to understand the risk of their communities, identify existing strengths and weaknesses, and establish resilience goals corresponding to the needs and capacities of communities
- Assisting local stakeholders to strengthen existing vulnerability reduction strategies

## 4.4 Case Study: Measuring Social Vulnerability for Hurricane Sandy (USA)

To illustrate the process of measuring social vulnerability, this section presents a case study based on Hurricane Sandy. The input data were collected from the US Census Bureau and the US Federal Emergency Management Agency (FEMA) to evaluate predictions from a social vulnerability model with a real-world disaster outcome. The analysis adopts multiple approaches to model pre-storm characteristics of affected populations from Sandy and to examine interactions among social vulnerability drivers within the respective models. We then compare the results against observed outcome data from Hurricane Sandy.

### *4.4.1 Hurricane Sandy*

"Superstorm Sandy" was the deadliest and most destructive event of the 2012 Atlantic hurricane season. At least 233 people were killed along the storm path, and direct economic losses were estimated at US$75 billion, a figure surpassed in the USA only by Hurricane Katrina in 2005. Hurricane Sandy affected twenty-four states, with severe damage in New Jersey and New York. The storm surge hit New York City, flooding streets, tunnels, subways, and cutting power around the city. The East River overflowed its banks, flooding Lower Manhattan with a water depth ranging from two to six meters (seven to twenty feet, Figure 4.7). The storm severely damaged or destroyed approximately 100,000 homes on Long Island with more than 2,000 homes deemed uninhabitable there. There were 71 Hurricane Sandy-related deaths in the state of New York. Damage in the state was estimated at US$42 billion. At the peak of the storm three million residents were without power. There were 43 Sandy-related deaths in New Jersey, with a direct economic loss estimated at US$37 billion.

Exposure and outcome data for New York and New Jersey were obtained from FEMA. The exposure data includes water depth over ground on October 31, 2012, as well as calculations of mean water depth and inundated areas during the storm. Sandy outcome data were used to develop a proxy impact measure at the census tract scale of geography, and includes loss estimates based on FEMA verified loss (HUD, 2015), total FEMA-approved applicants for flood compensations (individual assistance), total affected uninsured renters, and the number of affected nonseasonal housing units (Table 4.3). These variables were normalized using sociodemographic data from the 2008–2012 American Community Survey (ACS) from the Census (Table 4.4). The normalized data were used as outcome measures (dependent variables) to compare social vulnerability indicators with the real-world impact outcomes.

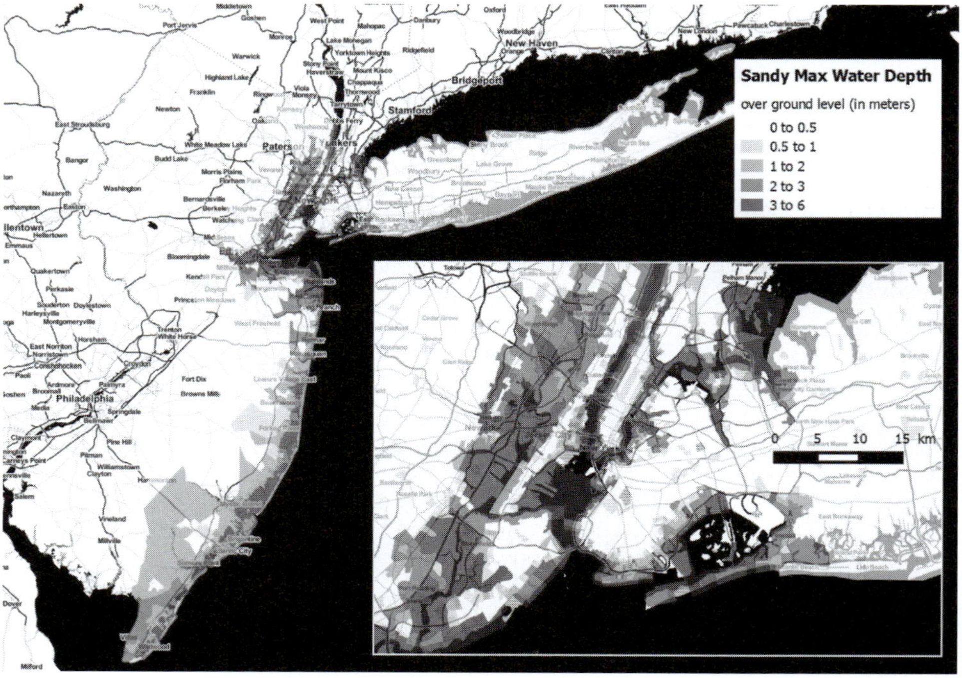


Fig. 4.7. Hurricane Sandy flood depths in New York and New Jersey After FEMA (2012) (A black and white version of this figure will appear in some formats. For the color version, please refer to the plate section.)

Table 4.3. *Sandy outcome indicators. After HUD (2015)*

| Empirical outcome | Normalization variable | Outcome variable |
|---|---|---|
| FEMA verified loss | Median house value | Relative loss |
| Applicants approved for FEMA individual assistance | Total population | Percent affected people |
| Nonseasonal housing affected | Total rental housing units | Percent affected nonseasonal housing |
| Uninsured renters affected | Total rental housing units | Percent affected uninsured renters |

### *4.4.2 Comparative Analysis*

We assessed social vulnerability using three different approaches for empirical measurement: (1) the SoVI, (2) a hierarchical model with variable selection and weighting based on expert knowledge, and (3) vulnerability profiling. For comparison, the same variable set was applied to each social

Table 4.4. *ACS (2008–2012) SoVI variables*

| Variable name | Description |
|---|---|
| POPDENS | Population density |
| AGE | Percent age dependency (under 5, over 65) |
| NURSRES | Percent nursing home residents |
| FEMALE | Percent female population |
| FHHOLDS | Percent female-headed households |
| FEMLBR | Percent female employment |
| FAMMARR | Percent children in married families |
| BLACK | Percent African-American population |
| NATAM | Percent Native American population |
| ASIAN | Percent Asian population |
| HISP | Percent Hispanic population |
| ESLANG | Percent English as a second language |
| EDU12LES | Percent adult educational attainment less than Grade 12 |
| PERCAP | Per-capita income |
| RICH200K | Percent annual income > US$200,000 |
| POVTY | Percent poverty |
| UNEMPLOY | Percent unemployed |
| SSBEN | Percent social security income |
| EXTRACT | Percent extractive sector employment |
| SERVICE | Percent service sector employment |
| NOAUTO | Percent no automobile |
| MOHOME | Percent mobile homes |
| VACANT | Percent vacant housing |
| PERPUNIT | People-per-housing unit |
| RENTERS | Percent renters |
| MDGRENT | Median rent |
| MHSEVAL | Median home value |

vulnerability model. The social vulnerability indicators were constructed for all affected census tracts in New York and New Jersey (n = 3,950; Table 4.4).

Two inductive models were created for flooded areas in New York and New Jersey. The first applies equal weights and the second was constructed with a weighting scheme derived from using weights derived from the PCA (OECD, 2008). Variables whose values increase as social vulnerability decreases (FAMMARR, PERCAP, RICH200K, MDGRENT, MHSEVAL) were multiplied by −1 to change their directionality. The data were then standardized and entered into a PCA. The first eight components represented 70 percent of variance. The remaining components with eigenvalues less than 1.0 were omitted. The first seven components corresponded to increasing social vulnerability within an index; whereas the contribution of the eighth component (representing high median rents) was reversed due to a

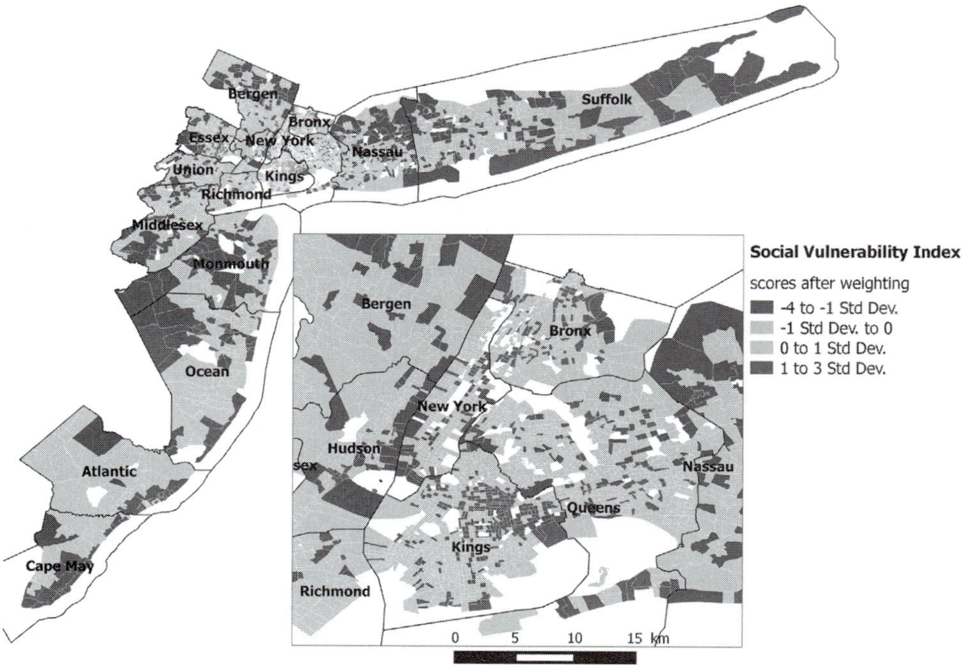


Fig. 4.8. Equally weighted SoVI scores (A black and white version of this figure will appear in some formats. For the color version, please refer to the plate section.)

negative association with the concept. The first seven factor scores for the equally weighted model were therefore aggregated by summation; the eighth component was normalized (subtraction by the mean and division by the standard deviation) and subtracted from the overall index to form the SoVI. The results were mapped in Figure 4.8 to highlight the outliers, with the highest vulnerability in red and the least in blue.

The social vulnerability scores for each census tract provide a comparative assessment within the study area's extent. Of special interest is the clustering of moderately high and high socially vulnerable tracts along the New Jersey coastline in Cape May, Atlantic, Ocean, and Monmouth counties. This zone of socially vulnerable tracts may entertain a particularly higher risk to damaging hazard events than inland tracts not only because of preexisting conditions within populations that foster risk, but also due to a myriad of infrastructure that is exposed to coastal hazard events. High levels of social vulnerability are also found throughout New York's counties, most notably in Kings (e.g., Brooklyn), New York (e.g., mid to upper Manhattan), and the Bronx.

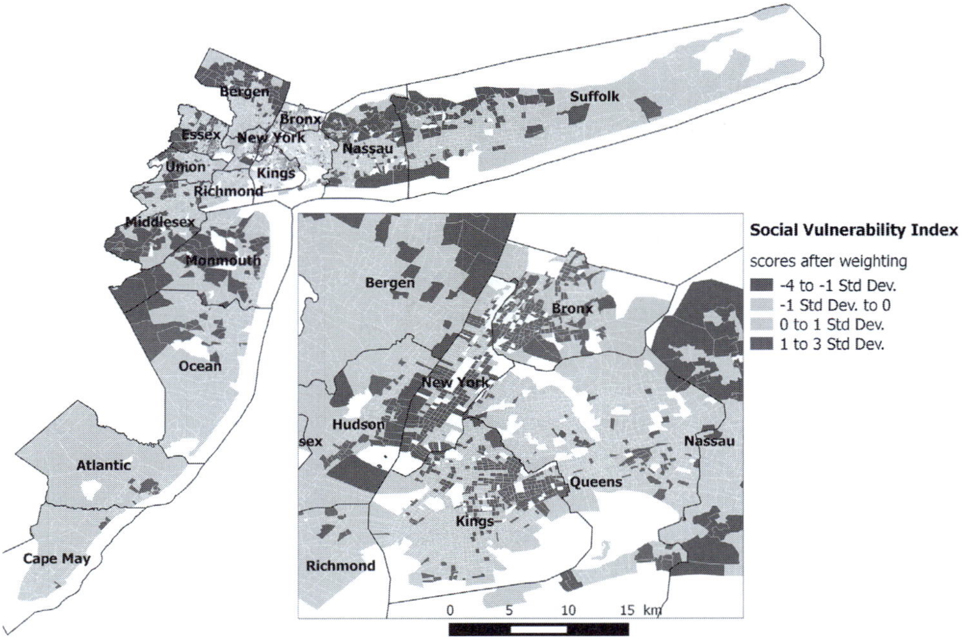


Fig. 4.9. SoVI scores after weighting by PCA eigenvalue. (A black and white version of this figure will appear in some formats. For the color version, please refer to the plate section.)

The second inductive model uses weights defined by the PCA. Here weights were applied to each factor score based on the initial variance explained by each component as represented by their eigenvalue. Seven of the eight components resulting from the PCA were weighted and summed to produce the final weighted SoVI score. The eighth component (Median Rent) was weighted and subtracted from the overall index due to its potentially negative contribution to the social vulnerability concept. The subsequent score was normalized and mapped following the same method for comparison (Figure 4.9). The results show a spatial clustering of moderately to highly socially vulnerable tracts that diverge significantly from the unweighted model. Along the New Jersey coast for instance, fewer tracts are classified as moderately high or highly socially vulnerable. In New York, tracts that are considered highly socially vulnerable in the unweighted SoVI model were reclassified to the other extreme of the measurement spectrum (i.e., to moderately low or low social vulnerability). This divergence demonstrates the sensitivity of the model to indicator weighting where changes in weights have profound implications on the final result.

For our second modeling approach, we constructed a hierarchical model of social vulnerability. To foster comparison with the inductive models, the same input variables (Table 4.2) were used. Individual indicators were aggregated into pillars and the pillars were then aggregated to create the index according to a weighting scheme based on expert knowledge. The expert weights were derived from a survey of hazards and disaster professionals to derive a level of importance for individual social vulnerability indicators (Emrich, 2005). The survey applied a budget allocation approach in which experts were asked to allocate a budget of 100 points among a set of indicators (Table 4.5).

Once selected, the variables were transformed using a Min-Max scaling scheme. The scheme is a method in which each variable is rescaled into a range of 0.0 to 1.0, with a score of 0.0 indicating the lowest social vulnerability rank and a score of 1.0 being the highest. All other values were scaled in between the minimum and maximum values. Similar to the inductive model, variables that rise as social vulnerability falls were adjusted by multiplying by −1 before aggregation. The 27 variables were then aggregated based on the weighting scheme. The final score was normalized by z-score (subtraction by the mean and division by SD) to create the output index score. The results are mapped in Figure 4.10, with high vulnerability outliers depicted in red, and low vulnerability outliers in blue. The spatial patterning is similar to that of the weighted SoVI where the greatest clustering of socially vulnerable tracts is found in New York's Bronx, Brooklyn, and Queens counties. Conversely, low levels of social vulnerability are found along the New Jersey shore in Cape May, Atlantic, Ocean, and Monmouth counties.

Vulnerability profiling is the third method we applied, using the same indicator set (Table 4.2). The starting point was the extraction of the most relevant vulnerability drivers for the case study. To limit collinearity, prevent implicit weighting, strengthen statistical power, and preserve a balance between the different dimensions of vulnerability, the association between variables was tested using a correlation analysis. Following the interpretation of the resulting correlation matrix, 18 variables were utilized in the final model in which all 18 variables (Table 4.6) were input into a PCA to derive components.

All PCA components with an eigenvalue under 1.0 were omitted leaving five components that represent 70 percent of the total variance in the data. The corresponding factor scores for each variable were used as a distance matrix for the classification (HAC), a divisive nonspatial clustering according to Ward's method, without any weighting. The optimal partitioning was determined without *a priori* expectations: a five-cluster partitioning was deemed most robust based on the Ward level index that minimizes the total

Table 4.5. *Expert-derived weighting scheme. After Emrich (2005)*

| Variable | Weight% | Description | Dimension |
|---|---|---|---|
| PerCap | 15.1 | Per capita income | Socioeconomic status |
| QPoverty | 14.3 | Percent families in poverty | Socioeconomic status |
| MedVal | 8.4 | Median value of owner occupied housing | Socioeconomic status |
| QOld | 7.5 | Percent elderly over 65 | Age |
| QKids | 5.7 | Percent children under 5 | Age |
| QCvlUn | 5.1 | Percent civilian labor force unemployed | Socioeconomic status |
| QBlack | 4.9 | Percent African-American | Race and ethnicity |
| NHRes | 3.8 | Nursing home residents | Special needs |
| QSSBen | 3.4 | Social Security recipients | Social dependence |
| QRich | 3.3 | Percent families earning <100K a year | Socioeconomic status |
| Immig | 3.2 | Percent international migration | Migrants |
| RentOcc | 3.1 | Percent renter-occupied housing units | Home ownership |
| QSpanish | 2.9 | Percent Hispanic | Race and ethnicity |
| PPH | 2.2 | Persons-per-household | Family structure |
| FLFP | 1.9 | Percent females in civilian labor force | Socioeconomic status |
| FHH | 1.9 | Percent female-headed households with no spouse | Family structure, gender |
| MedRent | 1.8 | Median gross rent | Socioeconomic status |
| Health | 1.8 | Percent healthcare employment | Economic sector |
| Service | 1.5 | Percent service occupations | Socioeconomic status |
| QIndian | 1.4 | Percent Native-American | Race and ethnicity |
| Primary | 1.3 | Percent employment in primary industry | Economic sector |
| NoHS | 1.3 | Percent no high school education for persons aged 25 years and over | Socioeconomic status |
| LFP | 1.2 | Percent civilian labor force participation | Socioeconomic status |
| Trans | 0.9 | Percent employment in transportation, communications, utilities | Economic sector |
| QAsian | 0.9 | Percent Asian | Race and ethnicity |
| Fem | 0.8 | Percent female | Gender |
| BRate | 0.4 | Birth rate | Population structure |

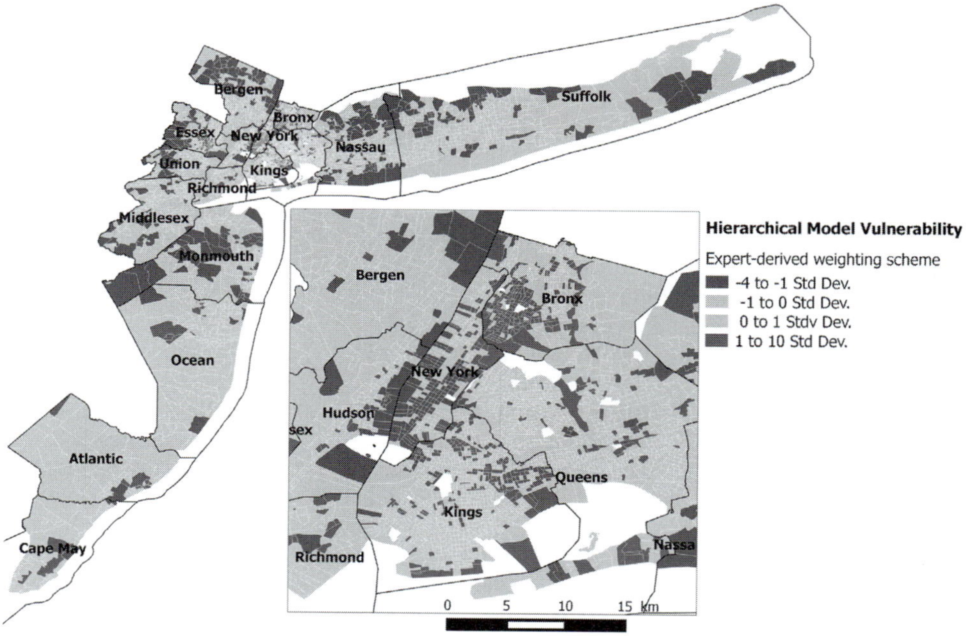


Fig. 4.10. Hierarchical model social vulnerability scores (A black and white version of this figure will appear in some formats. For the color version, please refer to the plate section.)

Table 4.6. *18 remaining variables available from ACS 2008–2012*

| Variable name | Description |
|---|---|
| POPDENS | Population density |
| AGE05 | Percent age under 5, over 65 |
| AGE65 | Percent age over 65 |
| NURSRES | Percent nursing home residents |
| FHHOLDS | Percent female-headed households |
| NWHITE | Percent population non-white |
| ESLANG | Percent English as second language |
| EDU12LES | Percent adults completed <Grade 12 |
| POVTY | Percent poverty |
| UNEMPLOY | Percent unemployed |
| SSBEN | Percent Social Security income |
| NOAUTO | Percent no automobile |
| MOHOME | Percent mobile homes |
| VACANT | Percent vacant housing |
| PERPUNIT | Persons-per-housing unit |
| RENTERS | Percent renters |
| MDGRENT | Median rent |
| MHSEVAL | Median home value |

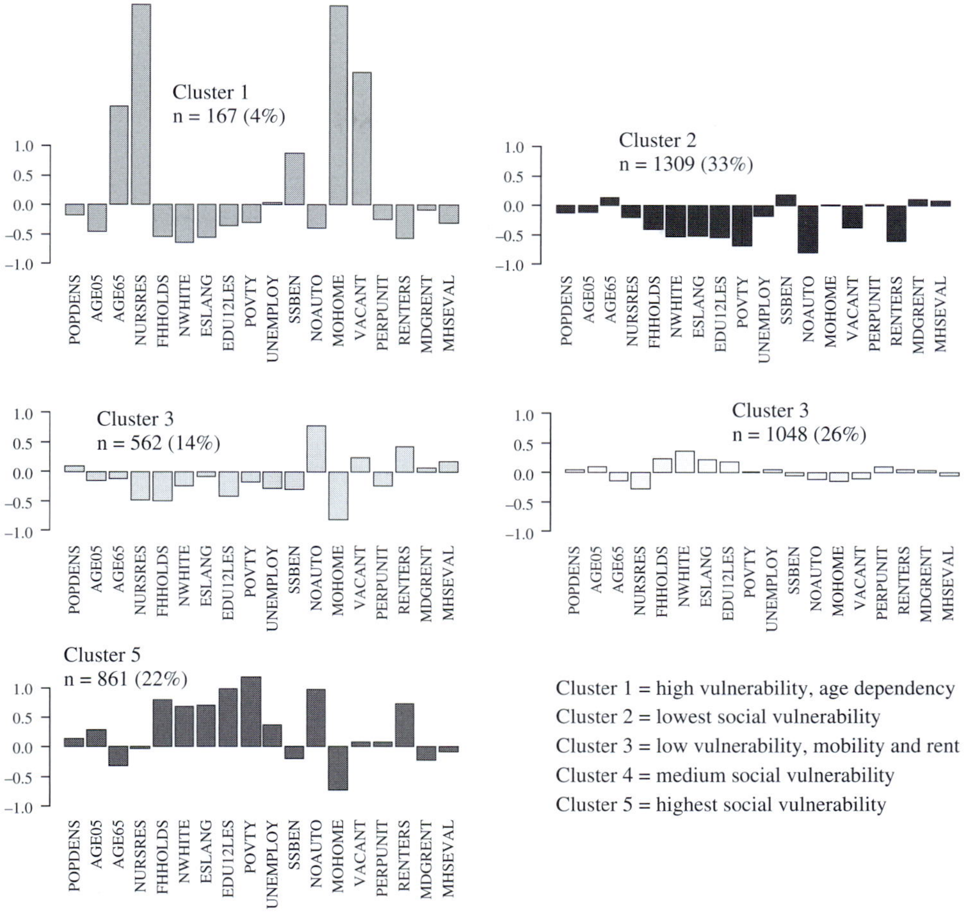


Fig. 4.11. Vulnerability clusters. (A black and white version of this figure will appear in some formats. For the color version, please refer to the plate section.)

within-cluster variance. For each of the five resulting clusters, the 18 initial variables were normalized, comparing mean and SD inside the cluster and on the total population (Figure 4.11).

The vulnerability profiles are interpreted as an over- or underrepresentation of the different variables. Their association or mutual exclusion in each cluster are interpreted regarding the resulting impacts on the strengthening (or reduction) of vulnerability (Figure 4.12). The first cluster (4 percent of census tracts) overrepresents the elderly and nursing home residents, as well as people living in unsafe or mobile homes, in vacant housing, and to a lesser extent people living on Social Security income. This profile is associated with the combination of dependency, housing and income insecurity or

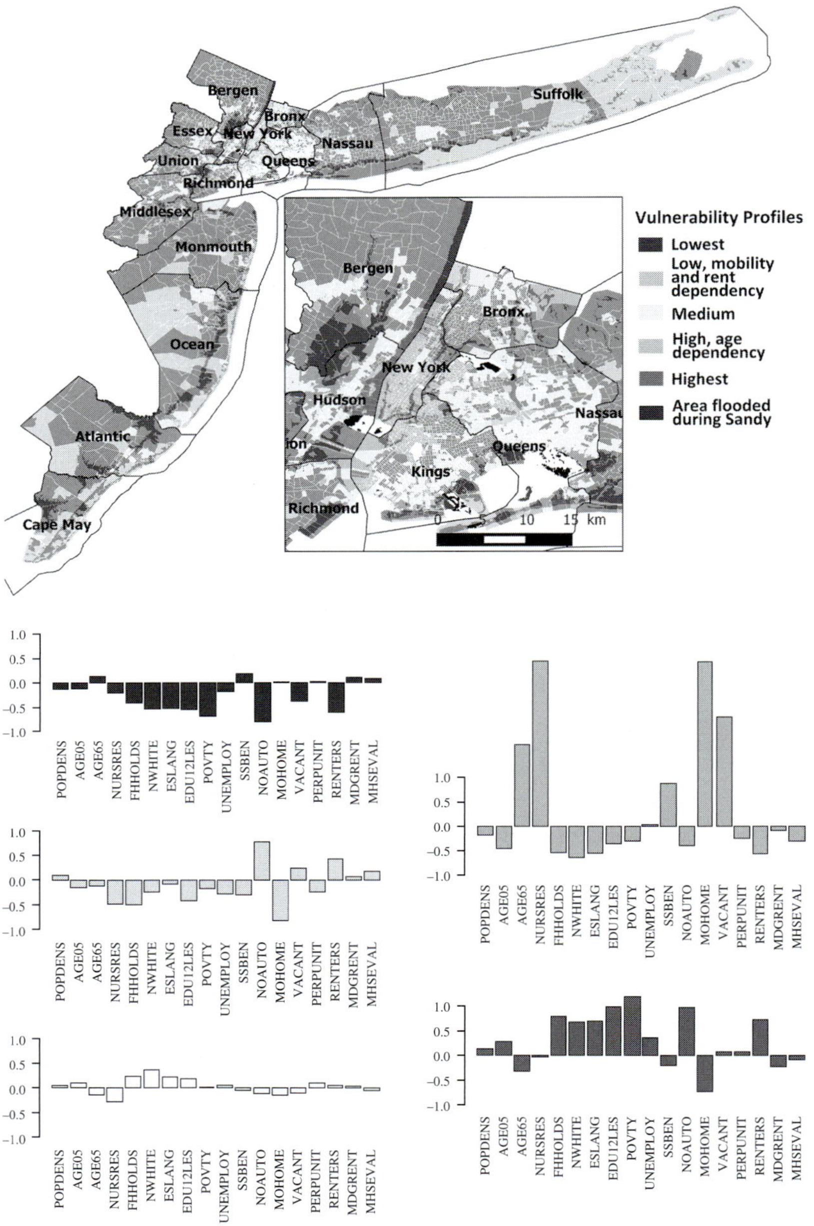


Fig. 4.12. Vulnerability profiles. (A black and white version of this figure will appear in some formats. For the color version, please refer to the plate section.)

deprivation, and was interpreted as high vulnerability marked by age dependency. The second cluster (33 percent of tracts) underrepresents most variables while staying close to the mean for the others. Hence, we interpreted cluster 2 as the lowest vulnerability profile.

The third cluster (14 percent of tracts) overrepresents people without an automobile, and to a lesser extent renters and vacant housing. Meanwhile, cluster 3 underrepresents all other vulnerability indicators. The interactions are limited and the cluster may therefore be interpreted as a low vulnerability profile with a public transportation and rent dependency. The fourth cluster (26 percent of tracts) is close to the average profile, with a minor overrepresentation of female-headed households, low educational attainment, low English proficiency, and a slight underrepresentation of the elderly, and nursing home, unsafe, or mobile home residents. We interpreted cluster 4 as a medium social vulnerability profile.

The fifth and final cluster (22 percent of tracts) reveals the strongest overrepresentation of poverty and transportation dependency, and a stark overrepresentation of renters and most other vulnerability drivers. Despite the low underrepresentation of the elderly, people living in mobile homes or on Social Security income, the interactions are important enough across all vulnerability dimensions to interpret it as the highest vulnerability profile.

### 4.5 Further Thoughts on Social Vulnerability Mapping

The results of the comparison between social vulnerability models are somewhat divergent. On the maps, places assessed as highly vulnerable by one model are characterized as less or least vulnerable by the others, and vice versa. The latter especially applies to the unweighted SoVI where the model's primary components dominate the social vulnerability scoring.

Up until this point, we have provided a rather spatially descriptive assessment of the distribution of the social vulnerability of New York and New Jersey's coastal zones by model type. To test the association between the models further we employed a series of regression analyses to examine the degree to which the social vulnerability models predict a "real world" and adverse natural hazard outcome (i.e., damage impacts from Hurricane Sandy). When associated with Hurricane Sandy outcome indicators using multiple regression, Pearson's product-moment correlation and Kruskal-Wallis non-parametric test, the statistical power for predicting an adverse impact varies by model. The statistical significance of those tests and regressions are summarized in Table 4.7.

When the inductive models are compared with outcome variables, the equal weights model has less statistical significance than the weighted scheme. Overall the inductive models have a weak relationship with observed Hurricane Sandy outcomes, even when controlling for exposure measured by water depth during the flooding. The hierarchical model performs better with an even greater statistical power. Only the vulnerability profiles have a

Table 4.7. *Vulnerability model versus Sandy outcomes (p-value)*

| Statistical significance | SoVI | Weighted SoVI | Hierarchical | Spectroscopy |
|---|---|---|---|---|
| Flood verified loss | | | * | ***** |
| Flood verified loss normalized by house value | * | * | ** | ***** |
| Share of affected people | * | ** | *** | ***** |
| Share of damaged houses | | | * | ***** |
| Share of uninsured renters affected | | | ** | ***** |
| Flood verified loss normalized by house value controlled by flood depth | * | ** | ** | ***** |
| Share of affected people controlled by flood depth | ** | *** | *** | ***** |
| Share of damaged houses controlled by flood depth | | | ** | ***** |
| Share of uninsured renters affected controlled by flood depth | | | *** | ***** |

p-value *0.001 **0.00001 ***$10^{-7}$ **** $10^{-10}$ ***** $< 10^{-15}$

highly statistically significant association with all Sandy outcomes, even when normalized by population or controlled by exposure. FEMA verified losses were normalized by the average house value in each tract; the affected people, damaged housing, and uninsured renters were normalized by their total population in each tract to express them as a proportion. The exposure was controlled with maximal water depth over ground during Sandy.

## 4.6 Conclusion

The measurement of social vulnerability as a tool for disaster risk reduction continues to gain importance in the face of increased exposure to risk from natural hazards. An improved understanding of how social vulnerability indicators and indices conform to real-world disaster outcomes could benefit both model development decisions, and how social vulnerability indicators are interpreted during decision making and public policy development processes.

This chapter has profiled the principal theoretical models of social vulnerability and the prominent construction approaches for defining social vulnerability models (inductive, hierarchical, profiles). Each model was harmonized to render them comparable, and statistically associated with normalized outcome measures from Hurricane Sandy. The findings highlight that the results of the model types are somewhat convergent when compared spatially.

However, when the models were statistically associated with the outcome indicators, a variation in model performance was observed. The inductive model, based on the factor analytic approach, displayed the least amount of explanatory power. This may be partially due to the data reduction process in which individual variables are reduced to create uncorrelated factors. The hierarchical model statistically performs slightly better. In terms of variable selections, both models have a strong statistical association that focuses on wealth (strong negative correlation) and poverty (strong positive correlation), gender, education, and race. The vulnerability profiles exhibit the highest statistical association with all outcomes measures.

Generally speaking, indicators can only explain the social vulnerability of a population to an extent and cannot explain the whole picture. Moving forward, a primary challenge is to derive a set of explanatory metrics (whether qualitative or quantitative) that is parsimonious yet explains the phenomenon as clearly as possible. Research needs for improving social vulnerability assessments include:

- Scale of analysis. Our case study chapter was conducted at the US Census Tract level of analysis. It is important to consider how changes in scale might influence the results. At minimum, research should be conducted to better understand the association between social vulnerability metrics and impacts at different scales.
- Variable selection. The variable selection process for our case study was based on the indicator selection for the SoVI because it has become a leading methodological approach. The SoVI variable set, however, was first the result of a multivariate analysis for which the data selection was extracted from the 1990 US Census. To test for consistency and the application of the SoVI variables to an ever-changing US landscape, it would be beneficial to derive a parsimonious data selection based on recent data.
- Hazard context. Context includes the geographic setting of the disaster, pre-existing social, economic and political conditions, hazard type, degree of exposure, scale of impacts and responses, and the disaster phase. More research is needed to incorporate context into social vulnerability measures.
- Conceptual framework. Another challenge to social vulnerability research is the end-to-end application of the conceptual frameworks designed to explain the topic. The application of frameworks like the PAR have been extensively documented. There is, however, limited available information of how to operationalize conceptual frameworks on social vulnerability. There is a demand for in-depth investigations of social vulnerability evidence, conceptual components, and measurement methods.

# 5

# Economic Vulnerability

THOMAS THALER AND BRENDEN JONGMAN

## 5.1 Introduction

Natural hazards have caused significant economic losses on a global scale in past decades; it is therefore urgent to find effective adaptation and mitigation strategies[1] to reduce future impacts (Hall et al., 2012; Noy, 2009). Disaster risk is understood to be a product of the occurrence of naturally occurring hazard events, such as a flood or an earthquake (hazard); the population, buildings and economic assets located in hazard-prone areas (exposure); and the susceptibility of the exposed elements to the natural hazard (vulnerability). Each of these three components is dynamic, and is influenced by human development, population and economic growth, natural climate variability (e.g., El Niño), and anthropogenic climate change.

Assessing economic vulnerability is crucial for the scientific community as well as for practitioners and private companies (mainly insurance and re-insurance businesses[2]) in terms of defining optimal risk management strategies and analyzing individual and societal risk levels (Dutta et al., 2003; Jongman et al., 2012; Merz et al., 2010; Meyer et al., 2013). Estimating economic vulnerability[3] has therefore received a great deal of attention in the past 20 years (Chang, 2001; Cole, 1995; Hallegatte et al., 2013; Jongman et al., 2012; Kreibich et al., 2014; Luino et al., 2009; Merz et al., 2010; Meyer et al., 2013; Penning-Rowsell et al., 2005; Shinozuka et al., 1998). The roots of such analysis stretch back to the 1930s and 1940s with the introduction of the 1936 US Flood Control Act, which requested the use of cost-benefit analysis in flood risk management (Congress, 1936; Hanley and Spash, 1994), and more specifically in terms of original works published by Gilbert White in 1936 and 1945, respectively.

White (1936) discussed the aim of an economic vulnerability assessment: it is supposed to ensure a more objective-technical perspective within the decision-making process, but especially to implement effective adaptation and mitigation measures. Moreover, he reasoned that the key task for natural hazards management is the assessment of benefits[4] from adaptation and mitigation measures, but also to compare the costs to determine optimal management. Therefore, stage-damage curves have been proposed, which basically showed the correlation between the impact of an event and the loss resulting from it (Green, 2003; Jongman et al., 2012; Messner and Meyer, 2005; Messner et al., 2007; Penning-Rowsell and Chatterton, 1977; Penning-Rowsell et al., 2005; Smith, 1981; White, 1945). The evaluation of the benefits is linked to natural hazards parameters (such as flow depth, pressure, etc.) to relative losses[5] (=loss ratio) or total damage losses (=monetary values; White, 1945)[6].

Stage-damage curves have been widely used and extended over time. However, there is no standard methodological agenda. Each country (and academic school) uses different approaches concerning how to assess losses (ex-post versus ex-ante approach) or which losses have to be included within the calculation (e.g. building envelope or content) etc. Additionally, the interpretation of the curves is differently used within the literature. For example, the US damage tool HAZUS predicts the potential losses as a proportion of the value of the element at risk (Green, 2003; Scawthorn et al., 2006a, b). Another approach includes the damage losses in relation to land use to distinguish damage between rural and urban areas (Amadio et al., 2016; Carrera et al., 2015; Green, 2003; Jongman et al., 2012; Kreibich et al., 2010; Messner and Meyer, 2005). The latest development foresees the application of 3D city models in the context of economic vulnerability assessment as a means to model the shape of the houses at a larger scale (Amirebrahimi et al., 2016; Biljecki et al., 2015; Varduhn et al., 2015). The greatest challenges can nevertheless be found in the lack of data (and data series) and the quality of that data (White, 1936). This causes high uncertainty in the results (Noy, 2009). Vulnerability and risk assessments are conducted at all spatial scales, from local to global. Such assessments have different levels of detail and accuracy, and serve vastly different purposes. Local assessments are often highly detailed. An example of such an assessment is the modeling of hurricane-induced flood risk in New York City (Aerts et al., 2014). This assessment uses detailed vulnerability functions on a building level, and analysed the effect of possible building-level mitigation measures on the reduction of risk. Such a local assessment can directly support policy decisions on flood protection and city planning. An example of a national-scale assessment is the cost-benefit analysis of nation-wide flood protection (Kind, 2014). The author analyzed potential flood damage using a national flood-damage model, which is based on vulnerability functions specific for different types of land-use, and computed the economically efficient standard of flood protection. Such a national assessment can be used for prioritization of investments and first-order budgeting efforts.

On a regional level, the World Bank has contracted risk assessments for a range of countries in Eastern Europe and Central Asia, using globally applicable flood and earthquake models in combination with local data where possible. These assessments are often based on individual and therefore hazard-specific vulnerability functions (depending on the hazard type), and are therefore less detailed. The purpose of a regional assessment is to create consistent risk profiles across various countries, and identify areas of high risk within the region.

On a global scale, the UNISDR publishes a biannual Global Assessment Report (GAR), which presents the results of global models on various hazards (UNISDR, 2015b). The global assessment is limited to exposure analysis, i.e. the assessment of the total population and assets at risk, without attempting to include vulnerability. The purpose of such a global assessment is to show the global trends and hotspots, and to highlight priority countries for international risk management efforts. Other global flood risk models, most notably GLOFRIS (Ward et al., 2013; Winsemius et al., 2016), have been attempting to include vulnerability functions in such global assessments, but uncertainties remain high.

The chapter has three main sections: (1) classification of damage types (indirect/direct/tangible/intangible), (2) state of the art of methods in damage assessment, and (3) a case study as example how to conduct economic vulnerability assessments.

## 5.2 Types of Damage

Damages[7] from natural hazard events can be classified (see Table 5.1): direct and indirect losses distinguished between tangible (monetary[8]) and intangible

Table 5.1. *Cost categorization in environmental hazard assessment*

| | Tangible (monetary) | Intangible (nonmonetary) |
|---|---|---|
| Direct losses | Physical damages to buildings, infrastructure, stocks etc. | Damages to physical and/or mental health, loss of life, loss of environmental goods, etc. |
| Indirect losses | Evacuation costs, production losses based on road disruption, emergency costs, etc. | Trauma/psychological inconveniences, loss of trust in public authorities, etc. |

Adapted from Green (2003), Meyer et al. (2013), and Penning-Rowsell et al. (2005).

(nonmonetary) losses (Carrera et al., 2015; Cochrane, 2004; Dutta et al., 2003; Jonkmann et al., 2008; Penning-Rowsell and Chatterton, 1977; Rose, 2004; Smith, 1981).

### *5.2.1 Direct Losses*

Direct-tangible losses result directly from the physical force of a hazard, such as damages to a building or household inventory items, infrastructure, agricultural crops, plants or even livestock (Merz et al., 2010; Meyer et al., 2013; Smith and Ward, 1998). On the other side, direct-intangible losses refer to losses that are difficult to measure, such as the loss of cultural heritage sites or environmental goods, damage to physical and/or mental health (psychological and physiology aspects), human-life, loss of community/neighborhood structure, loss of trust/confidence in public authorities dealing with risk management or loss of memorable and irreplaceable items, such as family pictures (Cammerer et al., 2013; Merz et al., 2010; Meyer et al., 2013; Penninng-Rowsell et al., 2005; Smith and Ward, 1998). Scholars categorize direct losses into primary and secondary direct losses. Primary direct losses result immediately from a hazard event (NRC, 1999). On the other hand, secondary direct losses are the additional losses resulting from cascading effects following a disaster (Gill and Malamud, 2014), such as economic losses through heavy rainfall events following a hurricane event as in the case of Hurricane Andrew (NRC, 1999).

### *5.2.2 Indirect Losses*

Indirect losses are losses caused outside of physical processes (Cochrane, 2004). Indeed, "indirect damages are induced by the direct impacts and may occur – in space or time – beyond the immediate limits of the flood event" (Hammond et al., 2015: 17). As an example, damages to a road network can cause interruptions to businesses without any damages to factories or office buildings. Employers become unable to reach the factory or a facility cannot supply a merchant, both of which affect business production levels (Rose, 2004). Besides negative supply side impacts, hazard events can also cause negative impacts on the demand-side. The temporary closure of a factory also reduces the request of input factors, meaning suppliers outside a road network or even on the other side of the world have to reduce their production (Rose, 2004). Indirect losses are therefore usually calculated as total damages less direct losses, but independent from the mechanism which triggered the impact (Cochrane, 2004).

Table 5.2. *Overview of economic impacts of environmental hazard event on property values*

| Authors | Place | Economic impact |
|---|---|---|
| Shabman and Damianos (1976) | Alexandria, Virginia (USA) | −22 percent |
| Skantz and Strickland (1987) | Houston, Texas (USA) | No effects on the property value, but strong increase of insurance rate |
| Tobin and Montz (1988) | Linda and Olivehurst, California (USA) | Property values decline around 4.2 percent; except high price selling houses – increased their value, because of renovation and repairs |
| Eves (2002) | Sydney (Australia) | −16 percent |
| Bin and Polasky (2004) | Pitt County, North Carolina (USA) | −5.8 percent |

Cochrane (2004) extended the term "indirect losses" to encompass a more holistic view of the damage assessment, rather than focusing exclusively on the negative effects and consequences of production interruptions. He introduced the term "systematic risk" which referred to the impacts on the financial system (shortages of credit loans and compensation payments from insurance sector). Therefore, he distinguished indirect losses between damages to businesses and lifestyle (real ripple effects), financial impacts within the banking sector (bank systemic risk), and damages to the insurance sector (insurance systemic risk). Examples for indirect losses include: permanent evacuation of individuals, disruption within the supply chain and production system (cause due to interruption of road network infrastructure), loss of income (based on spending reduction due to business closure), loss of leisure, lack of communication, and travel costs (NRC, 1999; Penning-Rowsell et al., 2005). Consequently, indirect losses tend to demonstrate a higher negative long-term impact in contrast to direct damages. An example of this would be the economic shrinking effects due to migration of people and/or businesses (Merz et al., 2010) or changes in property values as a result of past events (see Table 5.2; Lammond et al., 2010; Montz, 1992; Tobin and Montz, 1988, 1995).

## 5.3 State of the Art Methodological Approaches in Damage Assessment

### *5.3.1 Direct Losses*

Direct (tangible) loss estimation foresees a wide-range of different methodological approaches to assessing damages. The most common methodological concepts are stage-damage curves (Merz et al., 2010; Meyer et al., 2013), otherwise known as vulnerability or susceptibility function (Totschnig et al., 2011).

Damage assessment usually means to distinguish between ex-post and ex-ante damage evaluation (Molinari et al., 2014; Penning-Rowsell et al., 2005; Smith and Ward, 1998). Ex-post evaluation involves conducting data collection in areas with past flood experiences, where damage losses are collected directly after an event (empirical approach). In contrast, ex-ante evaluation is based on a synthetic approach, where what-if-questions[9] generate data (Merz et al., 2010; Molinari et al., 2014; Penning-Rowsell et al., 2005; Scawthorn et al., 2006a, b). Some damage impact assessments nevertheless combine both approaches, such as the HAZUS model used by US Army Corps of Engineers (Scawthorn et al., 2006a, b) or the RAM model used by Australian Government (NRE, 2000).

The advantages of using ex-post damage assessments include site-specific information on damage losses from past events as well as the assessment of the effects of local structural adaptation measures (Fuchs et al., 2007; Kreibich et al., 2005). However, the ex-post method requires detailed survey techniques and damage report information after an event; these are often impossible to gather or simply unavailable. Further, the information regarding severity or magnitude is also often lacking in loss reports. Lastly, comparability is in many cases also not possible due to the lack of standardization in methodology (Merz et al., 2010; Jongman et al., 2012; Papathoma-Köhle et al., 2015; Smith, 1994). On the other hand, synthetic damage functions have the highest advantage as a means for collecting the losses independently from a past event. The data collection demonstrates a higher level of transferability and standardization compared with other forms (Penning-Rowsell et al., 2005). Even so, the models require a large – and very detailed – data basis, which often is not feasible and sometimes even impossible particularly if the building type is heterogeneous.

Besides the choice of data collection (ex-post/ex-ante), further challenges refer to the use of relative or absolute losses in the assessment (Fuchs et al., 2007; Smith, 1994; Penning-Rowsell et al., 2005; Totschnig et al., 2011). Relative damage functions calculate the losses in relationship compared with the total value (in percentage), in contrast to absolute damage functions that assess the losses in monetary terms in relation to the unit area or building value (Fuchs et al., 2007; Merz et al., 2010; Penning-Rowsell et al., 2005). Relative damage assessments, for example, are common in Austria (Fuchs et al., 2007), in the HAZUS model (Scawthorn et al., 2006a, b) or in Germany within the MURL (2000) and ICPR (2001) models. Absolute damage losses have largely been used in the UK within the Multi-Colored Manual model developed by Penning-Rowsell et al. (2005) and in Australia, such as the Anuflood (NR&M 2002) or RAM models (NRE, 2000). One key advantage in using relative damage functions is the better transferability of the results; such results are independent from actual market and currency changes (Krzystofowicz and Davis, 1983). On the other side, relative damage functions request the value of the objective, causing greater uncertainties in the economic valuation of property value in particular (Merz et al., 2010). On the other hand, absolute damage functions have advantages in directly establishing the monetary value for a given scenario. Nevertheless, the results are dependent on the total value of the potential affected object as well as the difficulties in transferring the results based on currency fluctuations (Daniell et al., 2011; Merz et al., 2010).

A further challenge is the selection of input parameters for assessing losses (Cammerer et al., 2013; Merz et al., 2010), for which there is no comprehensive approach according to the literature (Gerl et al., 2016). Thieken et al. (2005) distinguished between impact (such as water level or velocity) and resistance factors (such as implementation of local protection measures schemes or use of an early warning system). Impact factors represent in this case the physical forces of a hazard event damaging a building. On the other hand, resistance factors involve the capacity of both the building and society in question to react to a hazard event (Merz et al., 2010; Thieken et al., 2005). Despite this, the key parameter in the impact assessment is still the depth of inundation (see an overview in Table 5.3).

Additionally, a few studies use susceptibility functions, especially within other types of natural hazards (like drought hazards, Horridge et al., 2005). The use of other economic models – besides the susceptibility functions – primarily functioned in the calculation of indirect losses.

### *5.3.2 Indirect Losses*

Estimating indirect (or secondary) losses caused by natural hazards necessitates economic models, such as econometric, input-output or general equilibrium models. The differences between methodological concepts are mainly

Table 5.3. *Examples of parameters in damage loss estimation*

| Parameter | Selected references |
|---|---|
| Inundation depth | e.g. White (1945), Smith and Tobin (1979), Handmer (1986), Smith (1994), Nicholas et al. (2001), Dutta et al. (2003), Penning-Rowsell et al. (2005), Scwathorn et al. (2006a, b), Fuchs et al. (2007), Jongman et al. (2012), and Papathoma-Köhle et al. (2015) |
| Flow velocity | e.g. Smith and Tobin (1979), Handmer (1986), Nicholas et al. (2001), Kreibich et al. (2009), and Pistrika and Jonkman (2010) |
| Duration of inundation | e.g. Smith and Tobin (1979), Handmer (1986), Nicholas et al. (2001), Dutta et al. (2003), and Penning-Rowsell et al. (2005) |
| Recurrence level | e.g. Elmer et al. (2010) |
| Contamination | e.g. Smith and Tobin (1979), Handmer (1986), Kreibich and Thieken (2008), Thieken et al. (2008), and Prettenthaler et al. (2010) |
| Timing (night/day; during week/weekend) | e.g. Smith and Tobin (1979), Dutta et al. (2003), and Keiler et al. (2005) |
| Use of building (such as businesses, residential etc.) | e.g. Dutta et al. (2003), Penning-Rowsell et al. (2005), Büchele et al. (2006), Totschnig and Fuchs (2013), and Papathoma-Köhle et al. (2015) |
| Building material | e.g. Nicholas et al. (2001), and Schwarz and Maiwald (2007, 2008) |
| Local protection measures schemes | e.g. Kreibich et al. (2005), Büchele et al. (2006), Thieken et al. (2008), and Kreibich et al. (2010) |
| Early warning | e.g. Chatterton et al. (1979), Smith (1981), Penning-Rowsell et al. (2005), and Pappenberger et al. (2015) |

data format and the level of analysis (time and space). Accordingly, each model has its own demonstrable advantages and disadvantages (Hammond et al., 2015; Meyer et al., 2013; Okuyama, 2009; Rose, 2004). Overall, the literature distinguishes between the five most commonly used methods (post-events survey, econometric models, input-output models, Computable General Equilibrium models and hybrid models) as follows.

First, using post-events surveys at the businesses level helps estimating the losses of past events based on surveys (event analysis). The advantage in this case is the scale flexibility when assessing indirect losses. While this advantage is apparent, the methodological approach often cannot assess and estimate a systematic view of the economic losses/impact of the event being analyzed. In particular, the method is usually unable to show the interdependences between different economic sectors (Kroll et al., 1991; Molinari et al., 2014; Pfurtscheller, 2014; Tierney, 1997).

A second option is contained in econometric models, which analyze the impact of past events within a time-series to quantitatively predict future developments (Herman, 1969; Hochrainer, 2009; Noy, 2009). The key advantage is the statistical robustness of the results. While useful in its robustness, the approach only estimates the total economic impact without attempting to know which sector is most damaged by an event (Rose, 2004).

The third option accommodates for this lack by including the use of input-output models, which assess the monetary losses for each economic sector based on input-output tables. The origin of using input-output models in the natural hazards debate dates back to Harold Cochrane's seminal 1974 work. A key advantage to this approach is the assessment of the interdependencies between the different economic sectors (producers and consumers), which allow estimation of losses from specific economic sectors, and their negative impacts on other sectors (Cochrane, 1974; Haimes et al., 2005; Hallegatte, 2008; Hallegatte et al., 2011; Okuyama, 2014; Okuyama et al., 2004). The main disadvantages here are the lack of flexibility in the economic system, which lacks the ability to assess behavioral changes during and after an event, and also includes a general inability to analyze the independences between price and output variability during and after an event. Moreover, there is a lack of focus on the demand-side of the economy (Carrera et al., 2015; Rose, 2004). Another drawback is the static view of the economy, mainly because of the lack of continuous and comprehensive data series within the input-output statistics (Lenzen, 2011).

Another approach is the Computable General Equilibrium (CGE) models mostly used in economics circles. Since the 1980s, the number of studies using CGE models has greatly increased; many analyze the external and internal macro-economic impacts for local, regional and national economies. CGE models calculate a partial equilibrium within the economy as a means of gaining some foresight. The idea in this case is to change/include new economic target variables based on political decisions or external shocks like natural hazards events, thereby allowing a more accurate assessment of the impact towards other economic variables (Berrittella et al., 2007; Boyd and Ibarraran, 2009; Carrera et al., 2015; Husby et al., 2016; Rose et al., 1997; Rose and Liao, 2005). CGE models allow simulation of a multi-market system based on optimizing individual behavior (Rose, 1995). A key advantage compared to other economic models is the flexibility this model presents to assess the supply changes through external and internal shocks as well as its flexibility in terms of the time horizon[10] (Carrera et al., 2015; Hallegatte, 2008).

CGE models thereby assume a perfect market situation and follow the idea of an economic equilibrium as an output after a shock event; reality demonstrates the opposite in practice, especially alongside natural hazards events (Rose, 2004). Another shortcoming is that CGE models generally underestimate the losses of natural hazards compared with input-output models, which usually overestimate the losses (Okuyama, 2003). Finally, in past years, there has been a stronger use of combined or hybrid models when estimating economic losses. Examples of hybrid models are the combination of CGE and input-output models (such as Rose (1995) or Rose et al. (2007)) or the combination of hydrological (or biophysical) with economic models (such as Jonkman et al. (2008) or Ciscar et al. (2014)). Hybrid-models have the goal to overcome limitations in preceding and dominant economic models, such as in the case of the Adaptive Regional Input-Output (ARIO) model (Hallegatte, 2008). The ARIO model aims to overcome the shortcomings in input-output models (like the lack of flexibility in the current economic system) but is not able to include shortcomings like the production system after natural hazard event. This shortcoming might be overcome with the help of CGE models (Hallegatte, 2008). Nevertheless, hybrid models include significant uncertainties within their approach to estimating indirect losses, as well as inconsistent data sets and a lack of data quality (Cochrane, 2004).

### *5.3.3 Nonmarket (Intangible) Losses*

The third group in the natural hazards evaluation includes the assessment of nonmarket (intangible) losses, such as losses in ecosystem services or human life (Meyer et al., 2013). Valuing nonmarket losses in monetary terms often results in conflicts and challenges due to a lack of a market on which the "goods" were able to be exchanged; without this, it is difficult to estimate the losses in monetary terms (Birol et al., 2011; Leiter and Pruckner, 2009; Thaler et al., 2014). Nevertheless, there are numerous techniques available to provide monetary data to assess nonmarket losses. In general, the economic framework exists broadly in two categories namely revealed preference (or conventional and surrogate markets) and stated preference methods (or hypothetical market). The choice of method for a study depends on several criteria, including the purpose of the study, availability of data, and particular economic values required (use and/or nonuse values; Hanley and Spash, 1994; Pearce and Moran, 1994).

Revealed preferences – introduced by Samuelson (1938) – include various options concerning how to assess economic losses. These include random utility (or discrete choice model), travel cost method, hedonic pricing, advertising behavior, costs of illness, replacement cost, production function or use, or market price (Birol et al., 2006; Hanley and Spash, 1994). Revealed preferences determine the demand for goods or services by examining the purchase of related goods in the private market (Meyer et al., 2013). This methodological concept is still generally preferred especially for certain types of nonmarket goods, such as impacts of natural hazards on house market or for analyzing the purchase of natural hazard insurances (Bin and Polasky, 2004; Daniel et al., 2009; Hamilton, 2007; Kousky, 2010; Michel-Kerjan and Kousky, 2010; Petrolia et al., 2013). In the last twenty years, there has been a dramatic increase in the number of academic papers and presentations related to the stated preference technique (Arrow et al., 1993; Bateman et al., 2011; Brouwer and van Ek, 2004; Luisetti et al., 2011; Turner, 2007). This is due in part to the availability of comprehensive reference texts on the subject (Arrow et al., 1993, for instance), and also to the growing interest in environmental problems and policies both nationally and internationally.

On the other hand, stated preferences develop a hypothetical scenario to analyze the losses based on willingness-to-pay (WTP) of individuals (Meyer et al., 2013). Economic scholars usually use standardized questionnaires based on stated preferences (=hypothetical market) so that they may assess the individual preferences – mainly with the help of WTP. The potential methods were conjoint analysis (such as choice experiments, contingent ranking, paired comparison or contingent conjoint rating), and the contingent valuation method (CVM) (Luechinger and Raschky, 2009; Hanley and Spash, 1994). Typically, these studies provide respondents with information about a hypothetical government program that would reduce the likelihood of future events; it would do this by implementing nonuse structural measures, such as creating retention pounds, wetland creation and restoration, rover dredging, removal of brush/debris along shorelines, and building dikes (Brouwer and Schaafsma, 2013). Respondents usually receive specific information about the exact nature of the damages that the program in question would prevent. This may take the form of an open-ended question (CVM), for example, asking the maximum amount they would be willing to pay for the program in question; it may also involve a series of questions that confront the subject with different prices for the program depending on their previous answers (choice experiment (CE); Hanley and Spash, 1994). On the other hand, life satisfaction analysis does not depend on individuals' choices. The methodology used in this case is an Ordinary least square (OLS) regression analysis that estimates the impact of past natural hazards events (Luechinger and Raschky, 2009). However, there are limitations in terms of not only the potential psychological biases based on the simplification of the complexity (Ryan and Spash, 2011; Vatn and Bromley, 1994), limited available information for the interviewees, and lack of transparency (Folmer and Johnasson-Stenmann, 2011; Vatn and Bromley, 1994), but also due to the fear of marketization of human life or environmental goods (Dempsey and Robertson, 2012).

## 5.4 Case Study: Economic Vulnerability Assessment for Flood Hazards (UK and Germany)

Challenges facing the assessment of economic vulnerability render mainly to the aspect of used method for the damage assessment and quality of data. However, there are thus many flood damage models around the world that make use of vulnerability functions for their damage assessment (see Table 5.4). This section makes use of data modeled by the study conducted by Jongman et al. (2012). The study modeled a comparative economic vulnerability assessment by using the maximum economic losses for two different communities (Carlisle in the United Kingdom and Eilenburg in Germany).

To assess the effect of the use of specific vulnerability functions, a hydrological model was created that uses long time series of daily rainfall to model potential flood-prone areas for all rivers in Europe. The result of this exercise is maps of possible inundation depths along all river stretches. The same analysis was conducted for scenarios of climate change, for which projections of daily rainfall are available. In addition, the team analyzed the correlation between the various river basins. This is important, because extreme weather

Table 5.4. *Vulnerability functions for residential housing from different flood damage models. After Jongman et al. (2012: 3738)*

| Model | Maximum damage per square meter of continuous urban fabric [€] |
|---|---|
| FLEMO | 268.00 |
| Damage Scanner | 852.00 |
| Flemish method | 718.00 |
| HAZUS-MH | 329.00 |
| Multi-Colored Manual | 286.00 |
| Rhine Atlas | 256.00 |
| JRC DE | 277.00 |
| JRC UK | 348.00 |

events often spread across different basins and countries at the same time. These flood inundation maps were then overlaid with the European land-cover maps from CORINE (EEA, 2016). This land-cover map classifies the different types of land-cover for each 100 m × 100 m grid cell, using classes such as residential areas, industrial and commercial areas, infrastructure, agriculture, etc. For each of these classes, stage-damage functions and maximum damage values were identified that link potential flood depth from the hazard model to the resulting economic damage for each of the grid cells. Using the correlations between the different river basins, the authors could then estimate the economic damage for the EU.

The next step in the assessment was the financial vulnerability. Some of the used models make use of empirical analysis of real-world flood damage to estimate the relationship between inundation depth and building damage (Thieken et al., 2008), whereas other models may use "expert judgement" to draw these relationships. The used models (see Table 5.4) vary strongly in their estimate of potential maximum damage per square meter (Table 5.5) and the shape of the vulnerability function. The difference in risk estimations as a result of the use of these different damage models can be up to a factor 50 (Jongman et al., 2012).

The used models show strong differences in the modeled damages. An important division that can be recognized in the used models refers to the aspect of object-based (such as HAZUS-MH or Multi-Colored Manual) and area-based models. The HAZUS-MH or Multi-Colored Manual show the advantages to include various building characteristics in their calculation instead of the more aggregated area-based models. Nevertheless, area-based models have the main advantages in their easily – rapid calculation over a large area. Another difference can be seen in the adapting regional differences in the economic vulnerability assessment. For example, the Flemish method, the Rhine Atlas or the Multi-Colored Manual, include local circumstances in their input parameters. The HAZUS-MH database, for example, is based on object-based asset values for a specific number of defined objectives, such as bridges, and standard values on census blocks (e.g. residential houses). The model uses a standardized stage-damage curve for the whole country, similar to the JRC-models, where the JRC develop a stage-damage curve on a NUTS 2 administrative level, based on existing national stage-damage curves and GDP per capita. A further dissimilarity relays on the used input parameters, which also show a strong impact on the results (see also Table 5.3). Finally, a central difference includes the damage calculation per se. The used models (FLEMO, Damage Scanner, or the Flemish model) value assets at replacements costs, where on the other hand, Multi-Colored

Table 5.5. *Results of the model runs for Carlisle and Eilenburg with modeled and reported damages. After Jongman et al. (2012: 3745)*

| | Modeled damages (million €) | | | | | | | |
|---|---|---|---|---|---|---|---|---|
| *Eilenburg* | | | | | | | | |
| Land use code | FLEMO | Damage Scanner | Flemish method | HAZUS-MH | Multi-Colored Manual | Rhine Atlas | JRC-UK | Reported [million €] |
| Continuous urban fabric | 0 | 0 | 0 | 0 | 0 | 0 | 0 | 77 |
| Discontinuous urban fabric | 130 | 252 | 494 | 128 | 174 | 67 | 165 | |
| Industrial or commercial units | 19 | 61 | 34 | 21 | 64 | 17 | 22 | 32 |
| Road and rail networks and associated land | 0 | 69 | 30 | 0 | 0 | 17 | 6 | 109 |
| Total | 150 | 383 | 558 | 149 | 238 | 102 | 193 | 218 |
| *Carlisle* | | | | | | | | |
| | Modeled damages (million €) | | | | | | | |
| Land use code | FLEMO | Damage Scanner | Flemish method | HAZUS-MH | Multi-Colored Manual | Rhine Atlas | JRC-UK | Reported (million €) |
| Continuous urban fabric | 2 | 8 | 7 | 19 | 5 | 1 | 14 | 321 |
| Discontinuous urban fabric | 38 | 56 | 121 | 172 | 73 | 18 | 214 | |
| Industrial or commercial units | 12 | 39 | 19 | 25 | 66 | 11 | 71 | 151 |
| Road and rail networks and associated land | 0 | 11 | 4 | 0 | 0 | 3 | 0 | 64 |
| Total | 52 | 123 | 152 | 216 | 144 | 34 | 299 | 535 |

Manual or Rhine Atlas uses depreciated values. All these significant differences are clearly translated in the model outcome.

## 5.5 Conclusion

In this chapter, we discussed a few of the aspects of economic vulnerability assessment and demonstrate how they are used by policy makers to support risk reduction policy. The measurement of economic vulnerability has become a more prominent role in our societies, mainly to compute the potential effectiveness of adaptation options in reducing losses as well as to understand the potential pressures on current financing systems that are in place to compensate losses from natural hazard events (de Moel and Aerts, 2011; Visser et al., 2014). However, a central challenge refers to used methodology for assessment of economic vulnerability. For example, local-level analyses in developed countries can often make use of high-resolution building-level datasets with a detailed definition of various building and infrastructure classifications (Jongman et al. 2014; Fuchs et al., 2017), enabling the application of a range of vulnerability functions. On a national or global level, such information is often not available, and more rudimentary approaches need to be applied (de Moel et al., 2015). Therefore, the vulnerability functions and the associated maximum damage values are directly linked to the land-use data that is used as input to the model. Consequently, economic vulnerability assessments in these are being derived using different methods, and can vary greatly (de Moel and Aerts, 2011; Jongman et al., 2012).

Apart from modeling efforts, there has been a recent effort to empirically estimate the level of vulnerability of different countries using reported loss databases. The idea of such studies is to separate reported losses ("risk") into the individual components of hazard, exposure, and vulnerability, and estimate whether the losses and fatalities as a percentage of the exposed assets and population is changing. Visser et al (2014) normalized a disaster loss database, correcting for population and GDP growth per country to find a trend in global vulnerability to weather-related hazards. Their results showed that exposure and hazard drove historical losses and that vulnerability seemed to be stable.

Jongman et al (2015), for example, showed that global vulnerability declined over the study period, meaning that the losses as a percentage of the exposed GDP is falling. As countries develop, the level of building standards, infrastructure, communication, healthcare, and education generally rises. Also, rich countries have a higher standard of flood protection (Scussolini et al., 2016) and better early warning systems. As a result, the population of low-income countries is often relatively hard hit when a flood strikes, causing a higher number of fatalities per capita than is seen in high-income countries (Jongman et al., 2015). People living below the poverty line have additional trouble in recovering from flood disasters, due to the low financial capacity to absorb shocks (Hallegatte et al., 2015). Key conclusions and further research needs for improving economic vulnerability assessment include:

- The economic vulnerability assessments are based on the following four steps: (1) the mapping of the location, magnitude and frequency of the hazard, such as flood depth with a certain probability of occurrence; (2) the mapping of the location and value of exposure, such as population density and building stock; (3) the assignment of vulnerability of the exposed elements to the hazard, such as building damage as a function of flood depth; and (4) the combination of these three elements in a probabilistic risk model, resulting in potential disaster losses for various scenarios or hazard levels.
- The methodology for the assessment of economic vulnerability varies, depending on the type of hazard considered, the scale of the assessment, the availability of data, and the required outcomes of the risk assessment.
- Going forward, the way in which global vulnerability is changing will therefore largely determine the economic and humanitarian impact of disasters. This emphasizes that the accurate estimation of vulnerability is key to understanding disaster risk.
- It also highlights that vulnerability, development, climate change, and risk management are all highly interconnected, and should be well understood in order to make effective decisions on risk reduction.

# 6

# Institutional Vulnerability

MARIA PAPATHOMA-KÖHLE AND THOMAS THALER

## 6.1 Introduction

When a natural or technological hazard strikes, the level of negative consequences on the society, which may be expressed as deaths, injuries, damaged buildings, and infrastructure, loss of product and business disruption, depends not only on the magnitude of the event but also on the vulnerability of the affected population and the built environment. Gilbert White focusing on flood hazards stated, "floods are an act of God but flood losses are largely acts of men" (White 1945: 2). These losses, however, depend not only on obvious drawbacks, such as poor quality housing, but also on factors that are not always apparent in the first place. Existing legislation frameworks supporting preparedness, response, or reconstruction efforts and their implementation are only few of them. Additionally, the level and efficiency of coordination and communication among actors that are responsible for disaster management, the degree of public participation and awareness, as well as available risk transfer mechanisms are factors that are actually embedded in fundamental socioeconomic, cultural, and political structures. These factors constitute the institutional setting of the society, which is responsible for the way a community prepares, responds and rehabilitates, before, during, and after the occurrence of a hazardous event.

Nevertheless, the institutional dimension of vulnerability is probably the dimension that has been, to date, the least explored. Most vulnerability studies focus on the physical or the social dimension, whereas research focusing on how institutions influence the vulnerability of the society to natural hazards is still scarce. The reason for this may be that institutional vulnerability often overlaps other vulnerability dimensions such as social, economic, and cultural. It is important to note that institutional vulnerability relates to the way institutions influence vulnerability to natural hazards rather than on the vulnerability of institutions themselves.

Recent extreme events revealed the importance of institutional vulnerability. Hurricane Katrina, a Category 4 storm that hit the southern coast of the United States in 2005 causing major devastation, revealed major institutional weaknesses at different scales (Elliott and Pais, 2006). However, the intensity of the storm cannot justify the extent of the disaster and the size of the consequences. Therefore, the devastation may be equally attributed to the inadequate response of the responsible organizations, the poor communication and coordination in different levels, as well as to actions that increased the vulnerability of the population and exposed social tensions (Baker and Refsgaard, 2007). The weaknesses of the institutional context revealed during Katrina combined with physical vulnerabilities (inadequate levees, lack of buffering) led to one of the worst natural disasters in US history.

However, even before Katrina, research on large-scale natural disasters between 1984 and 2004 worldwide clearly showed that countries with stronger institutions experienced lesser consequences in comparison with countries demonstrating institutional vulnerabilities (Raschky, 2008). Kahn (2005) investigating 4,300 catastrophic events from 1990 to 2002 in 57 countries concluded that nations with better institutions suffered fewer deaths from natural disasters. Kahn (2005) tested the correlation between the number of deaths and parameters such as the democracy level, income inequality, ethnic fragmentation, and the World Bank indicators of good governance for each country. The World Bank indicators included: voice and accountability, political stability and absence of violence, government effectiveness, regulatory quality, rule of law, and control of corruption (Kaufmann et al., 2010). Kahn (2005) concluded that institutions may contribute to the reduction of deaths from natural disasters.

Institutional issues are pertinent to all phases of the risk cycle (Figure 6.1). During the mitigation phase, policies, the legislative framework, as well as decisions for the implementation of structural measures are part of the pre-existing institutional context. In the preparedness phase, participation of the public, access to information, and the existence and performance of early warning systems are essential institutional components.

During an event, the institutional context and the decision-making process have an impact on the priorities of the emergency services and the handling of the vulnerable groups. Last but not least, in the rehabilitation and reconstruction phase, the mobilization and prioritization of resources, as well as the availability of insurance or state compensation, are institutional issues that directly affect the societal resilience.

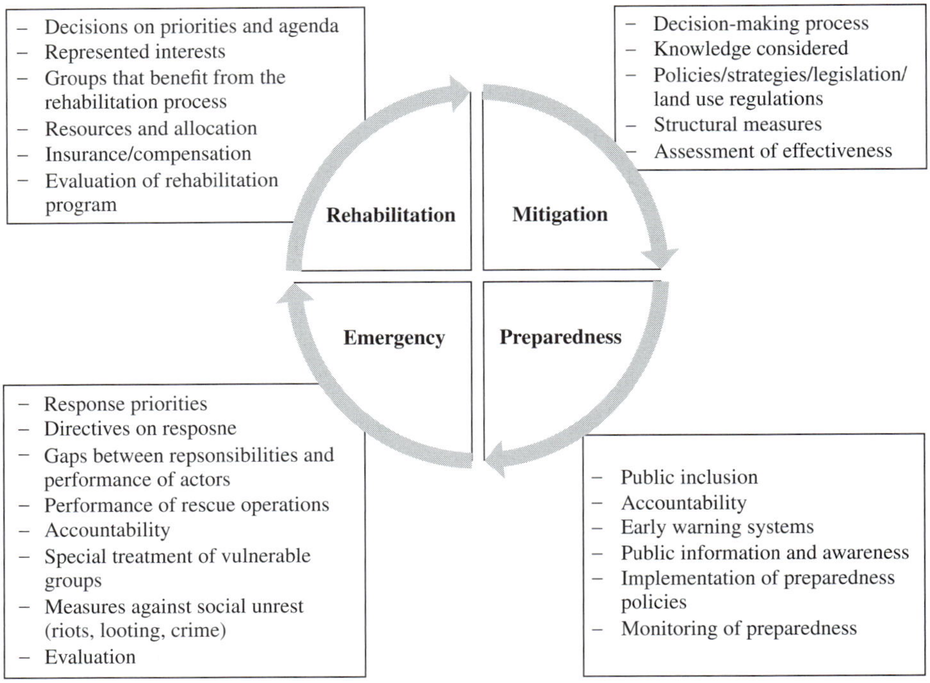


Fig. 6.1. Institutional issues related to the phases of the risk cycle

## 6.2 What Are Institutions?

Institutions regulate the social and economic behavior of individuals, playing a highly important role in the development of the society. Their main task is to organize structures that may optimize the social and economic behavior by reducing, for example, complexity, and therefore, uncertainty in our life (Hodgson, 1988; North, 1990; Vatn, 2009). The term "institutions" in the literature is perceived through different thoughts, meanings, and common understandings (Hodgson, 2006; North, 1990; Veblen, 1899). In general, institutions can be seen as a durable system, which includes social characteristics like culture, legislations, organizations, moral obligation, language, etc. (Hodgson, 2000, 2003) and creates economic and political stability over a certain period. According to Jessop (2001), an institution is the regulative framework of our society. Hodgson (2006) argued that institutions are seen as the framework for the social dimensions that consequently influence our individual life, which is imposed by social rules (Hodgson and Jiang, 2007). Institutions define how society can or should interact by forbidding, permitting, and assisting certain actions (Commons, 1931; Crawford and Ostrom, 1995; Vatn, 2009). Young (1996) suggested that institutions are consequently dependent on power relationships within the society (Lukes, 2005). They act as the rule makers for social actions, since they have the ability to define behavioral structures (opportunities and constraints), to specify procedures, to define the path of complex social relations, and to act as the "macrostructural matrices" of societies and social patterns (Jessop, 2001). According to North (1990), institutions consist in formal (laws, juridical decisions, constitutions, administration practices) or informal (social norms, traditions) norms and rules (Dutt, 2011; North, 1990; Raschky, 2008). Therefore, institutions refer to the regulative, normative, and cognitive aspects of the society (Moodysson and Jonsson, 2007; Zeller, 2004). Besides, the literature distinguishes between two main schools:

- The old institutionalism, which focuses on the idea that individuals are able to adapt to new developments within the society and institutions based on interaction (Commons, 1931; Veblen, 1914; Veblen, 1919).
- The new institutional economics which focus on the rational choice theory (maximization of the individual preferences) to analyze institutions (Hodgson, 2000; North, 1990).

It is clear that institutions have a direct influence on individuals, but also vice versa. Not only do institutions influence the behavior of individuals (top-down), but also their behavior and habitation act as a key driver for the development of new institutions or the change of the existing ones (bottom-up, Dequech, 2006; Hodgson, 2006). This is highlighted in the literature, by the identification of three hierarchical levels of social institutions related to the decision-making process: (1) the operational level (micro-level), (2) the collective choice (meso-level), and (3) the constitutional level (macro-level, Crawford and Ostrom, 1995; Gibson et al., 2000; Koontz, 2003; Turnpenny et al., 2008).

- The micro-level refers to individual agents involved in the policy discussion. Research in this level is oriented to the aspect of resources (human, time, and financial). In general, the decision process (choice) focuses on the input and output of used resources and decisions made by similar organizations (von Ciriacy-Wantrup, 1971). Evans et al. (2002) suggest that, at this level, choices are defined by policies made in the meso-level (collective choices).
- The meso-level covers the organizational structure (North, 1990). At this level, policy assessment includes the aspect of organizational norms and culture, formal and informal decision rules, leadership, and cooperation as well as coordination of the involved agents. Individuals and organizations are authorized to define collective choices, i.e., new policy strategies and

guidelines. The purpose is to define a general approach and control instruments (e.g., economic efficiency and effectiveness, social justice, and equity) for operational implementation of policy implementation process (von Ciriacy-Wantrup, 1971).
- Finally, the macro-level refers to the legal framework. Here, choices and policy discussion are related to "the authority of collective actors" (Paavola, 2008: 319). Constitutional rules or policy levels are set up for a long term, so their change is quite slow.

Institutional framework analysis includes the identification of the main actors and stakeholders, their interaction, and engagement. A key objective is to understand the role and interests of the actors and stakeholders involved in the policy arrangements as well as their legal limitations and responsibilities. Stakeholder and responsibility mapping enables the identification of the main actors and stakeholders in the policy discussion process, policy-making structures and decision-making practices. Furthermore, stakeholder mapping also allows us to clarify issues of responsibility and accountability. It is, therefore, clear that institutional and organizational arrangements influence the social structures and provide the framework of interactions and influences of the different stakeholders and actors in the decision-making practice (Hodgson, 2006).

## 6.3 Institutions Related to Disaster Management

As far as disaster management is concerned, institutions define and constrain our choices regarding decisions linked to the protection from natural hazards. Their role is to reduce the uncertainty generated by the complex set of variables that are involved in this field (Raschky, 2008). According to Frey (1990) institutions include: (1) formal and informal rules, and (2) organizations. In the following paragraphs rules and organizations related to the management of natural hazards are presented.

### *6.3.1 Rules*

#### *Formal Rules*

Activities leading to disaster risk reduction are mainly controlled by formal rules such as legal instruments, regulations, government guidance, policies and plans (Tompkins et al., 2012). These may include building codes, mandatory insurance, regulations regarding retrofitting of older buildings, land use planning regulations, and decisions regarding the balance between structural and nonstructural mitigation measures as well as risk mapping (Fuchs, 2009; Tompkins et al., 2012).

There are actually two approaches in reducing disaster risks for the built environment: the location approach and the design approach (Johnson, 2011). The location approach relates to land use planning that may be used in order to restrict or limit development to some areas and the design approach relates to the setting of standards for the construction of buildings and infrastructure that may limit damage.

Land use planning is an important nonstructural risk mitigation measure closely connected to and supported by a supportive legislative framework. Priority 2 of the Sendai Framework ("Strengthening disaster risk governance to manage disaster risk") also suggests that the legislative and the regulative framework of each country has to support land use and urban planning in order to manage risk efficiently (UNISDR, 2015a). Land use planning requires the identification of those areas that are at risk from specific hazards and the probability of occurrence in order to justify the land use regulations. This procedure is called hazard zone mapping. Methods for the development of the hazard zones vary according to the hazard type. For example, seismic hazard zones, like the ones developed in California, are based on peak ground acceleration, whereas coastal hazard zones for tsunami hazards are based on the expected tsunami height for different return periods (Smith, 2013). Hazard zones have been developed in many countries for a number of hazard types such as earthquakes, floods, rock falls, landslides, and snow avalanches.

Local population, landowners, developers, and local businesses often oppose the hazard zones, and for this reason their efficiency depends heavily on risk communication strategies, public awareness, and participation. Consequently, risk communication has become increasingly important in risk management. It represents a two-way communication between risk professionals and the public aiming at knowledge and views exchange in all the phases of the risk cycle (Höppner et al., 2012).

The legislative framework that accompanies hazard zoning may relate directly to its development or use, as well as its interpretation and associated restrictions. However, for buildings already built before the implementation of building codes, a retrofit program may be considered, which, in order to be attractive for the building owners, should be either funded by the government or combined with incentives given by insurance companies (Mason et al., 2013). The existence of building codes is not a guarantee for safety. First of all, building codes are often introduced following a disastrous event. For example, the first seismic code in Greece, which was introduced in 1959, was revised in 1984 following the devastation caused by two catastrophic earthquakes in Thessaloniki (1978) and the Gulf of Corinth (1981, Pomonis, 2002). Following the earthquake in Aeghio in 1995 the seismic code was modernized to its present form (Pomonis, 2002). However, a large proportion of the building stock in Greece (60–70 percent) was built before the enforcement of these codes (Pomonis, 2002). Moreover, in order to build robust buildings, the enforcement and implementation of the building codes have to be respected by homeowners and engineers. This has not always been the case. There are plenty of examples (e.g., in Turkey, El Salvador, Taiwan, India, Italy, and Japan) showing that due to corruption, indifference, or lack of qualified professionals, building codes are not implemented properly, leading to unnecessary loss and damage (Wisner et al., 2004). It is therefore clear that the introduction or reform of building codes is not enough to reduce disaster consequences without the adequate social, political, and economic context (Green, 2008).

The legislative framework related to disaster management may be local, regional, national, or supranational. At the EU level, for example, each member state has to comply with regulations that are binding legislative acts. Directives, on the other hand, are not legally binding. Each member state, however, has to develop or reform their own legislative framework to fulfil the objectives set by the directives. The EU Floods Directive, for example, came into force in 2007 and required that the member states produce flood hazard and risk maps until the end of 2013, and flood risk management plans until the end of 2015 (European Commission, 2007). The directive promoted not only participation of the relevant stakeholders and the public but also collaboration between member states in cross-border river basins. Moreover, the EU Guidelines for Risk Assessment and Mapping also prompted member states to conduct risk assessment and produce risk maps for a range of natural hazards (European Commission, 2010).

Apart from legislation related to hazard zones and land use mapping, legal frameworks may accompany all the activities related to the phases of the risk cycle in different scales (national, regional, local), which accompany all the activities related to the phases of the risk cycle. As far as emergency management is concerned, the legal framework may specify obligations and responsibilities, secure specific rights, and define the framework for civil protection (Alexander, 2002).

Nevertheless, the existence of laws related to construction, land use planning, disaster risk reduction, civil protection, etc., does not guarantee their implementation. Their implementation depends on a number of factors such as the level of public education, corruption, coordination of organizations, etc. Additionally, the quality and effectiveness of the existing legislation should be often evaluated and improved. For example, according to Vink und Takeuchi (2013), legislation related to disaster risk reduction for flood disasters in Japan, the Netherlands, and the United States does not consider the vulnerable groups (elderly, children, and disabled) in a sufficient way. In more detail, vulnerable groups are not clearly defined and there is no assessment regarding the potential increase of these groups in the future or a plan on how this increase could be handled.

Finally, legal frameworks, regulation, and acts may control the way that individual building owners or owners of agricultural land are compensated for the loss caused by natural hazards. Governmental catastrophe funds as well as private commercial insurance and the legislation that constitutes it mandatory (if any) are also formal institutions. Details about the risk transfer options and the compensation mechanisms are given later in this chapter.

### *Informal Rules*

Informal institutions include all these rules that determine human behavior and actions including customs, traditions, and unwritten laws. Informal institutions are related to the dissemination of information, response of vulnerable groups to an event, community involvement, risk perception, public awareness, and responsibilities (Tompkins et al., 2012). The role of informal institutions may be particularly significant in the efforts of risk reduction, especially in developing countries. During the tsunami of 2004 in Indonesia, a significant difference in the death toll among different cultural communities revealed the importance of culture as an informal institution. In more detail, people living on the island of Simeulue suffered 44 deaths in comparison with the 170,000 deaths that occurred in the cities of Banda Aceh and Meulaboh (Gaillard et al., 2008). According to Gaillard et al. (2008), the death toll reflects cultural differences that are responsible for the response of the people. Memories of past tsunami events and local knowledge regarding the precursory signs of tsunamis resulted in a better response of the Simeulue people that led to a higher survival rate (Gaillard et al., 2008). In a similar way, in Jakarta, "Kampung" people manage their risk in their own traditional way, which differs greatly from the governmental program. A study about the flood risk management at the community level showed that informal institutions due to their importance should be fully recognized and harmoniously connected to the formal institutions (Simarmata, 2015). Apart from cultural differences, religion may also act as an informal institution and affect the way people perceive and react to natural hazards. Fatalism is the downside of religion as an informal institution; however, there is a need to see religion as a "resource rather than a hindrance in the planning of disaster risk reduction" (Gaillard and Texier, 2010).

### *6.3.2 Organizations in Natural Hazard Management*

Organizations are defined as special institutions with clear boundaries and members including their accountability and responsibility (Hodgson, 2007). Organizations may include public administration, ministries, and national and international funding agencies, and donors as well as research institutes (Garschagen, 2013). Additionally, organizations may be NGOs, political parties, or groups of volunteers but even families, clubs, and clans (Fuchs, 2009). In many cases, nongovernmental organizations and volunteers may respond better to natural hazards than the official local authorities or the emergency services. In the case of Hurricane Katrina, which revealed huge deficiencies of local and regional authorities as well as of FEMA (US Federal Emergency Management Agency) and the emergency services, nongovernmental agencies, voluntary groups, and networks strongly affected not only the response but also the reconstruction phase (Baker and Refsgaard, 2007). It is noteworthy that some networks and organizations were developed following the aftermath of Katrina in order to enable and assist the rehabilitation of the local population. An interesting case is an organization called the "Common Ground Collective", which was involved in further activities including representation of residents in governmental hearings and participation of the local population in spatial planning, increasing in this way the resilience of the society but also the institutional capacities of the area (Baker and Refsgaard, 2007). A similar trend was witnessed during the reconstruction phase following the Kobe earthquake in 1995 in Japan where a clear increase in voluntary and nongovernment activities was observed (Shaw and Goda, 2004). In more detail, six years after the earthquake, the people of Kobe developed an action plan (the Kobe Action Plan) which aimed at strengthening links within the community, assisting the development of people's activities, and achieving a sustainable civic society (Shaw and Goda, 2004).

However, also less organized groups of people acting through traditional social networks may play a major role during the response and the recovery phase. For example, during the flood of 2000 in Mozambique, in remote areas where governmental aid was limited, fishermen assisted the evacuation process with their boats (Battista and Baas, 2004).

## 6.4 Framing Institutional Vulnerability

The institutional dimension of vulnerability is directly related to factors that may weaken or strengthen the institutional capacities of the society affecting directly the consequences of natural disasters. Institutional vulnerability has been investigated on a theoretical and mainly qualitative level; nevertheless, there is no thorough definition in the literature. According to Paul (2011), institutional vulnerability is determined by the capacity and commitment of the institutions to mitigate society's vulnerability and to achieve recovery from a disastrous event. On the other hand, the IPCC (2012) suggests that institutional vulnerability reflects the relationship between policymaking and implementation. It is, therefore, clear that institutional vulnerability relates to the sociopolitical and cultural settings that are responsible for the exposure of society, as well as the capacities to recover and adapt to future hazards (Fuchs, 2009). Since choices regarding the management and response to natural hazards are often assigned to the political decision maker, Fuchs (2009), based on data from the European mountain regions, stressed the fact that institutions take over responsibilities for hazard mitigation while the role of the individuals is at the same time diminished, highlighting in this way the significance of participation and accountability. Hochrainer (2006: 18) gave a very general definition of institutional vulnerability describing it as "the existence and robustness of institutions to deal with natural disasters." We understand institutional vulnerability as:

> the combination of the weaknesses embedded in institutions (purpose or nonpurpose built for disaster management) that reduce the capacity to resist/withstand/cope or recover from the impact of a hazardous event. These weaknesses are strengthened and shaped by the socio-political context and may be rooted in drawbacks of the legal framework and its implementation, deficiencies in communication and lack of participation of the relevant stakeholders and the public, as well as shortcomings of the available risk transfer mechanisms and poor land use planning.

Some of these institutional weaknesses are described in detail in the following subchapter. Despite overlaps and gaps in this description, light could be shed on the topic of institutional vulnerability contributing in this way to its better comprehension.

### *6.4.1 Socioeconomic Deficiencies and Corruption*

Political stability and established democracy may significantly affect the way society responds and recovers from a natural disaster. In Haiti for example, the political instability and violence of the past heavily influenced the consequences of the 2010 earthquake, the response, the management of the international aid and the reconstruction phase (van der Vink, 2007). The importance of the political system is supported by the fact that between 1964 and 2004 the vast majority of deaths due to natural hazards (more than 80 percent) occurred in just fifteen countries, 87 percent of which are below the mean democracy index. The democracy index depends on a number of variables including accountability, lack of violence, regulatory quality, rule of law, government effectiveness, and control of corruption (van der Vink, 2007).

Regarding the social structure, Cohen and Werker (2008) pointed out that fewer homogenous countries experience a higher death toll from natural hazards. Past disasters have shown that institutional policies may lead to further segregation of already segregated societies. For example, in New Orleans, racial disparities related to income and employment differences between black and white people led to black people residing in areas characterized by high flood hazard (Henkel et al., 2006). In addition to that, during Hurricane Katrina, most of the black population could not afford financially to evacuate. Finally, during the reconstruction period, mistrust from both sides and discrimination led to a delayed rehabilitation of the black population (Henkel et al., 2006). This phenomenon that became evident during the Hurricane Katrina is called institutional racism and has been defined as "a systematic pattern of inequality based on race which is embedded in social, business and governmental institutions" (Baker, 2014: 622).

Socio-economic deficiencies in society may also lead to corruption that has the capacity to significantly increase the institutional vulnerability. According to Transparency International (2010), corruption "is the abuse of entrusted power for private gain." Klitgaard (2010) proposes the following equation:

$$\text{Corruption} = \text{Monopoly} + \text{Discretion} - \text{Accountability}$$

Based on the above equation it becomes clear that in order to reduce corruption, competition has to be enabled to battle monopoly, rules have to become clear and transparent for the wider public to reduce discretion and accountability has to be enhanced (Klitgaard, 2010).

There is a strong link connecting corruption with underdevelopment and disasters (Lewis, 2011). Corruption is often observed in the construction industry relating to the failure of enforcement of building codes and standards usually revealed by earthquake events. For example, in Turkey, apartment blocks collapsed following the earthquake of 1999 mainly due to a failed system of building control and corrupt practices in the construction industry of the country (Lewis, 2011). Another reason why new buildings failed to comply with the building codes and regulations was mainly the pressure for more housing due to rapidly growing population (Escaleras et al., 2006). Escaleras et al. (2006) supported their claims regarding the role of corruption in the consequences of natural hazards by using evidence from 344 earthquakes in forty-two countries for the period 1975–2003, and showed that the level of public sector corruption is positively correlated with the number of deaths. Outside of the construction industry corruption may have a diverse effect on the vulnerability of the society. For example, in Pakistan, illegal and uncontrolled logging of forested areas contributed to a higher flood hazard potential leading to the death of 1,600 people during the 2011 floods (Lewis, 2011).

During the reconstruction phase, corruption may affect the trust of donors to the recipient country. In the aftermath of the 2010 earthquake in Haiti, the assistance for relief and reconstruction sent from international donors to the country was mainly received by UN agencies, international NGOs, and civil and military entities of the donors themselves. Only approximately 1 percent was received by the government of Haiti due to concerns regarding "systems of patronage, corruption and inefficiency" that previous Haitian governments have demonstrated in the past (Ramachandran and Walz, 2012). In such cases, disaster aid is often expected to fail due to the countries' legacy of corruption and lack of transparency.

### *6.4.2 Inadequate Risk Transfer*

In the aftermath of disastrous events, society faces the costly task of rebuilding destroyed housing and infrastructure under new land use planning and building regulations, in order to increase the resilience of the built environment and avoid future consequences. The costs are often too high for individuals or communities to bear, and for this reason, costs have to be transferred, shared, or spread among several parties. The costs of reconstruction and rehabilitation are normally shared between international aid, the public sector, and the insurance and reinsurance industry. A good share of the costs among these sectors can enable a promising and effective resilient reconstruction.

#### *International Disaster Relief*

A number of international bodies operate globally offering humanitarian aid following disasters everywhere in the world. One of the world's biggest donors is the EU through the European Community Humanitarian Office

(ECHO), the UN through the Office for the Coordination of International Affairs (OCHA) and the World Bank (Smith, 2013). The main aim of disaster aid is the medical care of the victims, the provision of shelter, and the avoidance of further mortality in the immediate aftermath of a disaster (Smith, 2013). The hours and days following a disaster are crucial not only because victims should be rescued and medically treated as soon as possible and dead bodies should be disposed of, but also because epidemics may cause additional health problems and deaths due to poor sanitary conditions. Most disaster aid is coming from charity initiated through international appeals for aid following a disastrous event. However, these appeals do not always have the same impact on donors as they depend on the media coverage, the place where the disaster occurred, and also political interests. Events that have an intensive media coverage (earthquakes, tsunamis) tend to receive more international aid than other events that have less media coverage due to their nature and onset (e.g., drought, famine, etc.). A good example is the 2004 tsunami in the Indian Ocean, which received a record aid estimated at US$13.5 billion (Smith, 2013). Although the record donation estimate is conservative, since it excluded private donations from nonTEC (Tsunami Evaluation Coalition) countries, private remittances and projects, private donations from the affected countries, communities and governments, it still exceeded by one-third the estimated losses (Telford, 2007). However, the funding decisions were not based on actual needs and not enough has been invested in order to "build back better" (Telford, 2007). Focusing on Sri Lanka, Korf et al. (2010) point out that although donations could have promoted peace and solidarity, the patronage rationale was supported instead and the opportunity was lost. Moreover, Khasalamwa (2009) investigated to what degree the humanitarian aid generated after the 2004 tsunami was adequately used to "build back better", only to find out that the pre-existing vulnerabilities in the area were not eradicated.

Besides, in some poor countries, governments invest deliberately too little in disaster prevention because they rely on external humanitarian aid (bail-out effect). In extreme cases, a population is deliberately ignored by the government (racket effect) so that humanitarian help will be received and abused (Cohen and Werker, 2008). On the other hand, relief organizations tolerate the abuse of disaster relief by governments because their priority is to comfort people that need help. Nevertheless, involving humanitarian relief organizations in prevention activities, providing disaster relief locally rather than centrally and supporting political development and proper disaster preparation activities in some countries may significantly reduce these effects (Cohen and Werker, 2008). However, relief organizations suggest that it is easier to collect donations for relief rather than for preventive activities (Cohen and Werker, 2008).

Raschky and Schwindt (2009) claim that foreign aid may have positive and negative effects on a country's protection against natural hazards, and call for redesigning strategies for international assistance. In more detail, Raschky and Schwindt (2009) conclude that ex-ante in-kind and conditional transfers (e.g., mitigation activities, local adaptation measures, and early warning systems) may reduce the "Samaritan's dilemma" challenge. Buchanan (1975) referred to the "Samaritan's dilemma" in order to stress that altruism of donors may lead to adverse behavior of the recipients should the donor not act strategically. Buchanan (1975) differentiates between a passive (the donor's help depends on the effort of the recipient) and an active (the donor will always help) "Samaritan's dilemma" (Raschky and Schwindt, 2009).

### *Commercial Insurance and Re-Insurance Industry*

Worldwide, governments and communities have to face the challenges related to the "natural disaster syndrome" which consists of the combination of limited protection before the event and the high costs following a catastrophic event (Kunreuther, 1996). On a global level, insurance penetration strongly varies among countries and continents. A global distribution of non-life insurance premiums per capita shows that although in the USA the insurance premium per year for the average citizen is US$500, in Africa or parts of Asia the premium is less than US$5. It is noteworthy that in many African countries there is absolutely no insurance coverage (Linnerooth-Bayer et al., 2011). Even when insurance coverage exists, this is limited to some wealthy households or to industrial and commercial assets that can afford to pay a premium (Arnold, 2008). However, elsewhere in the world, low-income households may have a priority with regard to potential financial assistance. For example, compensation in Germany is based on means testing, whereas in Romania, citizens that receive welfare payments are directly insured by the state (Priest, 2014). However, in the European Union, the market penetration rate of disaster insurance is generally low but it still depends on the hazard type. For example, penetration rates for insurance for storms are high in comparison to insurance for earthquakes or floods (European Commission, 2013). Nevertheless, in each member state the picture is different. In the UK, for example, flood insurance penetration was particularly low before the 1960s probably because of lack of product awareness. Following a series of changes in policy agreements between the government and the insurance industry, as well as technological advances that enable a more accurate assessment of risk, insurance penetration became increasingly higher (Penning-Rowsell et al., 2014).

An efficient insurance system should be involved in all the phases of the risk cycle to reduce economic loss and to assist reconstruction (European Commission, 2013). However, climate change and increasing development increase the pressure on the insurance market, which is called upon to cover high cost of extreme weather events more frequently than in the past. Insurance for natural hazards differs in many ways from other type of insurances due to the high amount of loss that is connected to the occurrence of natural hazards, and the high costs that are required to keep the market running (Raschky and Weck-Hannemann, 2007). Moreover, natural hazards affect at the same time a high number of concentrated properties in contrast to fire or theft, which normally affects individual properties (European Commission, 2013). In addition, the phenomena of moral and charity hazard and adverse selection affect the market for natural disaster insurances in a negative way.

"Charity hazard" is defined as "the tendency an individual at risk not to procure insurance or other risk financing as a result of a reliance on expected charity from others such as friends, family, community, nonprofit organizations or a government emergency program" (Browne and Hoyt 2000: 293). In this case, the individuals often rely solely on governmental funds that, in most of the cases, are not sufficient and in other cases are not fairly allocated due to corruption or political interest (e.g., forthcoming elections, Raschky, 2008). On the other hand, "moral hazard" concerns home or business owners who, having bought insurance, change their behavior by taking no precautionary measures whatsoever to protect their property from a potential natural hazard. In this case, the costs for the insurance companies, should a hazardous event occur, are particularly high leading to higher insurance premiums for the insured. Governmental relief not only is responsible for the low insurance penetration but also for the lack of incentives offered to homeowners to adopt local mitigation measures (Raschky et al., 2013). Although the need for public private partnerships (PPPs) has been pointed out very early in the past, the problem still exists. PPPs may encourage mitigation through low insurance premiums in exchange for mitigation and adaptation measures and enforcement of building codes (Kleindorfer and Kunreuther, 1999). In a different scale, "moral hazard" may be also observed when governments rely on external humanitarian help and underinvest in prevention measures (Cohen and Werker, 2008). Finally, "adverse selection" refers to the tendency of property owners to purchase insurance only if they are convinced that the risk involved is particularly high (European Commission,

2013). This results in only a small amount of individuals being insured, which increases insurance premiums and puts great pressure on insurance companies that cannot afford to pay for the disaster costs in the aftermath of the disasters. However, the correlation between risk and coverage varies significantly across insurance markets (Cohen and Siegelman, 2010).

Finally, poverty, especially in developing countries, forces household owners to self-insure and use their own savings to cover the costs of losses due to natural hazards (Arnold, 2008). Legislation supporting mandatory insurance schemes could spread the costs among more people and reduce high insurance rates (Messer, 2003). This is already happening in some European countries including Switzerland. In 19 Swiss cantons, insurance against fire and natural hazards (flood, windstorm, hailstorm, avalanches, snow, rock fall, and landslide) is mandatory and offered by public (cantonal) insurers, whereas in the rest of the country building owners are obliged to insure their properties with private insurance companies. The building contents of all buildings in Switzerland are also insured by private insurers (Hausmann et al., 2012). Moreover, various alternative insurance solutions are available; however, since most of them are running in form of pilot schemes and have not experienced a major natural disaster yet, their effectiveness cannot be assessed (Linnerooth-Bayer et al., 2011).

### *6.4.3 Deficiencies in Land Use Planning*

The trend in the 1960s related to the implementation of exclusively structural measures was followed by a clear shift towards nonstructural risk management solutions (e.g., land use planning regulations), which evolved into a combination of the two in the 1990s. However, as it has often been observed in the European Alps (Fuchs et al., 2017), the implementation of structural measures encouraged the development within protected areas, increasing vulnerability ("levee effect"). The term "levee effect" was introduced by Burton and Kates (1964) based on the case of US farmers described by White (1945) who farmed the land between the levee and the river since their land had been cut off by the levee. Giving examples from the Netherlands, New Orleans (USA) and the Po River (Italy), Di Baldassarre et al. (2015) refer to the "levee effect" as the increasing vulnerability related to the low frequency of flood events due to the existence of structural protection measures. The opposite effect, reflecting the decrease of vulnerability due to the frequent occurrence of flood events, is called the "adaptation effect". Di Baldassarre et al. (2015), comparing the two types of dynamics, suggest that societies following the adaptation effect tend to respond better to changes in the frequency and magnitude of floods and suffer less loss since they have moved their homes and their economic activities to higher ground.

Land use planning may lead to a decrease of vulnerability; however, communicating the use of hazard zones and associated restrictions and regulations to the public may be challenging. Problems may arise with buildings already located within hazard zones, as well as conflicts between hazard zone planning and development plans of the community.

## 6.5 Case Study: Institutional Vulnerability Assessment in Austria

### *6.5.1 Institutional Context*

As Austria consists of about 75 percent mountainous terrain the main hazard types that cause damages and disruption to communities are torrent processes, snow avalanches, and landslides. Mountain torrents (defined by Austrian Law as permanent or ephemeral streams, Republic of Austria, 1975) were responsible for thirty-nine deaths and 4,630 damaged buildings whereas 1,502 buildings were damaged during the period of 1994–2004 (Oberndorfer et al., 2007). Apart from mountain hazards, river floods are also a threat in Austria, as well as earthquakes that, although rare, may cause considerable damage (Embleton-Hamann, 2007). According to Fuchs et al. (2015), approximately 14 percent of the buildings in Austria are exposed to natural hazards. In more detail, 9 percent of the buildings are exposed to flooding and 5 percent are exposed to mountain hazards including torrential flooding and snow avalanches.

The institutions that set up the institutional context in Austria include Federal Ministries and other public authorities and organizations, private organizations and emergency services such as the police and the fire brigade (Table 6.1).

Some of the main Austrian institutions are presented in the following paragraphs including the legislative framework, the Austrian Catastrophe Fund and the procedure of hazard zoning.

#### *Legislative Framework*

Austria is a federal republic consisting of nine federal states that comprise most of the legislative and executive powers. In the case of natural hazards, however, and due to the localized nature of some events, the local governments and municipalities hold extensive competences in the field of disaster relief. On the national level, there is no national disaster management policy, but a national disaster management coordination (IFRC, 2012). Concerning

Table 6.1. *The Austrian institutions related to the management of natural hazards*

| Institutions | Description | Examples |
|---|---|---|
| Federal Ministries | Ministries involved in the disaster management in Austria include: Federal Ministries related to the natural hazards and the structural protection measures related to them, Federal Ministries responsible for issues of civil protection and ministries accountable for the financing of disaster management | Austrian Federal Ministry for Agriculture, Forestry, Environment and Water Management, Austrian Federal Ministry for Transport, Innovation and Technology<br>Austrian Federal Ministry of Science and Research and Economy<br>Austrian Federal Ministry of Interior<br>Austrian Federal Ministry for National Defence and Sport<br>Austrian Federal Ministry of Finance |
| Public authorities/ agencies | Public authorities at federal level responsible for the technical protection against natural hazards, preventive planning, monitoring, evacuation and the management of emergency operations | Water Authority, Spatial Planning Commission, Forestry Commission, Police, Civil Protection, Local Authorities |
| Public providers and organizations | Organizations involved in the collection of scientific data related to hazardous processes, the issue of warnings and monitoring of natural processes and hazard zone planning | Geological and the Meteorological Agency, the Federal Office for Water Management, Austrian Torrent and Avalanche Control Service |
| Private providers and organizations | Private organizations and corporations involved in disaster management in Austria | ÖBF: Austrian Federal Forests<br>ÖBB: Austrian Federal Railways |
| Emergency services | Services responsible for the response phase and the implementation of the emergency plan | The Austrian Armed Forces, the police, the fire brigade and the Red Cross as well as the crisis management group of each federal state. |

Adapted from Holub and Fuchs (2009) and Thaler (2015).

the Austrian legislation regarding natural hazards, relevant articles may be found in different acts at federal level such as the Austrian Forest Act (Republic of Austria, 1975), the Water Act (Republic of Austria, 2014) or the Disaster Fund Act (Republic of Austria, 1996). The structure of the legislation framework in Austria is demonstrated in Figure 6.2. It is clear that the Austrian Service for Torrent and Avalanche Control (WLV) and the Water Engineering Administration (BWW) have not only the policy but also the operational responsibility for a number of actions concerning the management of natural hazards such as the flood defense and the hazard zone mapping. However, their responsibility varies according to the location of the area of interest. The BWW is responsible for the management of the lower sub-catchment areas, whereas the WLV operates in the upper sub-catchment.

In order to understand the institutional framework of a country, an overview of the main actors, their tasks, and the policy related to them is necessary.

Due to the federal character of the country, federal governments and local authorities play the most important role in natural hazard management implementing a number of their own policies and laws. Every federal state has its own disaster management law. Additionally, building codes ("Bauordnungen") fall also under the responsibility of each federal state (IFRC, 2012). Last but not least, in Austria a communication and coordination mechanism called SKKM (Staatliches Krisen- und Katastrophenschutzmanagement) is responsible for the management of crises and catastrophes in Austria at national level.

### *The Austrian Catastrophe Fund*

In Austria, there is no compulsory property insurance regarding damages due to natural hazards. Compensations are mainly covered by the Austrian Catastrophe Fund. The Austrian Catastrophe Fund Act was launched in 1966 (Republic of Austria, 1966, 1996) with the main purpose to provide the financial means for the development of preventive measures and compensation to the affected population (Gruber, 2008). The Austrian Catastrophe Fund was revised many times in the following years becoming the main loss compensation instrument, since in Austria private insurance is not mandatory. The financing of the Austrian Catastrophe Fund is based on surcharges on taxes (income, wage, corporate, etc.). However, the compensation support is provided to the federal states that make decisions in supporting individuals (Holub and Fuchs, 2009).

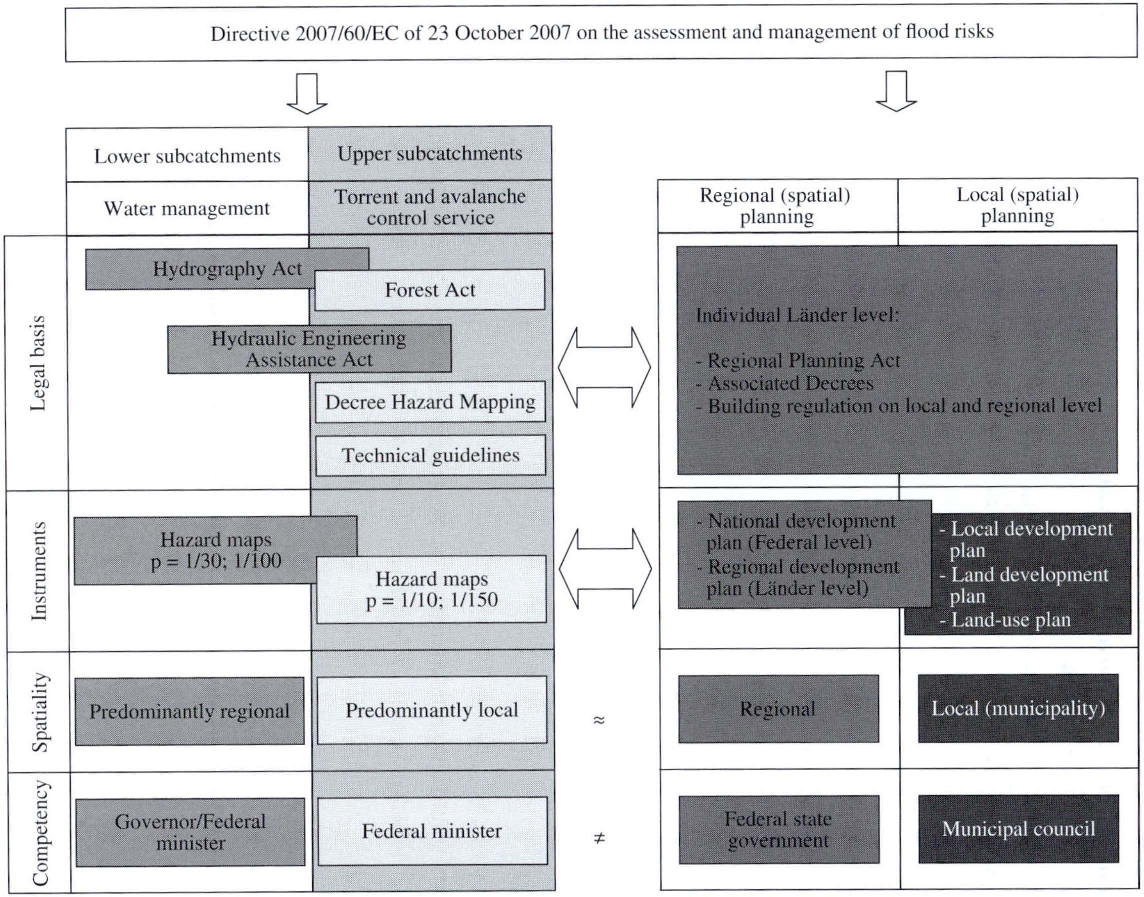


Fig. 6.2. Legislation structure in natural hazards management in Austria. After Holub and Fuchs (2009: 525)

### *Land Use Planning and Hazard Zoning*

According to the Austrian Forest Act 1975 (Republic of Austria, 1975), hazard zones should indicate the catchment areas of rivers and torrents as well as the areas that are susceptible to related hazards. The Austrian Service for Torrent and Avalanche Control is the responsible authority for the development of these zones that are provided for the whole of the country and mainly focus on torrents, avalanches, and erosion (BMLFUW, 2017; Republic of Austria, 1975). The hazard zones are based on modeling of several hazard types, records on characteristics and consequences of past events, and the interactions between different processes and the presence of elements at risk. They are based on two-design events, an infrequent (1-in-150 years) and a frequent one (1-in-10 years, Holub and Fuchs, 2009). The hazard zones are given to the municipalities for public review four weeks before they are officially approved, in an effort to encourage participation of the public but also to avoid possible conflicts that are often related to hazard zone mapping (BMLFUW, 2017). The hazard zones are related to recommendations for development within the zone; however, they are not legally binding and serve more as an expert opinion or a forecast unless they are referred to by the individual federal spatial planning law (Holub and Fuchs, 2009). Additionally, an internet-based platform, HORA Flood Risk Zoning Austria (www.hora.gv.at), is available in Austria presenting risk zones for floods, earthquakes, and hail individually for the whole country (Stiefelmeyer and Hlatky, 2008). HORA was developed by the Federal Ministry of Agriculture, Forestry, Environment and Water Management (BMLFUW) in collaboration with the Austrian Insurance Association (Stiefelmeyer and Hlatky, 2008). Its main aim was the risk awareness of all interested parties, and for this reason, it is provided free of charge online. It is based on a hydrological and a hydraulic model that contributed to the representation of the flood extent and water depth for three return periods (1-in-30, 100, and 200 years) for 25,500 km of river network within the country. HORA clearly showed that more than 10 percent of the building stock in Austria is located within flood risk zones without considering the effect of existing flood protection (Stiefelmeyer and Hlatky, 2008). HORA was later extended to include more hazard types, namely, earthquake and hail. As far as floods are concerned, HORA has been recently complemented by the new system WISA (www.wisa.bmlfuw.gv.at) that fulfills the requirements of the EU Floods Directive (European Commission, 2007).

### 6.5.2 Institutional Vulnerabilities in Austria

*Compensation*

According to Gruber (2008), although there is a wide range of insurance options in Austria, the market penetration of private property insurance is very low. This may be due to underestimation of risk by the homeowners and due to the nature of some hazards (e.g., flood) that tend to have extended costs. This leads to difficulties for the insurance companies to compensate the loss without involving reinsurance companies (Url and Sinabell, 2008). However, the main reason for the low popularity of private insurance in Austria seems to be the Austrian Catastrophe Fund. Nevertheless, the Austrian Catastrophe Fund covers 20–30 percent of the damage costs, although in some cases up to 80–100 percent may be covered for private households (Raschky et al., 2013). Private insurance, on the other hand, pays better but still only up to 50 percent of the insured amount for private households and more than that for businesses (Gruber, 2008; Holub and Fuchs, 2009). On the contrary, in the case of the Austrian Catastrophe Fund, the premiums paid are not dependent on exposure or location leading to social injustice. Federal states that are often impacted by floods pay the same premium (as percentage of their share from income taxes) as federal states that are not affected so often by natural hazards (Holub and Fuchs, 2009). An additional disadvantage of the Catastrophe Fund is that it focuses more on the ex-post phase of a disaster. The Catastrophe Fund compensates for losses and damages but it does not invest ex-ante in raising risk awareness or promoting local structural adaptation measures that may reduce significantly consequences and costs. A significant drawback is that the remaining budget is accumulated through the years until it reaches an upper limit amount (€ 29 million) and then it is used for other national expenses. Hence, in the case of extreme events with unexpected consequences, such as the floods of 2002 and 2003, the financial means available for the compensation may not be enough (Holub and Fuchs, 2009). Furthermore, the responsibility for the compensation of the homeowners lies with the federal state government and the owners have no legal right for financial assistance. Thus, property owners that have additionally an optional private insurance do not receive the adequate compensation by the state, leading to the paradox that risk-aware homeowners are actually worse insured than others (Holub and Fuchs, 2009). Last but not least, the fact that in Austria homeowners have a low uncertainty related to if and to what degree they will be compensated by the state may lead to crowding our effect on private insurance (Raschky et al., 2013).

### *Challenge of Exposure*

Hazard zones in Austria strongly support the efforts for prevention; however, partly for technical reasons and partly due to the fact that hazard zones are not a legal binding instrument, a very high number of buildings are still located within red hazard zones (Url and Sinabell, 2008). In addition, although the number of buildings within the hazard zones is known, their value and the potential economic consequences are not known due to inadequate damage documentation and incomplete inventories of elements at risk in Austria (Url and Sinabell, 2008). Despite the existence of hazard zones, development has not been prevented within them, since at least 32 percent of all the exposed buildings were built after 1976 when the Directive on Hazard Mapping was amended (Fuchs et al., 2017). Building in the hazardous areas may be explained by administrative particularities (land use plans are often interpreted locally), by overestimation of the protection offered by engineering structures, and by conflicting priorities (economic growth versus ecological concerns, Fuchs et al., 2017).

Moreover, although hazard zones in Austria are available in advance for public inspection, conflicts between authorities and property and business owners located within the hazard zones still exist. The main problems involve the effect of the hazard zones on the prices of properties and land within high-risk zones, existing property and development within the high-risk zones, and conflicts regarding future development plans. Furthermore, people located within high-hazard zones and especially vulnerable groups (e.g., children and elderly) are not adequately informed on the measures they have to take (response behavior, local adaptation measures, etc.) in case a hazardous event occurs. Adequate risk communication strategies and encouragement of participation may solve these problems. However, the problem of residual risk remains.

### *Local Adaptation Strategy*

Raschky (2008) suggests that the construction of protection measures and other examples of resources allocation for the management of natural hazards should be based on the concept of "public goods". Public goods are characterized by nonrivalness and nonexcludability. This means that their consumption or use by one person does not reduce its quantity and availability for the others and that no one is excluded from using them (Burrell, 2011). The responsibility for public good projects should not be centralized to ensure that local needs are met (Raschky, 2008). An alternative to structural protection measures for a large number of buildings are the local adaptation measures that are normally located within or are adjacent to an element at risk, may reduce the costs of the impact of a natural hazard significantly (Holub and Fuchs, 2008). These measures may include sealing plates for doors and windows, deflecting structures, deposition areas, drainage works, stabilizing structures, etc. (Egli, 1996). Holub and Fuchs (2008) conducted a cost-benefit analysis for local adaptation measures and showed that the benefit-cost ratio is higher than the one for local structural measures. However, the homeowners need incentives in the form of reduced premiums or reduced bank credits in order to invest in local adaptation measures and to succeed in a significant reduction of risk. Additionally, enhancing awareness and participation of the public would also encourage people to invest in these types of measures although enforcement by inclusion of local adaptation in building codes would also be possible (Holub and Fuchs, 2008). Finally, the cost benefit analysis of local adaptation measures should include not only the damage costs but also the relocation costs of the homeowners for the time that a building is not functional. Local adaptation measures and adaptation design may reduce the costs of the impact but also the physical resilience of the building, which may return faster in its initial state.

## 6.6 Conclusion

It is noteworthy that although governments, scientists, and individuals have a growing improved knowledge and understanding regarding natural hazards, the resulting consequences and losses are increasing. White et al. (2001) recognized that research on natural hazards should include integrated research of different dimensions of the problem, which should be included in broader environment and development goals. Institutional capacity is a prerequisite for the achievement of this goal. Improvement of the institutional setting and promotion of specific features (such as participation, equality, justice, communication, etc.) may lead to successful natural hazard management and more resilient communities. This was also pointed out by Baker and Refsgaard (2007) that, investigating the institutional gaps of hurricane Katrina, also concluded that institutional vulnerability decreases when institutions promote collaboration between individuals and the community, make resources available promoting education and awareness and training as well as, enhance resilience by investing in participatory planning, integrated command systems and evaluation of plans and practices following catastrophic events. Furthermore, the importance of participation of all stakeholders in disaster risk reduction has been stressed also in the Sendai Framework which clearly emphasizes that for successful DRR an "all-of-society engagement and partnership is required" in order to share responsibilities (UNISDR, 2015a). However, there may be several degrees of participation. As the ladder of Arnstein (1969) suggests, citizens may not participate at all or they are even manipulated (bottom of the ladder), there may be a level of participation where citizens participate but without influencing decisions actively (tokenism) and, finally, at the top of the ladder, there is "citizen power" characterized by "partnership", "delegated power" and "citizen control" which represents a full control of the decision-making process by the citizens. Additionally, the importance of informal institutions should not be underestimated, and local characteristics and particularities should be considered in all phases of the risk cycle. Successful participation can be supported by adequate risk communication strategies. Höppner et al. (2012) suggest that risk communication should be based on a two-way rather than one-way communication between stakeholders in risk management and to contribute to the building of social capacity through the building of trust between the stakeholders. According to the same study and based on a review of risk communication practices across Europe, there is a need for comprehensive long-term communication strategies. In other words, raising awareness is not enough if it is not sustained through the years to come. Moreover, risk communication strategies vary for different hazard types, and for this reason, they should be tailored accordingly. However, the main challenge in risk communication is apparently the change from one- to two-way communication. People at risk should be involved in participatory projects during the early phases of the risk cycle (preassessment, appraisal phase, and prevention). Last but not least, risk communication strategies should be evaluated in terms of effectiveness; but also design, language, content and timing, and should be improved accordingly (Höppner et al., 2012).

It is widely accepted that decentralization may have positive and negative effects on disaster risk reduction. According to Twigg (2015), the positive effect of moving responsibilities from the central to the local level is that decisions will be made taking into consideration local needs. On the other hand, in this way, DRR may become isolated from the central decision-making and sometimes fragmented into smaller initiatives, although they may not have the capacities (financial and technical) to manage DRR. Additionally, local actors cannot deal with the deeper roots of vulnerability since they are not capable to see the whole picture but also they are not entitled to address it (Twigg, 2015).

Political stability, transparency, and impartiality are paramount features of resilient societies and may significantly reduce institutional vulnerability. As already mentioned above, marginalized communities and segregated societies may amplify disaster consequences. According to Alexander (2002), there is no room for racial and ethnic insensitivity in the way we manage natural hazards. Emergency plans should consider cultural norms and local individual characteristics. The degree of segregation of a community and the existence of marginalized groups should be taken into consideration in the designing of DRR.

Focusing on the reconstruction and recovery phase, institutional vulnerability is often related to risk transfer mechanisms that are often translated into availability of private or public insurance schemes. Alternative insurance solutions, such as micro-insurance schemes, should be further developed; however, additional research should be carried out regarding the value of such schemes for the benefit of the poor. A multi-stakeholder approach has to be ensured including community groups and the implications of climate change scenarios should also be investigated (Arnold, 2008).

Furthermore, insurance companies and authorities should promote the implementation of local adaptation measures. In the case of floods, these may include replacement of vulnerable floors with water-resistant ones, moving of electrical appliances to higher floors, creating water barriers with sandbags, etc. A survey conducted in the Netherlands focusing on the willingness of homeowners to undertake and finance mitigation measures in their properties revealed that many would undertake mitigation measures in exchange for an insurance premium discount. By including these types of mitigation measures in cost-benefit analysis the benefits associated to them will become evident (Botzen et al., 2009).

To conclude, the most important closing remarks of this chapter are outlined:

- Decentralization as well as multilayer and polycentric institutions reduce the institutional vulnerability of communities and enhance the accountability of authorities.
- Participation and multi-stakeholder involvement improve risk communication as long as the participation level is high enough to ensure involvement in the decision making, racial and ethnic sensitivity and consideration of informal institutions.
- Political stability, transparency and impartiality may safeguard the adequate implementation of strategies, policies and legislation.
- Alternative risk transfer options may ensure the satisfactory coverage of damage costs and the timely rehabilitation of the built environment as well as offer incentives for local adaptation measures.

# 7

# Resilience: An Introduction

CHRISTOPHER T. EMRICH AND GRAHAM A. TOBIN

## 7.1 Introduction

Previous chapters in this book have examined hazard vulnerability through its various themes, including physical forces, social issues, economic pressures, and institutional organizations. The authors have presented an extensive series of arguments and analyzed various models that together highlight the complexity associated with understanding vulnerability. Indeed, there is extensive literature on vulnerability, from both academic and gray sources, that seeks to address community hazardousness (see for example, Cutter et al., 2003; Emrich and Cutter, 2011; Gaillard et al., 2009; Kelman and Mather, 2008; Montz et al., 2005; Pelling, 2003b; and Wisner et al., 2004; among others). For a far-reaching treatise on vulnerability measurement, see the edited collection by Birkmann et al. (2013).

For the next section of this book, we introduce the concept of resilience, a dynamic often regarded (wrongly, as we shall see) as simply the converse of vulnerability. We incorporate several topics, including a review of definitions of resilience, a discussion on the evolution of resilience frameworks, an examination of temporal and spatial scales as they relate to resilience, an overview of approaches and thematic areas under the umbrella of resilience, appraisal of the shortcomings of a resilience focus in hazards research, and an assessment of future trends. Furthermore, it is difficult to discuss resilience without some consideration of sustainability; the two are closely tied, and it is to this that we turn our attention as it applies to community (Tobin 1999: 14),

> Sustainable and resilient communities are defined as societies which are structurally organized to minimize the effects of disasters, and, at the same time, have the ability to recover quickly by restoring the socioeconomic vitality of the community.

Nevertheless, as we shall see it is difficult to establish a precise definition of resilience as demonstrated by the titles of these recent scholarly articles: "Resilience: A bridging concept or a dead end?" (Davoudi et al., 2012) [Note this latter title has been shortened]; "The geographies of community disaster resilience" (Cutter et al., 2014); "The paradox of social resilience: How cognitive strategies and coping mechanisms attenuate and accentuate resilience" (Shaw et al., 2014); "The geographies of resilience: Challenges and opportunities of a descriptive concept" (Weichselgartner and Kelman, 2015); "Resilience to what? Resilience for whom?" (Cutter, 2016a).

## 7.2 Definitions of Resilience

The definition cited above provides a generalized picture of what community resilience entails; to be resilient communities must have the wherewithal to mitigate the impacts of disasters and at the same time the ability to recover. This is useful as an introduction but it does not cover all the facets of what we now understand constitutes resilience. Is it, for instance, a process through which impacts of disasters are minimized; or is it an outcome, the actual reduction in losses? Dovers and Handmer (1992: 270) differentiated three levels of societal resilience: (1) resilience through resistance to change; (2) resilience through incremental change at the margins; and (3) resilience through openness and adaptability, which suggests that resilience needs to be addressed at several levels. Fortunately, the literature has blossomed since the turn of the century as researchers have explored in great detail what makes a community resilient, although it should be stressed that there is no single definition accepted by all (Gaillard, 2007; Klein et al., 2003).

Resilient cities have been the focus of some important research. Tobin (1999) argues that sustainability goals may be somewhat unattainable in the light of significant disaster impacts, whereas Godschalk (2003) takes a more optimistic view citing the need for research into hazard mitigation. In reality both authors suggest that outcomes will depend in part on levels of vulnerability and amount of resources available in particular to the communities under threat. As Godschalk (2003: 138) states "the ability to accommodate change gracefully and without catastrophic failure – is critical in times of disaster." These three studies are useful as guidelines but they do not provide details on exactly what we understand as resilience.

Bahadur et al. (2010: 2) point out that while the relationship between vulnerability and resilience is contested, resilience is often seen as the opposite of vulnerability, that is, low resilience equates with high vulnerability and high resilience corresponds with low vulnerability. To a degree and under certain circumstances this can be true, but there are many other outcomes and it may be that the emphasis on vulnerability in the earlier literature perhaps detracted somewhat from evolving concepts of resilience. In one sense if vulnerability constitutes the level of risk and exposure, then resilience counteracts this; but just as vulnerability means far more than risk and exposure (see preceding chapters in this book), so resilience encompasses more nuanced relationships among recovery, adaptability, and preparation. For instance, in measuring and operationalizing resilience to natural hazards, definitions have tended to incorporate three common themes: multidimensional concerns and composite indices; systems approaches and equilibrium arguments; and temporal/spatial scale concerns. Weichselgartner and Kelman (2015: 249), in their excellent review and assessment of the resilience literature, suggest that "resilience has been replacing 'vulnerability' and 'sustainability' as a guiding principle in development planning." The very nature of a general or simplistic definition, they argue, could render the term almost meaningless.

> While an all-encompassing multi-interpretable idiom has attractions as a unifying concept and political vision, there is an inherent danger that the term becomes an empty signifier that can easily be filled to justify a specific goal as happened to 'sustainability' to a large degree.

Weichselgartner and Kelman go on to cite Stumpp's (2013) article in support of this contention. Nevertheless, it is worth exploring some of the definitions proposed by authors from various disciplines because they have added new perspectives to understanding resilience.

Taking a different approach, for example, Cutter (2016a) rightly asks in her introductory commentary, "Resilience to what? Resilience for whom?" In other words, what guiding principles should we use to address resilience especially if society is to cope with environmental change, brought about by both acute and chronic conditions? Cutter (2016a: 110) adds,

> The discourse on disaster resilience and vulnerability entails diverse research and policy communities each assigning different meanings to concepts, which in turn influences their measurement and implementation in decision-making contexts.

While definitions of resilience abound from a variety of disciplines (see Bec et al., 2016), several terms stand out: bounce-back and recovery concepts; adaptive capacity; coping and the ability to absorb change; and acceptance of levels of risk. The concept of resilience meaning to bounce back is widespread. For example, Saunders and Becker (2015) refer to the ability to bounce back or recover from short-term phenomena, but also address other evolving aspects of resilience. Similarly, Paton et al. (2015: 37) applies the concept of resilience, in this case to Christchurch following the earthquakes of 2010 and 2011, stating the need to understand "attitudes and actions that increase people's ability to cope with, adapt to and recover from earthquake hazard consequences that can occur suddenly and can last for extensive periods." This idea that communities must be able to recover, that is bounce back, from disaster is intrinsic to most definitions of resilience.

Several definitions emphasize the capacity of communities to absorb and accommodate change. For instance, the UNISDR (2009: 24) definition states the "ability of a system, community or society exposed to hazards to resist, absorb, accommodate to and recover from the effects of a hazard in a timely and efficient manner, including through the preservation and restoration of its basic structures and functions." Similarly, the National Academies (2012: 1) describe resilience as "the ability to prepare and plan for, absorb, recover from, and more successfully adapt to adverse events." There is an inherent weakness in such definitions, however, for "the ability to bounce back" does not necessarily address issues of vulnerability. A community that returns to the status quo after a disaster will merely perpetuate the same conditions that existed prior to the event, and do nothing to mitigate the exposure of vulnerable populations.

The question arises as to individual responsibility and how to address hazard impacts so that resilience is an ongoing concern of society. Crowley and Elliot (2012) point out that resilience is essential because populations need to have the ability to live with hazards and must accept a certain level of risk and as such need to be informed about what that level of risk is. This is where both local and national governments need to play a more pivotal role, while others have demonstrated that, in addition to public action, individual responsibility is also necessary (Queensland Government, 2016; Tobin, 2015; UNISDR, 2012a). Mileti (1999: 32–3) supports the need for at least local level responsibility and the necessity for local communities to weather the disaster without too much outside support.

> Local resiliency with regard to disasters means that a locale is able to withstand an extreme natural event without suffering devastating losses, damage, diminished productivity, or quality of life and without a large amount of assistance from outside the community.

However, Boon et al. (2012) using Bronfrenbrenner's bio-ecological framework (2005) point out that there is limited research linking individual characteristics to community resilience. In this context, they argue for the study social factors such as,

> [1] personal attributes such as social competence, problem solving, autonomy, self-efficacy, and purpose and [2] contextual environmental influences such as peers, family, work, school, and local community
>
> *(Boon et al., 2012: 402).*

Nevertheless, some authors have stressed the need for research into the individual and social networks in understanding community resilience (Faas et al., 2014; Jones et al., 2015). Zautra (2009:1–2) argues that "people who are resilient display a greater capacity to quickly regain equilibrium physiologically, psychologically, and in their social relation following stressful events [and] equally important, is sustainability, that is, the capacity to endure and continue forward in the face of diversity." Such findings suggest that there is a positive relationship between resilience and individual traits, which foster helpful outcomes to disaster. For instance, Masten and Obradovic (2008: 9) in discussing development science, point out that "individual resilience refers to the processes of, capacity for, or patterns of positive adaptation during or following exposure to adverse experiences that have the potential to disrupt or destroy the successful functioning or development of a person." To understand resilience fully, therefore, we should consider both individual and societal concerns.

There are multifaceted dynamics that can determine the level of community resilience, described by Crowley and Elliot (2012: 209) as "a complex web of social interactions, characteristics and capacities that enable a community to live with the hazards they face." In fact, resilience is multidimensional and spreads across assorted systems, exhibits an inverse relationship when referenced/compared with vulnerability, but still lacks a contemporary definition that is accepted across multiple disciplines (Bahadur et al., 2010). Consequently, it is apparent that the play between vulnerability and resilience is decidedly complex, and that both the human and natural environments must be evaluated if resilience is to be fully determined (Adger et al., 2005).

Other researchers have addressed multidimensionality through a systems approach, questioning whether a human-based system can absorb impacts and adapt to change (Gunderson, 2010). It is through a systems perspective that we can provide a comprehensive framework for examining resilience, identifying not only the vast array of variables that exacerbates or ameliorates disaster impacts over time and place, but we can also focus on the linkages among the many variables.

Godschalk (2003: 137) says that a resilient city is a "sustainable network of physical systems and human communities" both of which must function during times of crisis; a failure in either system can result in high levels of vulnerability. Godschalk (2003: 137) goes on to point out that "Resilient cities are constructed to be strong and flexible, rather than brittle and fragile." Welsh (2014: 15) supports this necessity for linkage between the two systems,

> Resilience is primarily conceived as the property that captures the capacity of the entity to anticipate, adapt to and recover from the event such that it resumes its original configuration, shape, functional relationships or trajectory afterwards. The

linking of social and ecological systems and integration of complexity theory produces a model of interlinked systems in continual adaptive cycles of growth, accumulation, restructuring and renewal.

Gunderson (2010) talks of the ability of systems to absorb change and yet still persist, of the stability of systems, and the ability of human systems to adapt and predict. Still others refer to resilience as "the capacity of a system to absorb and recover from the occurrence of a hazardous event" (Timmerman, 1981: 21). Wilson (2010) suggests that while resilience is a contested term, it is seen generally as the capacity of a system to absorb disturbance and reorganize while undergoing change to retain essentially the same function, structure, identity, and feedbacks. Resilience is then measured by the size of the displacement the system can tolerate and yet return to a state where given functions can be maintained. Walker et al. (2004: 2–3) describes four aspects that have a bearing on the resilience of systems: latitude, resistance, precariousness, and panarchy. Latitude refers to the extent (defined as a threshold) to which the system can be stressed before losing its ability to recover; resistance is the ease or difficulty with which the system can be changed; precariousness refers to how close the system is to its threshold; and panarchy is the scale of interaction with other (sub)systems. Walker et al. (2004: 3) provide examples of panarchy as external oppressive politics, invasions, market shifts, or global climate change, which can trigger local surprises and regime shifts. Thus, thinking of resilience in systems terms provides a valuable framework through which measurements might be taken to evaluate overall sustainability. However, from a practical perspective, an emergency manager or hazard planner would still need to know the goals and outcomes to which communities might be striving.

Benson and Craig (2014: 779) focused on resilience in the context of sustainability and suggested that,

> The pursuit of sustainability inherently assumes that we (a) know what can be sustained and (b) have the capacity to hold onto some type of stationarity and/or equilibrium. In contrast, resilience thinking acknowledges disequilibrium and nonlinear change in SESs [social-ecological system].

To paraphrase Carpenter et al. (2001), Folke et al. (2010), and Benson and Craig (2014) the concept of resilience must be considered as a system that is capable of accommodating change and retain its function and organization while increasing its capacity learning and adaptation.

Another issue noted in the literature is adaptive capacity or the ability of society to change during and after moments of stress. According to Bahadur et al. (2010: 2) there is a lack of consensus on the precise definition of adaptive capacity with respect to resilience; sometimes adaptive capacity is viewed as the "ability to be resilient" and at other times as a learning process in response to disturbance in systems. Furthermore, Robinson and Carson (2016: 118) stress that much of what we know about resilience is "not accompanied by clear statements" and instead see resilience as evolving into "resourcefulness" as that gives local communities a chance to address overlooked aspects. This lack of clarity on precise definitions can lead to confusion within disciplines and no doubt leads to steep learning curves when entering into interdisciplinary research and applications.

## 7.3 Evolution of Resilience Frameworks

The development of resilience concepts in hazards research can be seen as a progression especially during the second half of the twentieth century. Initially, hazard research generally sought explanation of disasters through studies of the physical environment. While this work provided valuable lessons, particularly through diverse case studies, it did not fully explain the rise in losses and the increase in costs of mitigation, nor did it develop any sound theoretical underpinnings (Alexander, 1997; Hewitt, 1997). Thus, attention was turned to the human dimension of hazards and with this so emerged a better understanding of human vulnerability and community resilience and appreciation of complex events (Cutter et al., 2003; Montz et al., 2005; Smith and Petley, 2009; Tobin and Montz, 1997). The inclusion of vulnerability and resilience principles has fundamentally changed hazard research and we can point to a variety of works that have furthered our knowledge (see for example, Birkmann, 2006a; Paton and Johnston, 2006; Weichselgartner and Kelman, 2015; Wisner et al., 2004). Drawing closely from this literature, Tobin and Montz (2015: 2) state that,

> Disasters, therefore, are now understood to have systemic root causes embodied in socioeconomic and political systems that place people, often marginalized groups, in hazardous locations, in unsafe structures, and with few resources available for them to respond with security to events. There has indeed been a paradigm shift, turning attention to more general principles of hazardousness, risk, vulnerability, and resilience.

## 7.4 Temporal and Spatial Concerns of Resilience

One aspect clearly standing out from the definitions discussed above is that resilience operates at many scales, and must be addressed both temporally and spatially. We can identify, for instance, multiple levels within a society whereby resilience can be affected either negatively or positively, and simultaneously there are multiple sectors functioning on different time frames.

If we first look at the temporal components of resilience, we can see that the different stages of disaster can play fundamentally different roles. Saunders and Becker (2015) for instance discuss two timescales, in the context of the Christchurch earthquakes, the short-term period immediately following a disaster, where the emphasis focuses on survival and personal concerns, and the second long-term where the attention turns to recovery and rebuilding. In essence, however, we need to address a continuum along the time frame of disaster from the preimpact phase to impact and emergency operations, to the immediate post-impact period and recovery efforts, and finally to reconstruction and rehabilitation (which may include consideration of new and innovative mitigation measures). Of course, different socioeconomic and political forces are operating during each of these periods, which can confound how we address resilience.

Spatial scale also plays a role in determining resilience levels. Should attention be given to the community level or the individual? What is the responsibility of different levels of government? To what extent are there linkages between levels of government? Resilience has different applications at different scales and for different types of communities. Scale, therefore, is important because resilience takes on different meanings when initiated at the local scale, and it can be difficult to scale upward as the goals and data become too large to provide any meaningful insight (Robinson and Carson, 2016). In other words, the multidimensionality and dynamic nature of society make it difficult to fully understand and ultimately to address resilience.

In effect, of course, temporal and spatial components operate hand-in-hand. Barrios (2014) for example, looks at community resilience from an anthropological perspective and suggests that resilience stems from the ability of national and local political organizations as well as NGOs to work with hazard-prone communities and not draw dividing lines among them. Barrios (2014: 329 and 347) concludes by arguing that "definitions of community resilience and disaster mitigation programs must take the emergent and relational nature of communities into account in order to address the long-term causes and impacts of disasters" and that current definitions "of disaster resilience that implicitly theorize 'community' as static, unchanging and geographically bounded miss the relational and emergent nature of social groups." Ignoring the dynamic forces within society and over time, especially in post-disaster environments, may prove costly in terms of understanding community resilience and thus flexibility and change must be incorporated into community resilience planning.

We know that every sector of society is impacted differentially and hence has potentially greater or lesser resilience to disaster. In this sense, resilience is directly related to vulnerability; highly vulnerable areas invariably suffer disproportionately greater losses and face magnified challenges during the recovery phase. As put by Masten and Obradovic (2008: 2) a broad conceptual umbrella should include,

> (1) achieving better than expected outcomes in high-risk groups of people, sometimes referred to as overcoming the odds against healthy development; (2) sustaining competence or maintaining effective functioning under highly adverse conditions; and (3) regaining or attaining effective or normal functioning following a period of exposure to traumatic experiences or conditions of overwhelming adversity often described in terms of recovery, bouncing back, normalization, or self-righting.

Many attempts fail to capture the range of data from the household to the national level and subtleties are lost due to factors that cannot be measured by simple indicators. Quantitative attempts also fail to capture the larger social picture. To some extent research into social networks has added to our understanding of resilience but the ties are still somewhat tenuous. We know that certain characteristics of individuals' social networks lead to greater support and recovery. It is argued, for instance, that community recovery depends in part on individuals feeling that they are part of a strong network and can thus overcome adversity (Hall and Kerr, 1991; Hobfoll et al., 2002; Jones et al., 2014; Reissman et al., 2005; Tobin et al., 2014). For example, gender has been shown to play a role in recovery and hence foster or hinder resilience. In disaster environments, women are frequently the decision makers and active agents for change (Akçar, 2001; Enarson and Morrow, 1998; Enarson and Chakrabarti, 2009; Phillips and Hearn Morrow, 2008). Further work by Faas et al. (2014: 223) found that "women and men used different types of informal social support networks, with men receiving significantly more material, emotional, and informational support than women." It is apparent though that men and women often experience different challenges and advantages/disadvantages when seeking support through local and extra-local institutions.

Similarly, others have examined social networks highlighting: individual help-seeking, well-being and coping ability (Casagrande et al., 2015; Jones et al., 2015; Tobin et al., 2011); social capital (Aldrich, 2011, 2012; Hawkins and Maurer, 2010; Weil et al., 2012); interorganizational collaboration (Butts et al., 2012; Doerfel et al., 2013; Drabek and McEntire, 2002); health, stress, and anxiety (Kaniasty, 2012; Norris et al., 2002; Whiteford and Tobin, 2009); ethnicity (Messias et al., 2012); and impacts of relocation and resettlement (Faas et al., 2014; Oliver-Smith, 1991). Each of these attributes of social networks undoubtedly affects the nature of resilience, but it should be stressed that there is no simple relationship such that strong social capital always leads to higher levels of resilience, or that intergovernmental collaboration consistently fosters positive outcomes and recovery. A simple framework does not exist. Take for example resilience and personal social networks – any scientific relationship between them remains elusive and we do not know what network structures work best to foster resilience in disasters. Jones et al. (2013) have addressed this, examining personal networks to determine the connections between risk perception and number of relations, network density, various forms of network particularly degree of centralization, and the presence of subgroups in the network. They argue that dense networks of strong ties might create redundant feedback loops not conducive to the introduction of new information regarding evolving risk conditions. In addition, a number of subgroups within social networks would seem to be important because this increases choice options for individuals to seek help and support (Avenarius, 2003; Avenarius and Johnson, 2004; Unger and Powell, 1980). Similarly, density of social networks may be important because it is associated with trust within the network (Buskens, 1998); in contrast individuals from different networks could be expected to have a negative association. Perhaps a meta-analysis of these many studies would reveal prevailing patterns? In effect, networks are not necessarily resilient, even though they might have survived a disaster, because intrinsic vulnerability may not have been ameliorated.

## 7.5 Thematic Aspects of Resilience

Resilience science is seen by some as a rebranding or repurposing of existing systems theory although the debate around socioecological resilience has been ongoing for more many decades. From Holling's (1973) work on resilient ecosystems, to Dovers and Handmer's (1992) definitions of resilience, through Tobin's (1999) work connecting the concept of resilience to sustainability planning, to Alexander's (2013) etymological overview of prior work on the topic, there is seemingly endless academic rhetoric surrounding not only the term's definition, but also its genesis and application within different fields. What is certain is that the term resilience and its core meaning date back many years (see for example Folke, 2006). Alexander (2013: 2710) states:

> resilience (resiliency, resile) has a long history of multiple, interconnected meanings in art, literature, law, science and engineering. Some of the uses invoked a positive outcome or state of being, while others invoked a negative one. In synthesis, before the (…) [twentieth] century, the core meaning was "to bounce back." Mechanics,

aided by politics, had already started to change that: in both the literal and the figurative sense, under the aegis of the resilience concept, ductility had been added to elasticity.

Its emergence as a key term, and indeed a field of study within multiple disciplines, is a testament to the fact that science is only interested in gaining a higher-level understanding of individual and combined processes. In global terms, this concept is captured well by looking at the "Spaceship Earth" concept, which suggests that we must learn to see those characteristics uniting us or linking us together in order to understand the complex trends and critical needs of the world. Viewing the world as an interdependent system of relationships is fundamental to understanding how resilience of single systems can also lead to resilience across, between, and within systems of systems (Fuller and Snyder, 1969) or across multiple levels of functioning (Bronfenbrenner, 2005). The planet is one system – a dynamic combination of an infinite number of subsystems – and a resilient system when left to operate independently. However, the presence of other human activities and the accumulation of greater knowledge have brought about significant changes to many systems, changes that left unchecked and unchallenged will alter the functions of all other systems. In this taxonomic view, lower resilience or lack of resilience in one system can and does influence the resilience of other systems (Drabek, 2012). The concept of resilience comes into play within this theory as "the oil" within the system, ensuring that slips in one or more gears can be compensated for across the system and do not lead to catastrophic failure across all systems. Being resilient in the short term enables appropriate adaptation and adjustment strategies to be invented, vetted, and put into place to avoid larger scale calamities. This type of open-loop system theory, although complex, dynamic, and iterative, allows the identification of causes and takes into account improvements or interventions through a feedback mechanism which influences other systems or subsystems (Birkmann, 2006a; Cote and Nightingale, 2012; Norberg and Cumming, 2013). For a valuable synthesis of the literature on general resilience see Carpenter et al. (2001).

Stepping away from the umbrella that is the basic structure of the systems approach allows for a more pragmatic view into disentangled versions of resilience concepts as they are applied to its various subdomains. Here we can expound on insights from diverse actors within resilience science to understand in more detail some recent resilience themes stemming from environmental and ecological perspectives combining into the larger resilience field, namely: physical (infrastructure), social-ecological, economic, and organizational systems.

### 7.5.1 Physical Resilience

Engineering has always been interested in designing and building the most efficient and effective structures to suit human needs. The basic path of exploration and research in this field has largely remained unchanged since its inception and has led to amazing growth and findings in what we know today as "infrastructure resilience" or ability to reduce the magnitude and/or duration of disruptive events (Resilience, 2009). Just as in other areas of resilience science, however, physical resilience manifests itself differently depending on the context. What has become clear in recent years is the need to promote inclusivity in decisions around the built environment, its function, and its form. Bosher et al. (2007) found that inclusion of architects in all phases of the (building) process is vital to ensuring that the final products are not only resilient but also palatable to those who would utilize them. What is becoming more apparent in the race to build climate-ready communities is the fact that we cannot afford total infrastructure investment across the board. Some communities cannot afford to move and must invest in strengthening existing infrastructure such as softened coastlines, breakwater towers, artificial islands, and island array techniques (Mallory-Hill et al., 2012). Understanding that each country, city, or town faces a unique set of challenges in becoming resilient, entities such as UNISDR (2012b: 25) provide guidelines for becoming more resilient that provide a menu-like set of selections – each effecting resilience in a slightly different way are routinely updated,

1. Put in place organization and coordination to understand and reduce disaster risk, based on participation of citizen groups and civil society. Build local alliances. Ensure that all departments understand their role in disaster risk reduction and preparedness. 2. Assign a budget for disaster risk reduction and provide incentives for homeowners, low-income families, communities, businesses and the public sector to invest in reducing the risks they face. 3. Maintain up-to-date data on hazards and vulnerabilities, prepare risk assessments and use these as the basis for urban development plans and decisions. Ensure that this information and the plans for your city's resilience are readily available to the public and fully discussed with them. 4. Invest in and maintain critical infrastructure that reduces risk, such as flood drainage, adjusted where needed to cope with climate change. 5. Assess the safety of all schools and health facilities and upgrade these as necessary. 6. Apply and enforce realistic, risk compliant building regulations and land use planning principles. Identify safe land for low-income citizens and upgrade informal settlements, wherever feasible. 7. Ensure that education programmes and training on disaster risk reduction are in place in schools and local communities. 8. Protect ecosystems and natural buffers to mitigate floods, storm surges and other hazards to which your city may be vulnerable. Adapt to climate change by building on good risk reduction

practices. 9. Install early warning systems and emergency management capacities in your city and hold regular public preparedness drills. 10. After any disaster, ensure that the needs of the affected population are placed at the centre of reconstruction, with support for them and their community organizations to design and help implement responses, including rebuilding homes and livelihoods.

Strides continue to be made in the physical resilience area. Scientific research into new and more resilient materials and construction practices never ceases and new engineering techniques are discovered, tested, and deployed to build more resilience physical systems (see Chapter 8 on physical resilience).

### *7.5.2 Social-Ecological Resilience*

As detailed above, many of the main thrusts of resilience science utilized today have roots in environmental and ecological analysis and modeling. Here, Holling (1973, 1978, 1986, 1996, 2001; Holling et al., 2002) and Gunderson (Gunderson, 2000, 2003; Gunderson and Light, 2006; Gunderson et al., 1995) have laid the groundwork for appreciating the dynamic and interconnectedness present in each system of the natural world. Research and analysis within and across environmental and ecological resilience shows us first hand, and often in relative short order, the real cause and effect relationships between shocks and stresses, adaptation, adjustment, and at times critical failures. Findings from pioneers studying environmental and ecological linkages have provided invaluable learning opportunities for many other disciplines, which have only recently begun to explore the topic.

Social resilience, itself, is a relatively new field of inquiry stemming from the work of environmental scientists. While social resilience began as a study of individuals and the effects of shocks, stresses, and interventions of outcomes days, weeks, or years after an event, the term social resilience has grown to encompass such ideas as social sustainability or the long-term existence of certain groups, customs, or anthropological characteristics (Magis, 2010). Scholars such as Peek (2008) and Fletcher et al. (2016) aim to develop capacities in children affected by disasters, while others like Crane (2010) and Lucini (2014) focus on a grander concept – cultural resilience as part of a larger environmental/ecological approach. Shaw et al. (2014), for example, conclude that social resilience responds to vulnerability in an independent manner due to the inherent positive or negative nature of the concept. For instance, a person could be internally resilient, but externally vulnerable due to a misguided belief about their status. Most recently, the concept of community resilience, or the characteristics of a community that help it to absorb, deflect, and rebound quickly from system pressures and shocks.

Much work has been done in the area of community resilience (see above). The most influential scholarly work in this area stems from Gilbert White and the Chicago School of natural hazards research. This field of inquiry incorporated human dimensions into its research questions acknowledging that socioeconomic factors were important to consider when addressing adjustments to disasters. Built on the ideas, adapted from Burton et al. (1993) that hazards are acts of God, but hazard losses are largely acts of man, the concept of community resilience suggests that individual and collective actions all contribute to resilience outcomes. Following this line of thinking, Bourdieu (1984) suggests that social resilience is a multilayered concept where human agency and capital play an increasing role in outcomes. Bourdieu separates cultural capital into three forms: embodied (personal dispositions and habits), objectified (knowledge and tradition stored in material forms), and institutionalized (educational qualification). Cultural capital, in other words, to a large extent shapes human agency through social experience and practice as well as education. Cutter et al. (2008) and Norris et al. (2008) both target community resilience as a concept with measurable, and (most importantly) modifiable characteristics and drivers. Others expand on the community concept to tease out more nuanced connections between community resilience and environmental transitions (Wilson, 2010) and present a view of resilience communities linked to resourcefulness (Robinson and Carson, 2016).

### *7.5.3 Economic Resilience*

Economically resilient systems are often central to community survival in hazardous conditions. In most instances, economic resilience to disasters manifests from policy decisions either attenuating or amplifying a community's ability to withstand or recover from the effect of exogenous shocks and stresses such as those produced by disaster events (Briguglio et al., 2009). As indicated by Rose (2009: 9) economic resilience should be addressed through a lens of adaptability; in order to survive the uncertainties of nature whereby catastrophic losses can lead to the unraveling of the social fabric, communities must be able to adapt. Two components stand out: static economic resilience and dynamic economic resilience. On the one hand, Rose (2009: 8) defines static economic resilience as the "ability of an entity or system to maintain function (e.g., continue producing) when shocked" such as by disaster. The occurrence of a disaster, then, can aggravate the challenge of efficiently allocating resources. On the other hand, Rose defines dynamic economic resilience as "the speed at which an entity or system recovers from a severe shock to achieve a desired state." The former might relate to short-term impacts, while the latter more to recovery of the economic engine (see also, Rose, 2007).

Bahadur et al. (2010: 3) take a broader perspective to resilience and highlight some important economic concerns. They stress the need for diversity in groups, in economic opportunities, and in planning, response, and recovery. They point out the significance of scale in understanding resilience, resilience "is built through social, political, economic and cultural networks that reach from the local to the global scale." At the mid-level scale, Hill et al. (2008: 3) consider economic resilience by defining such regions as economically resilient if they return to or exceed their previous growth path within a short time, as shock-resilient if the growth path does not change, and as nonresilient if the region does not rebound. However, they stress the need for further case studies to ascertain exactly what constitutes resilience that includes organizational and institutional systems at the regional level.

### *7.5.4 Organizational and Institutional Resilience*

The resilience of organizations and institutions has been the focus of attention for other researchers. Illustrative of this is the National Academies (2012: 55) treatise documenting the needs for building resilience and integrating an all-hazards approach: "Effective resilience-focused private-public collaboration will often span geographic and political jurisdictions, include multiple agencies and levels of government, and cross other social, economic, and cultural boundaries." The National Academies (2012: 55) also emphasize the need for community collaboration and suggests that, such "coalitions can be charged to assess the community's exposure and vulnerability to risk, to educate and communicate risk, and to evaluate and expand the community's capacity to handle such risk." Thus, coordination and integration of organizations and institutions in hazard planning are seen as essential to enhance resilience. MacKinnon and Derickson (2013: 235) break this down further and point to the need for integration of ecological, social, and organizational concerns to shape resilience:

> First, the ecological concept of resilience is conservative when applied to social relations. Second, resilience is externally defined by state agencies and expert knowledge. Third, a concern with the resilience of places is misplaced in terms of spatial scale, since the processes which shape resilience operate primary at the scale of capitalist social relations. In place of resilience, we offer the concept of resourcefulness as an alternative approach for community groups to foster.

Mackinnon and Derickson, therefore, place greater emphasis on the political economy rather than simple concerns of spatial scale. Additional research on the role of organizations and institutions in fostering resilience is needed encompassing studies of different political systems within different local contexts.

## 7.6 Measuring Resilience

To date, while many have posited definitions, theories, quantitative examples, and conceptual drivers of resilience, few have taken the next step towards quantification and measurement of resilience. Those who have attempted to turn scientific inquiry in empirically derived, testable, and comparable indicators open themselves up to endless scrutiny. Interestingly, this vanguard also lays the groundwork for developing real and actionable decision support tools and informatics. Among the most notable of these quantitative revolutionaries are those situated in fields like geography and sociology, where comprehensive comparative analysis is strengthened by detailed fieldwork and case studies. Cutter's (2016b) work on summarizing the field of resilience indicators stands as a testament to the volume of work already in circulation and provides a warning of potential overdependence on quantification.

## 7.7 Resilience Across Disciplinary Fields

Standing in sharp contrast to advances in theory, research, and applications in resilience science over the past decades are the numerous critiques of where we are now and where we should be moving into the future. What are we doing right, and what can we do better or different? These challenges can be combined into two points for growth in science and practice: (1) a lack of a unique academic identity and extension into all fields of academic inquiry; and (2) a general disconnect between research on the topic and applications for decision-making.

Some recent publications point out that although resilience concepts and applications have matured over the years they still do not have firm footing across many sub-disciplines and that many indicators used in some specific fields do not easily apply in other contexts (Beccari, 2016). Bec et al. (2016) point out that current resilience studies in relation to tourism are weak due to the majority of studies focusing on reaction to natural disaster response and that an increase in studies focusing on the strength of tourism for communities is needed. Furthermore, Boon et al. (2012), linking Bronfenbrenner's bioecological theory to community disaster resilience, indicate that academia still needs to assess community resilience in relation to disaster studies and also needs to develop multiple methods that take into account multi-scale and temporal variations to include predisaster, community data, and post disaster (Boon et al., 2012). Others recognize that more work needs to be done on the livelihood, psychological, and community-agency categories of resilience (Paton et al., 2015). Some believe resilience should remain more tightly coupled to the vulnerability concepts and ideas where the concept of resilience should be a question of social transformation (Weichselgartner and Kelman, 2015). Conversely, a set of scholars deeply entrenched in real world applications of scientific findings in disaster management, as we have seen, advocates that the concept of resilience is not the flip side of the vulnerability coin and that the characteristics of a resilient community are not simply lower levels of vulnerability (Cutter et al., 2008, 2010).

## 7.8 Resilience in Practice

The gap between scientific findings and the creation of knowledge and when these important realities are employed in everyday decision-making can be long and painful, especially for those who strive to make science applicable to the real world. Even after decades of research into the concepts, interactions, and outcomes around resilience, the concept and its application is still viewed as a new buzzword in science. Weichselgartner and Kelman (2015) argue that there is a lack of scientifically reliable and context-appropriate data that lead to the broad usage of it. Burby (2006: 172) adds that increases in resilience would be moot if such changes do not influence "uninformed local government decision making about urban development that results in millions of household and businesses occupying at-risk-structures in vulnerable locations."

In other works, current policies are not creating resilience, but rather often exacerbating risk and vulnerability. Still, societal learning and adjustments are coming closer in line with what we know about resilience. In some cases, local governments that have adopted resilience considerations within flood planning have seen a decrease in per capita losses. Such practices, perhaps required by higher levels of government (state/national) and enacted in real and actionable local planning, might enhance resilience through sound mitigation practices (Burby, 2006). Torres (2017) examines some of these concerns in Florida in relation to sustainable policies.

Some believe that the academic literature is replete with examples and significant evidence proving a core concept that "resilience is a biopolitical response to the contemporary idea that life itself is less governable than we previously thought on some deep and fundamental level" (Pugh, 2014: 316). Entities, be they communities, towns, cities, regions, or nations with more money are less vulnerable, more resilient, and typically exhibit a much higher adaptive capacity than countries that lack financial capital (Goklany, 2007). Furthermore, Pugh (2014: 318) asserts that the resilience concept promotes consequences "brought about by liberalism and encourages us to embrace the idea of living dangerously in an increasingly complicated and dangerous world." Progress is being made in the area of disaster resiliency at the nexus of public administration and real world decision making. Research into and findings related to the influence of interdisciplinary efforts on resiliency planning, response, and outcomes are taking root in the real world and decision makers are beginning to take notice (Kapucu et al., 2013, Torres, 2017).

## 7.9 Conclusion

The ability to cope with or accommodate environmental change, therefore, would seem to be fundamental to resilience; if communities are sustainable then a level of coping while under stress is essential. However, this might be achieved by preparation, perhaps through implementation of mitigative measures, or by successful emergency response and long-term rehabilitation and reconstruction practices. For the most part, the preceding definitions do not articulate such details and so we return to Cutter's (2016a) remarks regarding resilience to what and to whom, and to the National Research Council (2010: 36) which identified four basic principles:

1. Disaster resilience correlates strongly with community resilience, including economic, environmental, health, and social justice factors.
2. Private-public collaboration is based on collaborative relationships in which two or more private and public entities pool and coordinate the use of complementary resources through joint pursuit of common objectives.
3. Effective collaboration ideally encompasses the full fabric of the community and is representative of all walks of life – including minorities, the impoverished or disenfranchised, children, and the elderly – so community engagement is essential for the success of resilience-focused collaboration.
4. Principles of comprehensive emergency management, incorporating an all-hazards approach, guide resilience-focused collaboration-building efforts.

Scholars and practitioners, having peered into the crystal ball or reflected in the magic mirror, provide us with a set of future directions around resilience, its definition, measurement, and the future of resilience research and applications. While many comments about the future of any subject area are rhetorical in nature, the literature is clear on a few points related to resilience. These can be broken up into two main related groups: theoretical/ conceptual, and application-oriented. Advocates for more and better future research around resilience require that the resilience field needs to move from a descriptive concept (what is done) to a normative agenda (what ought to be done) as argued by Weichselgartner and Kelman (2015). In addition, recognizing the role of diversity in contributing to resilience will increase resilience opportunities among and between systems (Gunderson, 2010). Those who argue that we need to move from theory into practice would like to see the following extensions and growth areas around resilience.

- The focus and goal of urban hazard mitigation, a particular branch of hazard mitigation practice, should be on developing resilient cities (Godschalk, 2003).
- Creating a resilient society will greatly depend on land use planning measures (Saunders and Becker, 2015).
- Individuals should learn to become more adaptable to the complex economic, social, and environmental forces that they face (Pugh 2014).
- Rather than relying solely on short-term resilience to guide a community through a disaster period, advancements in preevent recovery planning can greatly increase the long-term sustainability and overall resilience of any area (Saunders and Becker, 2015).
- Resilient communities should be sustainable communities built on a need to meet changing legislative requirements, and to ensure the needs of future generations are made through economic, social, cultural, and environmental factors (Saunders and Becker, 2015).
- A real and meaningful understanding of the connection between how society cares for its environment and how this care can influence the devastation from disasters will increase community resilience (Adger et al., 2005).
- Communities need the creation of the proper social conditions (safety, housing, community, etc.) to experience recovery, therefore resilience. To create those conditions, national and local political and NGOs need to create an open dialogue with all players in the process instead of making decisions without consultation with the local affected populations (Barrios, 2014). Furthermore, proper coordination is required to successfully create a built environment that is resilient (Bosher et al., 2007).

We end this chapter with the comprehensive recommendations made by Bahadur et al. (2010: 2–3) to operationalize resilience:

- High level of diversity: A high level of diversity in groups performing different functions in an ecosystem; in the availability of economic opportunities; in the voices included in a resilience-building policy process; in partnerships within a community; in the natural resources on which communities may rely; and in planning, response, and recovery activities.
- Effective governance and institutions: which may enhance community cohesion. These should be decentralized, flexible, and in touch with local realities; should facilitate system-wide learning; and perform other specialized functions such as translating scientific data on climate change into actionable guidance for policymakers.
- Accept uncertainty: The inevitable existence of uncertainty and change is accepted. The nonlinearity or randomness of events in a system is acknowledged, which shifts policy from an attempt to control change and create stability to managing the capacity of systems to cope with, adapt to, and shape change.
- Public participation: There is community involvement and the appropriation of local knowledge in any resilience-building projects; communities enjoy ownership of natural resources; communities have a voice in relevant policy processes.
- Preparedness and disaster management: Preparedness activities aim not at resisting change but preparing to live with it; this could be by building in redundancy within systems (when partial failure does not lead to the system collapsing) or by incorporating failure scenarios in disaster management plans.
- Social justice: A high degree of social and economic equity exists in systems; resilience programs consider issues of justice and equity when distributing risks within communities.
- Social values: The importance of social values and structures is acknowledged because association between individuals can have a positive impact on cooperation in a community, which may lead to more equal access to natural resources and greater resilience; it may also bring down transaction costs as agreements between community members would be honored.
- Not return to the status quo after disaster: The nonequilibrium dynamics of a system are acknowledged. Any approach to building resilience should not work with an idea of restoring equilibrium because systems do not have a stable state to which they should return after a disturbance.

- Updating and learning: Continuous and effective learning is important. This may take the form of iterative policy/institutional processes, organizational learning, reflective practice, and adaptive management; and may merge with the concept of adaptive capacity.
- Different scales: Resilient systems take a cross-scalar perspective of events and occurrences. Resilience is built through social, political, economic, and cultural networks that reach from the local to the global scale.

In this chapter, we have discussed some varying concepts and applications of resilience in disasters, and have cited some of the most recent literature in the field. However, the second part of our chapter title (Humans and Nature) reflects the debt we owe scholars from the past. Marsh (1869), in his work "Man and Nature", built on the works of Alexander von Humboldt to emphasize the importance of understanding the interconnectedness of nature and the significance of humans in interacting with the environment. Resilience is a logical extension of this thinking, a holistic examination of humans and nature in disaster environments. Time will tell how well society has addressed resilience, especially in the context of global climate change.

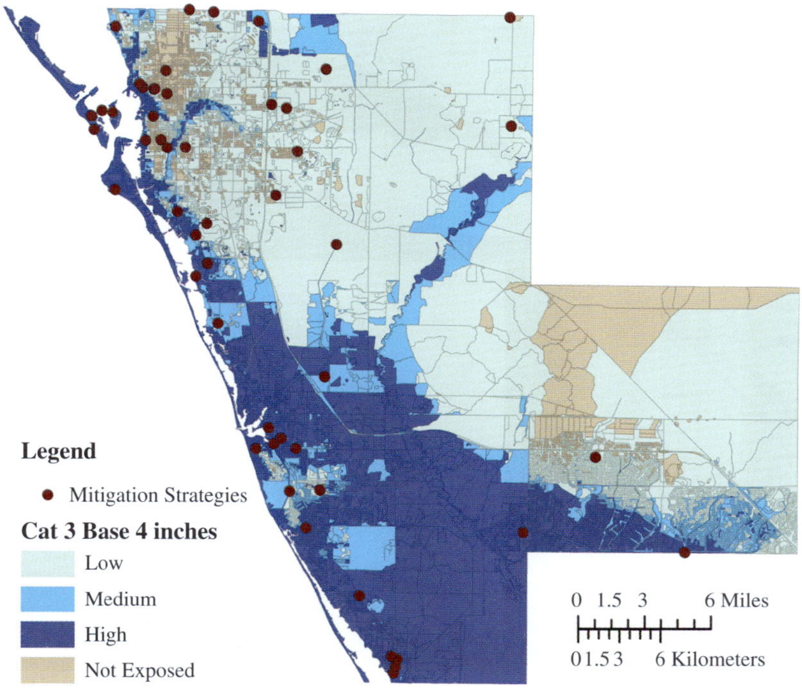


Fig. 3.4. Physical exposure from Category 3 storm surge with inland precipitation

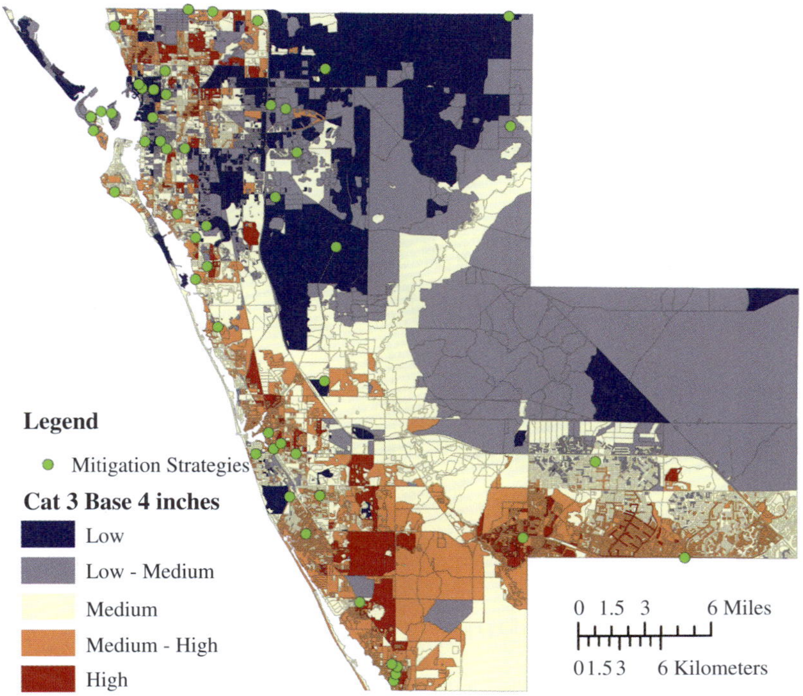


Fig. 3.5. Physical vulnerability to Category 3 storm surge with inland precipitation

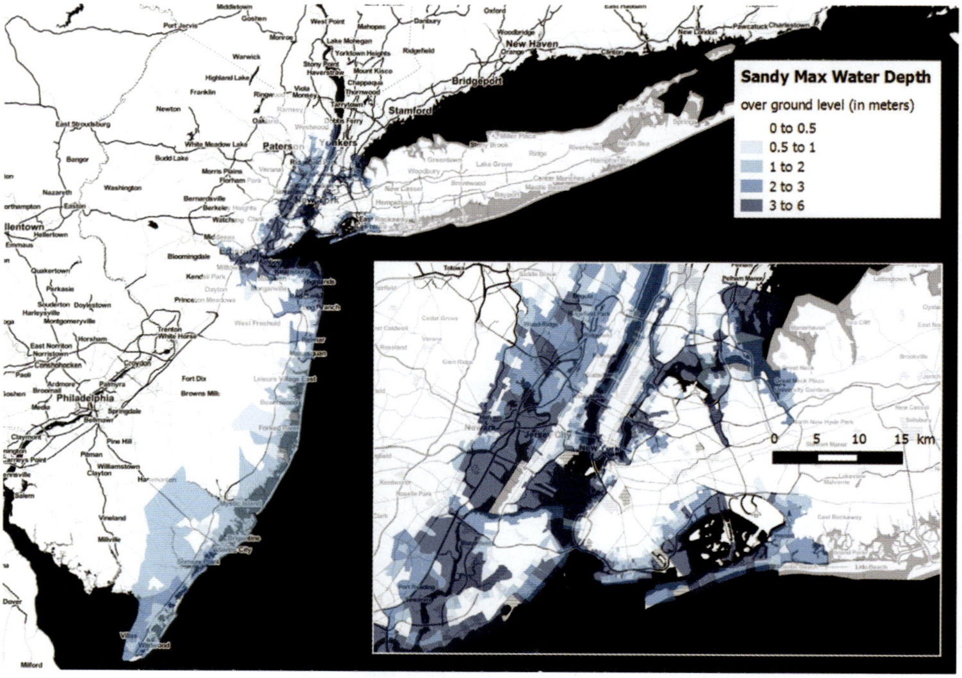


Fig. 4.7. Hurricane Sandy flood depths in New York and New Jersey
After FEMA (2012)

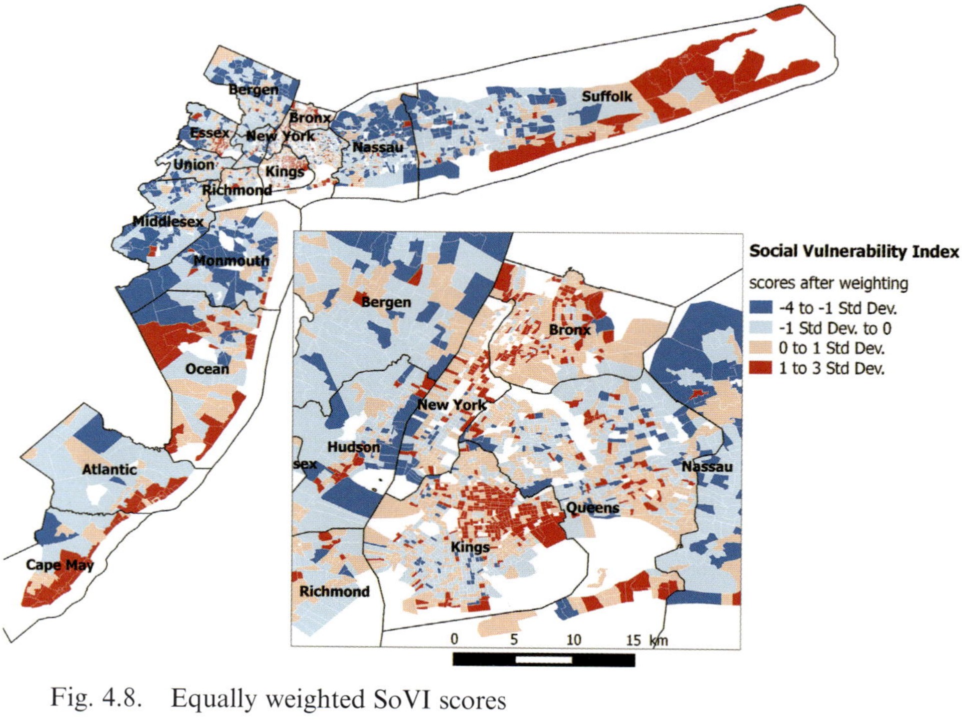


Fig. 4.8. Equally weighted SoVI scores

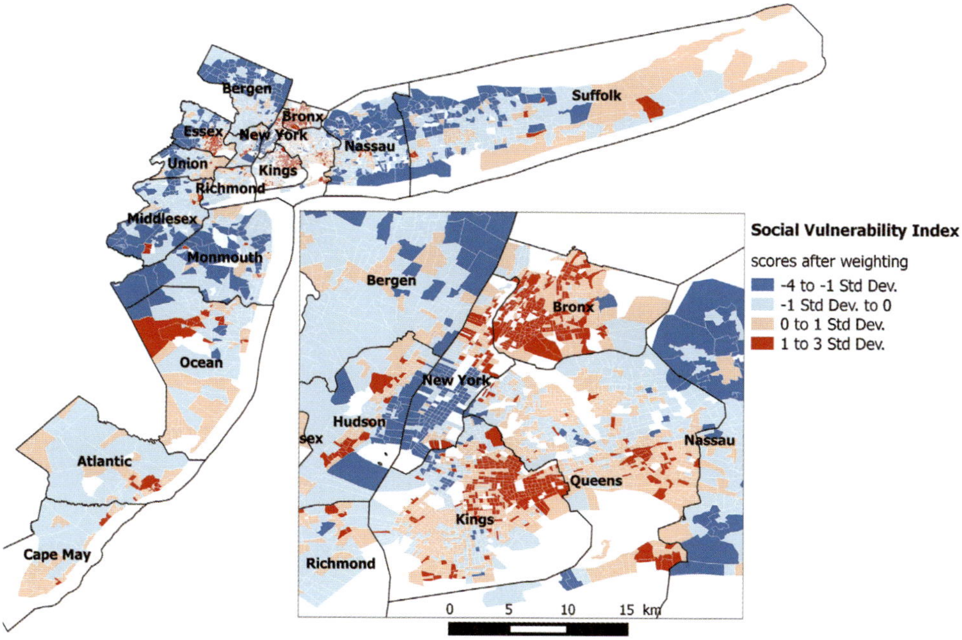


Fig. 4.9. SoVI scores after weighting by PCA eigenvalue

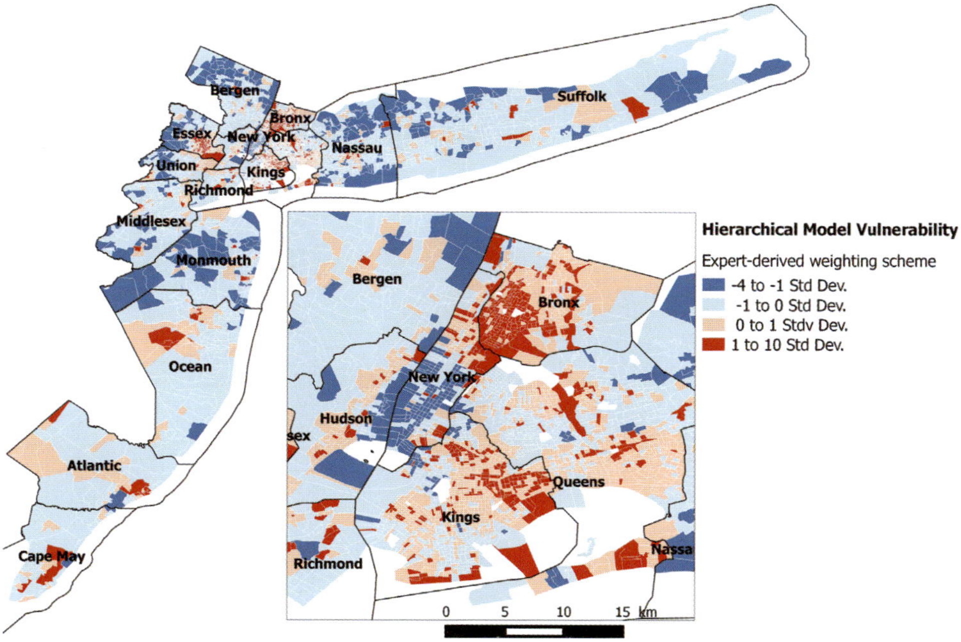


Fig. 4.10. Hierarchical model social vulnerability scores

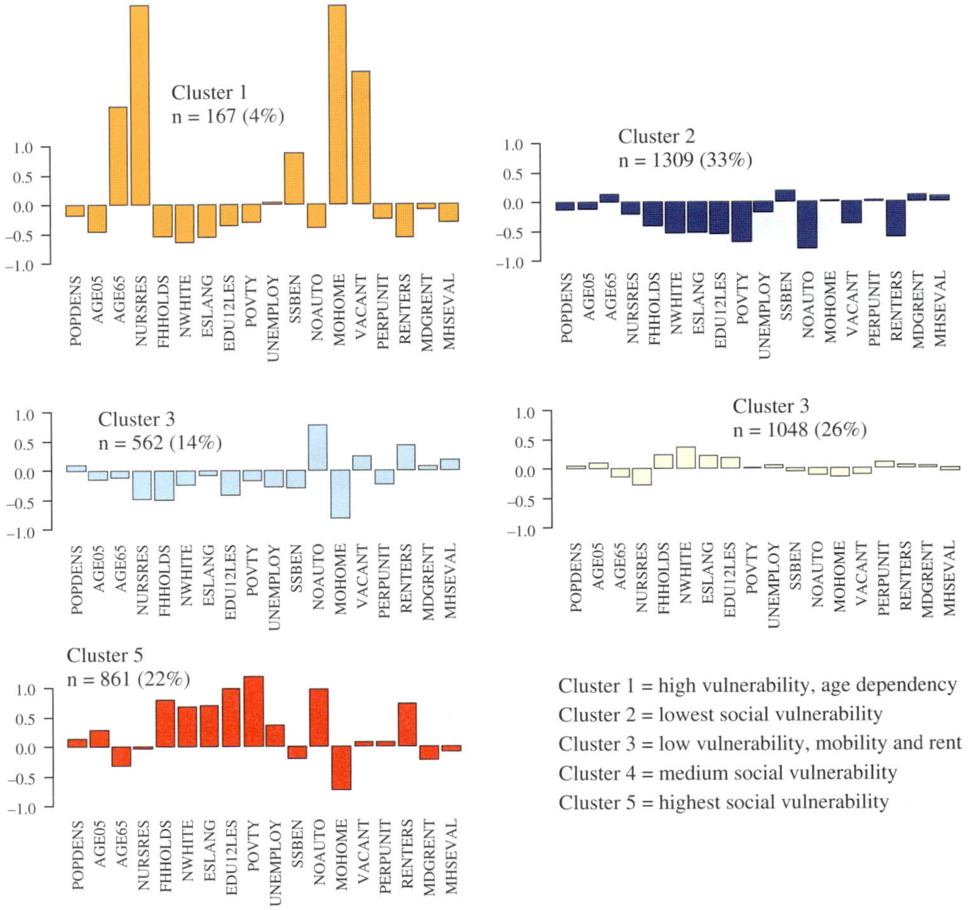


Cluster 1 = high vulnerability, age dependency
Cluster 2 = lowest social vulnerability
Cluster 3 = low vulnerability, mobility and rent
Cluster 4 = medium social vulnerability
Cluster 5 = highest social vulnerability

Fig. 4.11. Vulnerability clusters

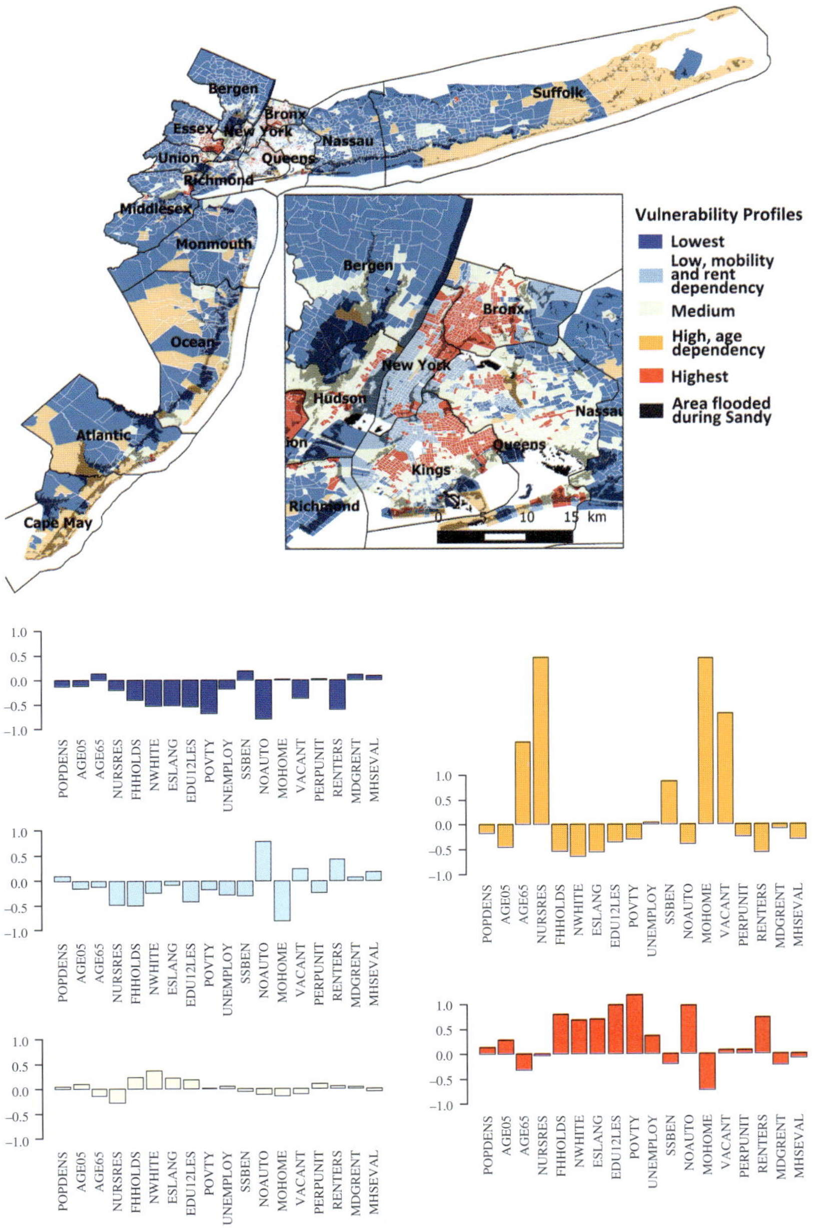


Fig. 4.12. Vulnerability profiles

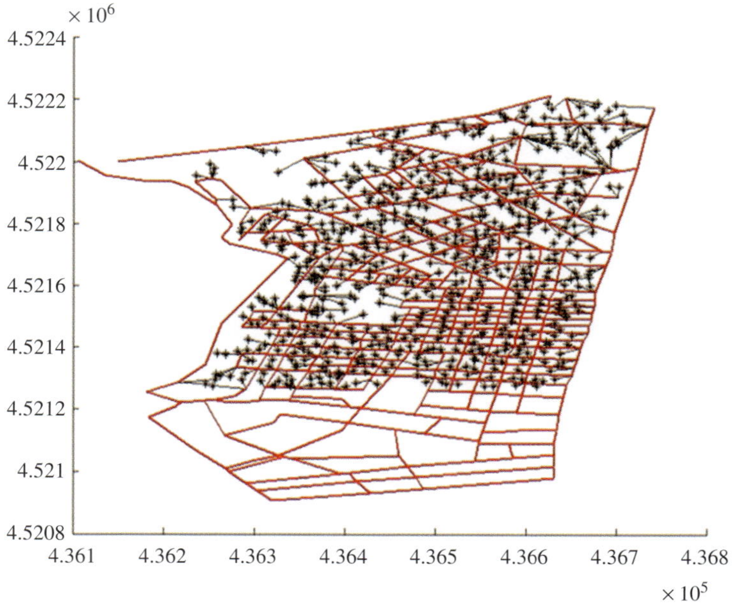


Fig. 8.10. HSPN of the historical center of the city of Naples (the Quartieri Spagnoli area) where only residential buildings have been modeled

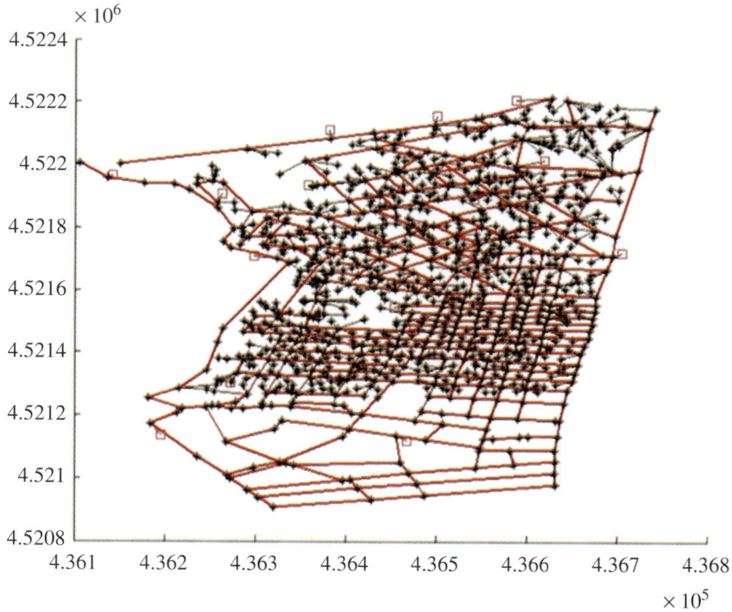


Fig. 8.11. HSPN of the historical center of the city of Naples (the Quartieri Spagnoli area) where both school (grey-squared markers) and residential buildings (black-starred dots) have been modeled

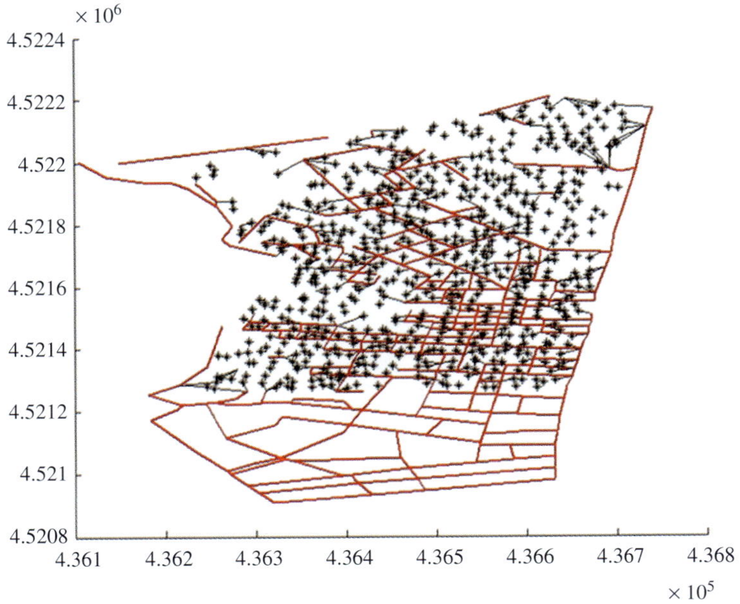


Fig. 8.12. Residential HSPN of the historical center of the city of Naples after the occurrence of a seismic event that caused 15 percent of buildings to collapse

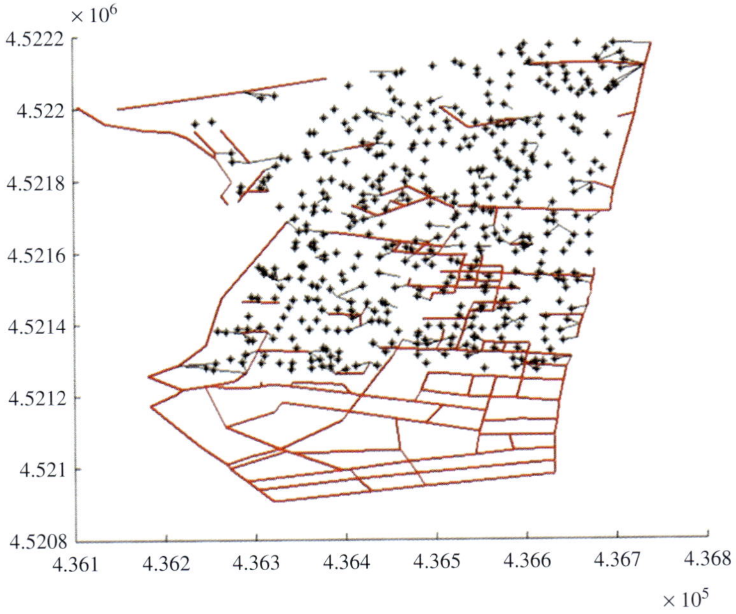


Fig. 8.13. Residential HSPN of the historical center of the city of Naples after the occurrence of a seismic event that caused 30 percent of buildings to collapse

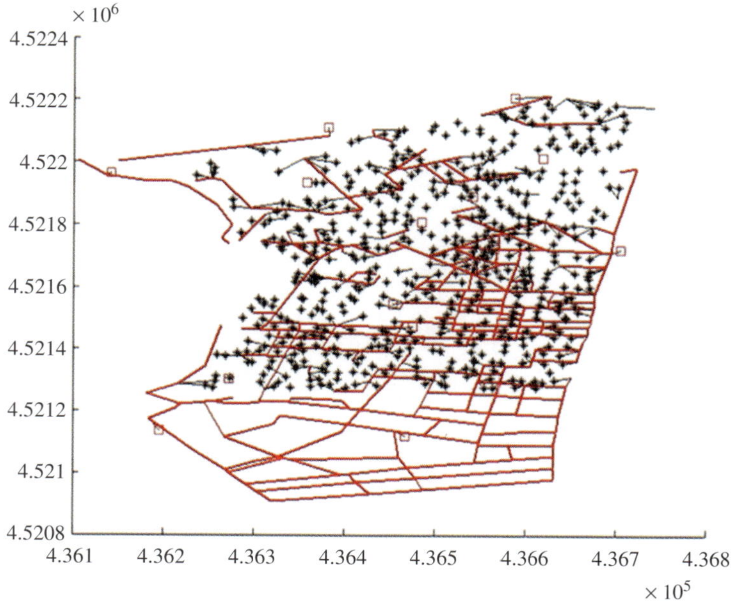


Fig. 8.14. School HSPN of the historical center of the city of Naples after the occurrence of a seismic event that caused 15 percent of buildings to collapse

Fig. 8.15. School HSPN of the historical center of the city of Naples after the occurrence of a seismic event that caused 30 percent of buildings to collapse

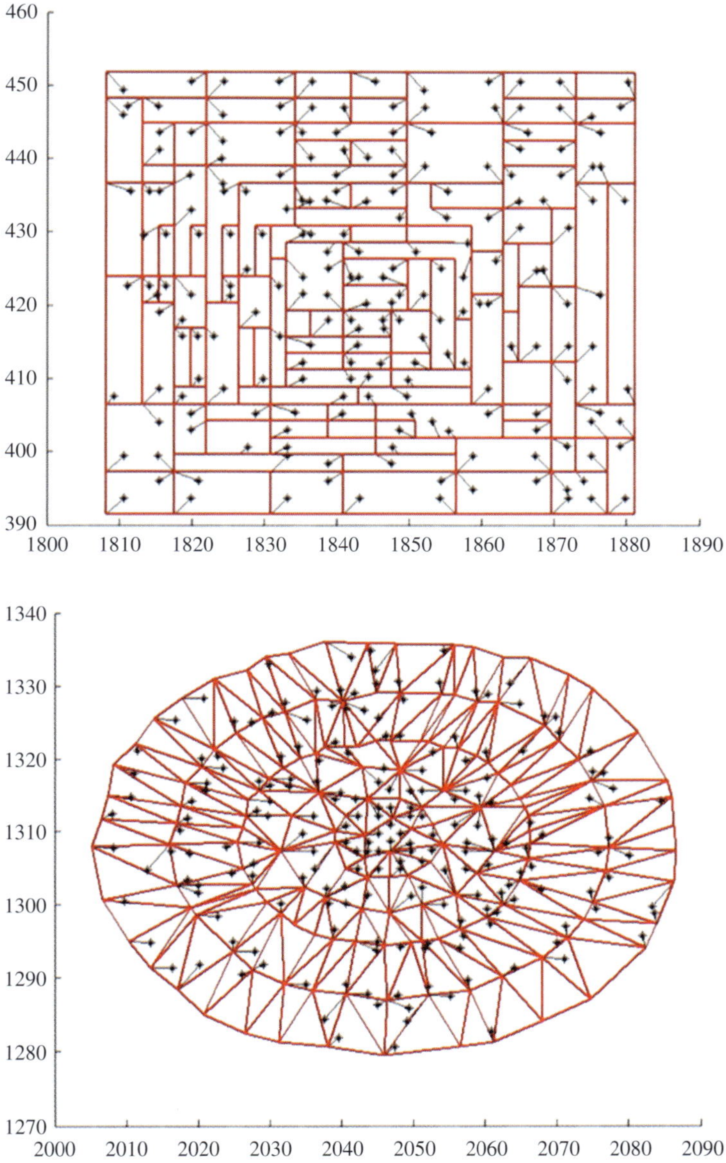


Fig. 8.16. City shapes: rectangular (A), circular (B), hexagonal (C), and star (D) modeled for the case analysis of 200 buildings

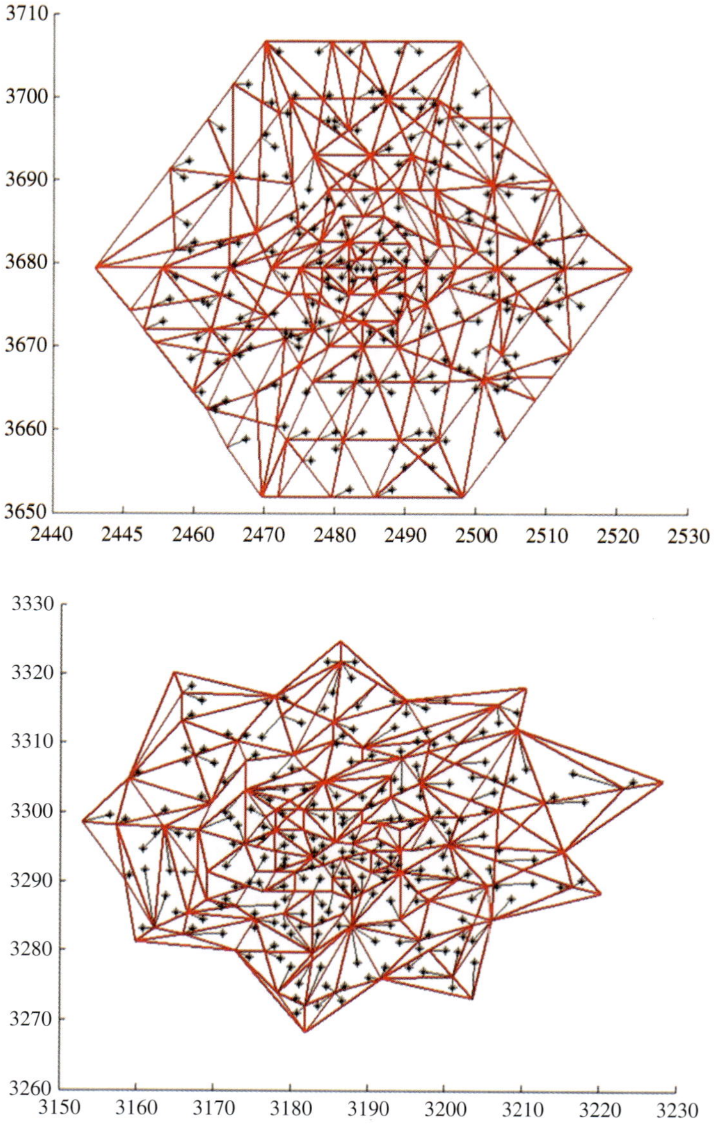
3710
3700
3690
3680
3670
3660
3650
2440
2445
2460
2470
2480
2490
2500
2510
2520
2530
3330
3320
3310
3300
3290
3280
3270
3260
3150
3160
3170
3180
3190
3200
3210
3220
3230

Fig. 8.16. (*Cont.*)

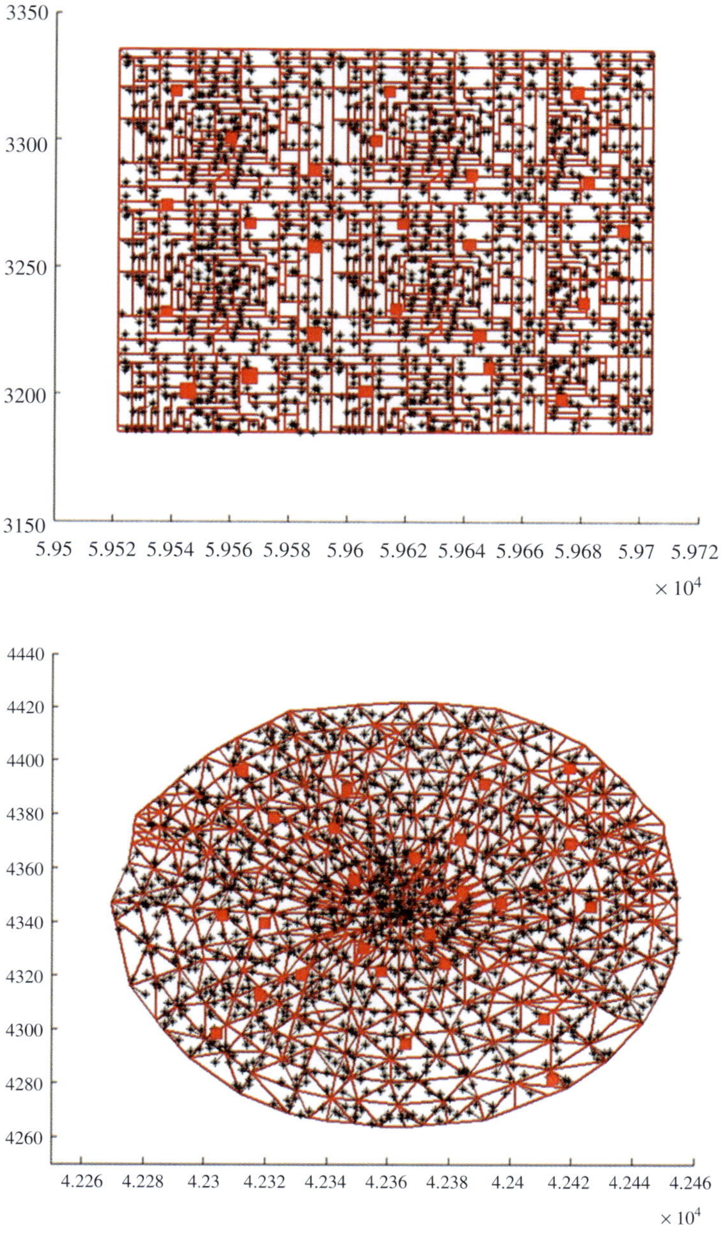


Fig. 8.17. City shapes: rectangular (A), circular (B), hexagonal (C), and star (D) modeled for the case analysis of 1250 buildings (25 of which are considered to be schools)

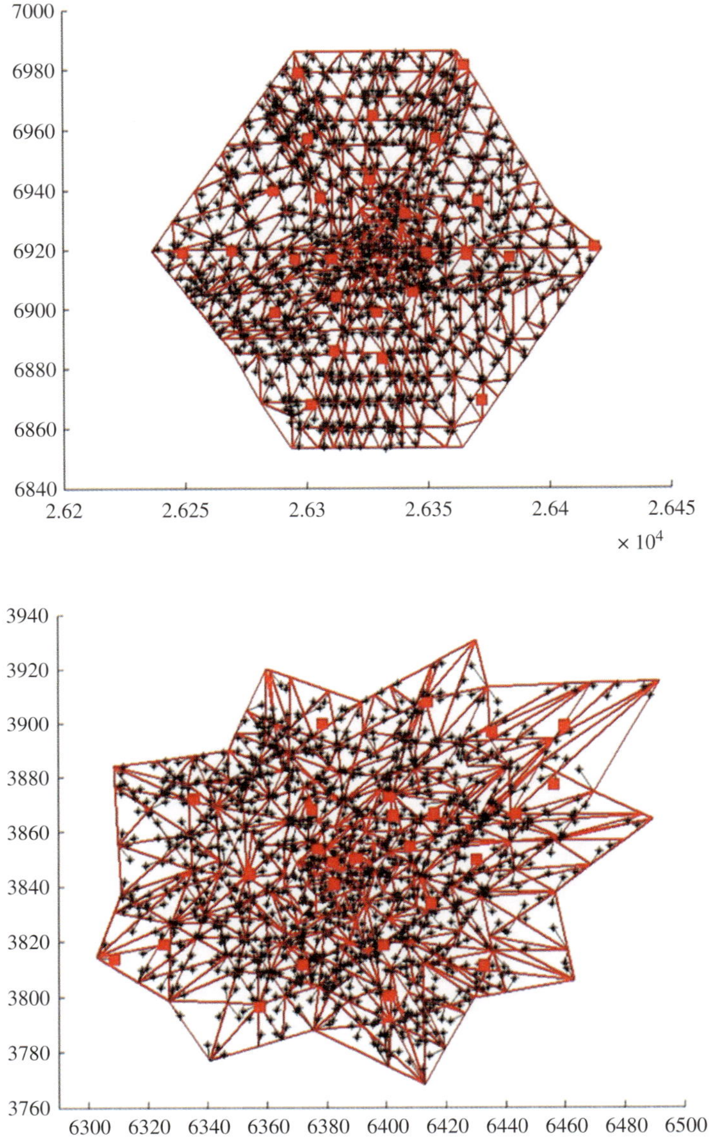

7000
6980
6960
6940
6920
6900
6880
6860
6840
2.62
2.625
2.63
2.635
2.64
2.645
× 10^4
3940
3920
3900
3880
3860
3840
3820
3800
3780
3760
6300
6320
6340
6360
6380
6400
6420
6440
6460
6480
6500


Fig. 8.17. (*Cont.*)

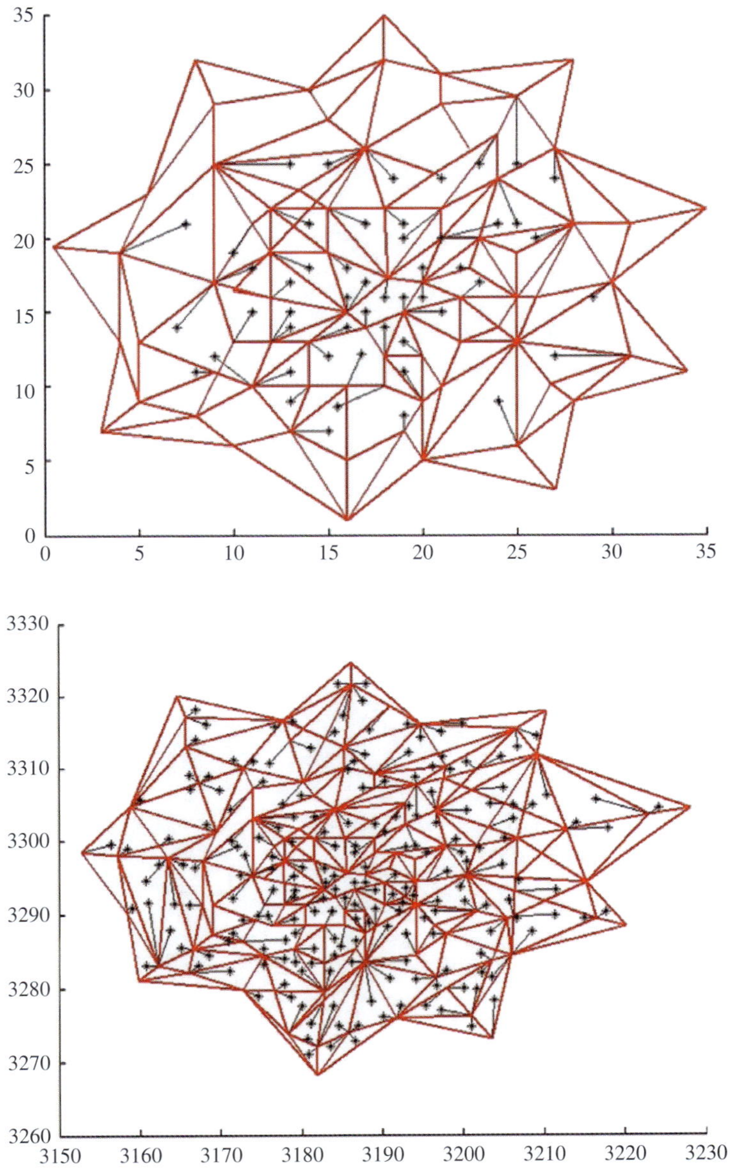


Fig. 8.18. Example of scaling in the case of a star-shaped city with 50 (A), 200 (B), 1250 (C) and 5000 (D) residential buildings

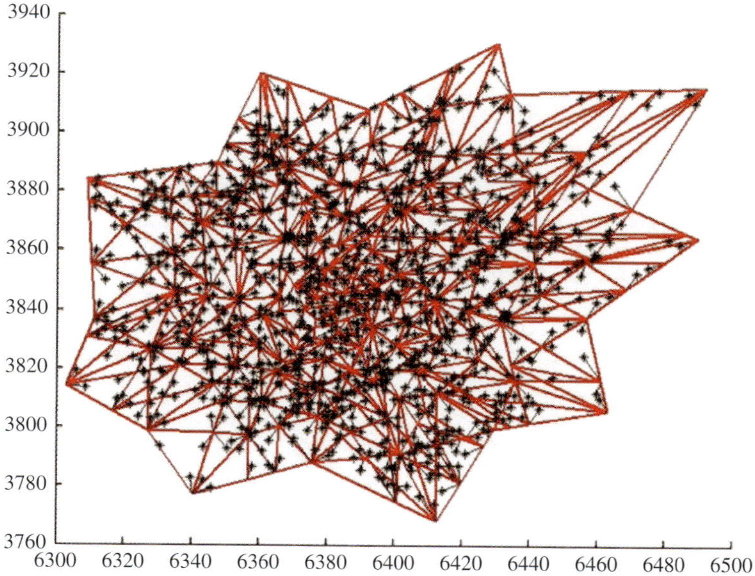
3940
3920
3900
3880
3860
3840
3820
3800
3780
3760
6300
6320
6340
6360
6380
6400
6420
6440
6460
6480
6500

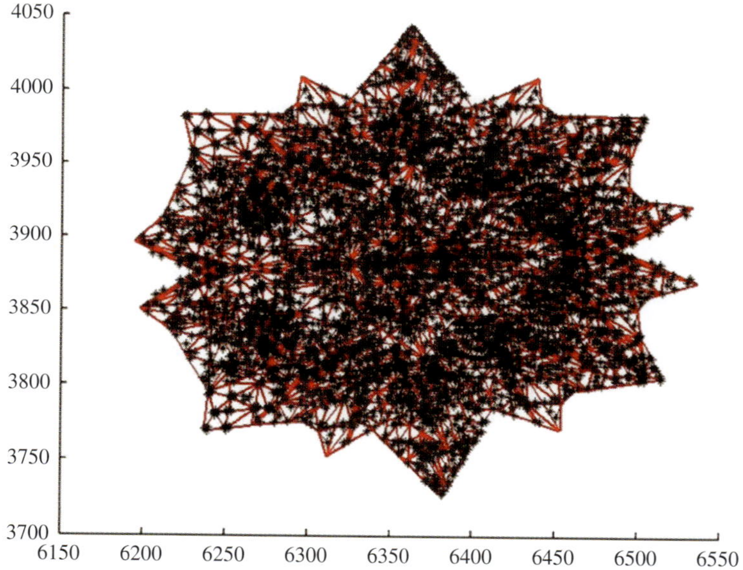
4050
4000
3950
3900
3850
3800
3750
3700
6150
6200
6250
6300
6350
6400
6450
6500
6550

Fig. 8.18. (*Cont.*)

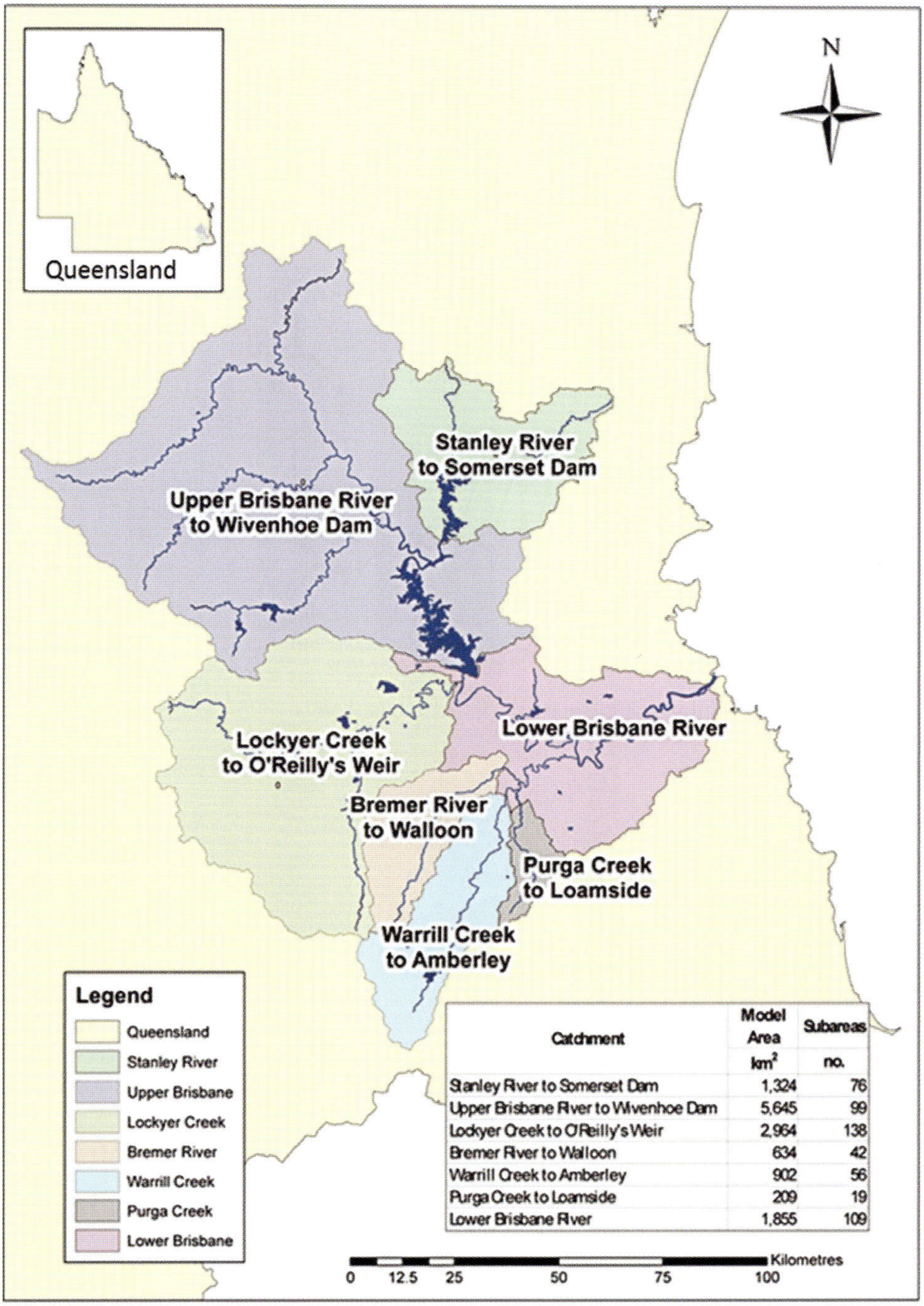


| Catchment | Model Area km$^2$ | Subareas no. |
|---|---|---|
| Stanley River to Somerset Dam | 1,324 | 76 |
| Upper Brisbane River to Wivenhoe Dam | 5,645 | 99 |
| Lockyer Creek to O'Reilly's Weir | 2,964 | 138 |
| Bremer River to Walloon | 634 | 42 |
| Warrill Creek to Amberley | 902 | 56 |
| Purga Creek to Loamside | 209 | 19 |
| Lower Brisbane River | 1,855 | 109 |

Fig. 10.2. Southeast Queensland water supply grid

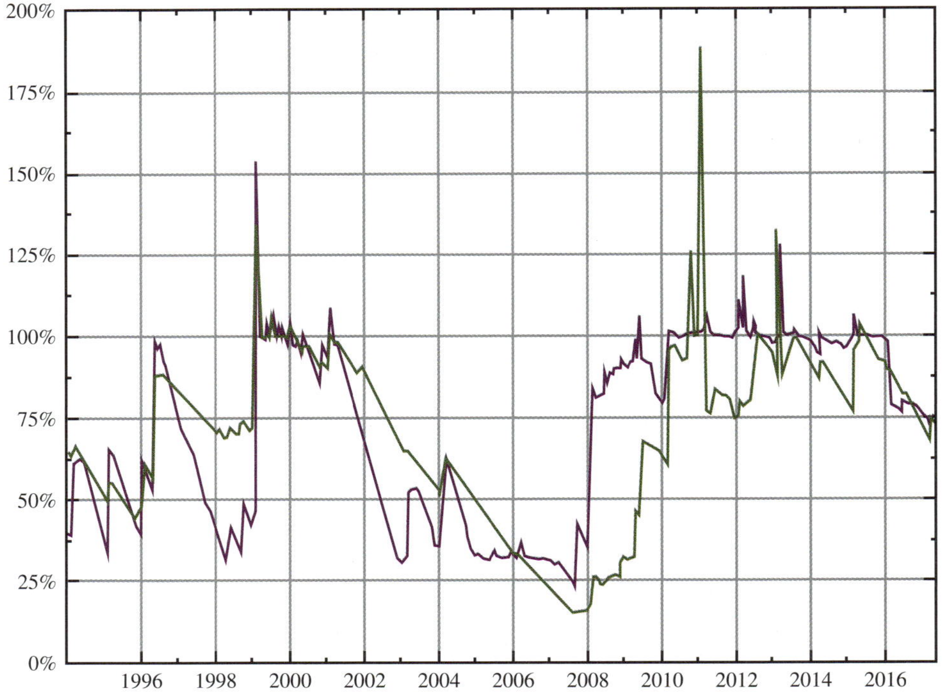


Fig. 10.3. Southeast Queensland dam storage levels 1995–2017

# 8

# Physical Resilience

ANNA BOZZA, DOMENICO ASPRONE, AND GAETANO MANFREDI

## 8.1 Introduction

Increasingly complex urban dynamics rule modern societies worldwide, marking an important landmark in human history and characterizing the urbanization era. This inexorable trend leads to contemporary cities being the cornerstone of both social and technological developments, and even more exposed and vulnerable. Opportunities and challenges arise from such phenomena, meaning that novel approaches to urban management are required. In particular, a major issue is natural hazards, which need to be accounted for according to pioneering engineering and human-centric visions in order to build sustainable and resilient cities. In this chapter, urban resilience against disasters is understood as the engineering vision according to ecosystem theory, and an engineering-based methodology for resilience quantification is presented. This enables any urban context to be modeled as a complex network, and resilience to be assessed as regained efficiency after a catastrophe occurs and for each stage of the recovery process.

Due to the high rate of occurrence and huge economic losses caused by seismic events in the last few decades, earthquake scenarios are simulated to endorse the methodology. A real case study is developed for the historical center of the city of Naples (Italy). Furthermore, according to the more widespread city configurations, urban contexts are modeled with diverse shapes and sizes to study the urban resilience trend in the face of the city's geographical configuration and scaling in relation to its size. The social and physical city subsystems are individually modeled as complex networks and then overlaid to account for their interrelations as a hybrid social-physical network (HSPN) (Bozza et al., 2015; Cavallaro et al., 2014). The vulnerability of infrastructures, namely buildings and streets, is accounted for through the integration of probability-based models. Efficiency measures are also performed, enabling the local and global loss of functionality to be assessed and urban damage to be evaluated in a systemic manner. Finally, diverse recovery strategies are simulated and resilience is calculated for each studied city context.

In this chapter, the developed framework for the resilience quantification is presented in Section 8.2. Hence, each step of the methodology is analyzed, and specific features are highlighted. Section 8.3 presents the case study of the Quartieri Spagnoli area (Naples, Italy), whose analysis results are analyzed and discussed. Analogously, Section 8.4 presents the case study where synthetic HSPNs with different sizes and shapes are analyzed.

## 8.2 An Engineering Framework to Quantify Urban Resilience

Contemporary patterns of urbanization lead to cities being the cornerstone of human activities and technological development. Indeed, according to the 2014 revision of the world urbanization prospects by the United Nations 54 percent of the world's population lives in cities, with this figure expected to rise to 66 percent by 2050 (United Nations, 2014). As a consequence, infrastructures and community assets are also increasing in number, causing urban areas to be exposed and vulnerable now more than ever. As a result, urban management is now one of the most important challenges when it comes to guaranteeing the development of local and global communities and the building of sustainable and resilient cities. A major issue in this is disaster risk management, which is in step with the growing awareness of challenges related to climate change and the alarming increase in the rate of occurrence of natural hazards worldwide. One of the most threatening of these problems is seismic risk. Indeed, since 2004, six of the ten costliest catastrophic events have been earthquakes, which have led to huge human and economic losses (Munich Re, 2011).

Seismic hazards are, in fact, a common issue that many local authorities have to deal with. As a result, the scientific community and urban stakeholders have to pay particular attention to the search for innovative solutions. New paths of prevention, emergency management, mitigation, and adaptation actions such as response and recovery strategies are continuously being developed. Nonetheless, many doubts remain about how to identify the best of these interventions. So, which are the relevant criteria for selecting the most appropriate strategy after a catastrophe? How do local administrators choose to adopt one particular action over another? The key is to build resilient cities.

The term resilience dates back to the nineteenth century, when it was used in physics to indicate the ability of materials to withstand loads without suffering damage. In this light, physical resilience was understood as strictly related to the Hooke's law from 1678 concept of elasticity. Linking resilience to the concept of elasticity led to an ambiguous understanding of the word, which was interpreted as a property owned by a system.

Recently, the notion of resilience has triggered a renewed interest when referring to urban physical systems; hence it has been reinterpreted. In keeping with this, Ayyub (2014) has highlighted the importance of understanding resilience as "an emergent property of what an engineering system does." Hence, resilience is now identified as the outcome of a wider process that includes statics and dynamics variables. In particular, resilience represents the capacity of a system to withstand and recover from external stresses and achieve an equilibrium state through mitigation and adaptive management of complex engineering systems. Consequently, each potential action is as effective as the contribution it makes to increase resilience. Such a feature is strictly related to the capacity of the studied system to also be sustainable. Indeed, the more a system is efficient in using its own resources to recover from a shock, the more it can strive for future sustainable development. This is even more evident when dealing with natural disasters that affect urban areas, where a more efficient recovery process is guaranteed from the greater sustainability of the reconstruction phase within the lifecycle of a city. A city is, in fact, as resilient as it is sustainable during the occurrence of a hazardous event, which is when it suffers an external stress and makes an effort to reconfigure its equilibrium (Asprone and Manfredi, 2015; Bozza et al., 2015).

On the other hand, the concept of resilience is multidisciplinary, as it is also used in psychology and social sciences (Garmezy, 1991; Werner and Smith, 1982), medicine (Lotka, 1924), and engineering (Bruneau et al., 2003). In this chapter, resilience is understood as the engineering definition (Bruneau et al., 2003; Holling, 1973; Pimm, 1984) according to ecosystem theory. As a consequence, resilience is defined as the capacity of a city system to withstand external stresses and bounce back to an equilibrium state that can be both the same as the preevent condition and also a new, different state (Asprone and Manfredi, 2015; Bozza et al., 2015).

In this study, the external stress considered is an earthquake and the city system is modeled as a complex network that is assimilated to an ecosystem (Holling, 1973, 1986, 2001), therefore merging within the geographical space in which all the interacting urban components are embedded, i.e., the social and the physical. However, modeling urban contexts must be conducted with great care because complex city systems can differ greatly from one another. On the one hand, urban environments typically differ in size and geographical shape, and so, when attempting to model a city, it is essential to account for its territorial extent and the spatial distribution of its components. On the other hand, these environments share similar socioeconomic dynamics and topological features (Cardillo et al., 2006), which enables them to be compared.

So, despite the huge diversification among urban environments worldwide, some general rules can be observed that typically scale with a city's size (Bettencourt, 2013; Bettencourt and West, 2010, 2011; Bettencourt et al., 2007). The way such scaling arises greatly depends on the type of phenomena being observed. According to Bettencourt and West (2011), Bettencourt et al. (2007) and Lobo et al. (2013), the thick network of interrelations that develops within urban contexts is based on very diverse underlying mechanisms. These authors observed changes in a huge number of indicators with respect to city size and also recognized two predominant trends, which can be associated with social or economic dynamics. In particular, indicators related to economy of scale mechanisms exhibit superscaling with a city's size. Meanwhile, subscaling is observed for indicators related to social processes (Bettencourt et al., 2007).

In order to characterize the urban efficiency trend, damage indicators, and resilience against a city's size and shape, this chapter proposes an integrated framework that simulates a seismic event and enables expected damage to be assessed and resilience to be quantified.

According to Batty (2008), since contemporary cities are typically fractals, the most effective way to model and simulate their behavior is to deconstruct the rules that have been used in the past and design ideal cities. On the other hand, most of these realizations rarely reflect the quality of life of inhabitants, as they are too simple when it comes to the real workings of the development process. Accordingly, and keeping this in mind, synthetic city models are developed in this study within the current framework, but typical features of actual urban contexts are also accounted for. A real case study is also performed to validate the proposed metrics for the Quartieri Spagnoli area in the inner city of Naples (Italy).

Cities are modeled according to graph theory as complex networks. The infrastructure and social networks are modeled separately and then overlaid and included in the related geographical space to finally obtain a HSPN (Bozza et al., 2015; Cavallaro et al., 2014). Georeferencing is performed through a geographic information system (GIS), which enables specific information on the built environment and a wide range of other data to be integrated.

In particular, synthetic HSPNs are modeled based on the geometric shapes that are most common worldwide and have historically formed contemporary cities. As a result, diverse cities are modeled with rectangular, circular, hexagonal, and star shapes. Each of these shapes is then increasingly scaled and a seismic scenario is simulated for each of them.

Scenario analyses are also performed that account for the vulnerability of the built environment through a probability-based methodology. Two diverse recovery strategies are modeled and simulated, with the former focused on social dynamics and the latter on economies of scale related to a city service. Efficiency, as it is a robustness metric of the network and represents urban connectivity, is evaluated before the event, soon after its occurrence, and for each stage of the recovery process. Consequently, urban damage is assessed in a systemic fashion as the city's efficiency decays in the aftermath of an earthquake. Finally, urban resilience is quantified as the capacity of a city to bounce back to equilibrium. An alternative resilience metric is also used to evaluate the city's performances according to the level of damage.

The results are then analyzed and compared to identify the most efficient city shape and the resilience trend against the city's size.

### *8.2.1 Contemporary City Ecosystem Modeling*

To date, cities have always been studied as entities with a well-defined functional structure. So, when dealing with disaster management, each urban context is usually understood as a unique system, which is then placed and considered in a wider national framework. Nowadays, this perspective can sound too restrictive, as it does not account for complex dynamics and the interdependencies that arise from typical self-organizing processes in each city.

With this, each city is characterized by underlying mechanisms that are governed by the people living in it. City inhabitants live, follow rules, and make choices that can influence the urban structure from both a human and a topological point of view. These social dynamics have different outcomes depending on the city's geographical configuration and its sociological, cultural, and economic background. On the other hand, citizens are fed by urban services, and so always act through a dense infrastructure network in which they are strictly interrelated. It is clearly evident that each urban context should be modeled starting by considering the link between each citizen and each physical structure. The main reason for this lies in the opportunity to assess urban performances according to the peculiarity of each studied city. So, in this study, the focus is on the local scale. This forms the basis for a multiscale approach, which can finally lead to an upper-scale approach, namely the global, according to modern, bottom-up thinking.

Cities are modeled as spatial networks. These are a particular kind of complex network that is embedded in a two-dimensional space, and whose typical metric is the Euclidean distance. A system of typical street patterns is created to model each type of urban geometry in a GIS environment, with the topology inspired by major European and US cities and ancient city centers.

The robustness of the proposed approach is examined in a real case study, namely the historical center of the city of Naples. Furthermore, to study differences in terms of the size and shape of the studied urban center, four different city geometries are modeled as follows:

- Circular (e.g., Rome, Figure 8.1; L'Enfant's plan for Washington DC; Regent's park in London; Karlsruhe).
- Rectangular (e.g., Savannah, Regensburg on the southern bank of the Danube river from Roman times; Milgram, 1967). This is better known as the typical structure of modern US cities, e.g., Orlando, New York, and Philadelphia (Figure 8.2). These cities typically have T-shaped crossings as self-organized urban networks. Venice and Cairo also have similar geometric shapes, although they are not just rectangular, but also self-organized cities.
- Star, the ideal city renaissance model. An example is the Italian city of Palma Nuova (Figure 8.3), outside Venice, which was originally credited to the architect Scamozzi.

Each urban center investigated in this chapter is modeled as an HSPN (Bozza et al., 2015; Cavallaro et al., 2014). The HSPN is a novel approach based on complex network theory, and enables us to account for all of a city's components. Moreover, interrelations between urban physical (buildings and infrastructures) and social components (citizens) can be characterized to understand the city's physiological behavior from a human-centric perspective.

In particular, the infrastructure and social network are first individually modeled as graphs and then overlaid in the global network, namely the HSPN (Bozza et al., 2015; Cavallaro et al., 2014).

In this study, the infrastructure network is represented through the modeling of the street network. This is because most urban services are typically arranged according to urban street patterns. As a consequence, this simplification enables us to study the interactions between the city's inhabitants and services by modeling only two planar graphs. A complex network is always represented by a graph $\Gamma = (N, L)$, which is composed

Fig. 8.1. Picture of Rome and a map of the city showing an analogy with the circular shape. After Google Maps (2017a)
*Source*: Getty Images. Credit: Norbert Rehm

Fig. 8.2. Picture of Philadelphia and a map of the city showing its typically rectangular shape. After Google Maps (2017b)
*Source*: Getty Images. Credit: Harold Stiver / EyeEm

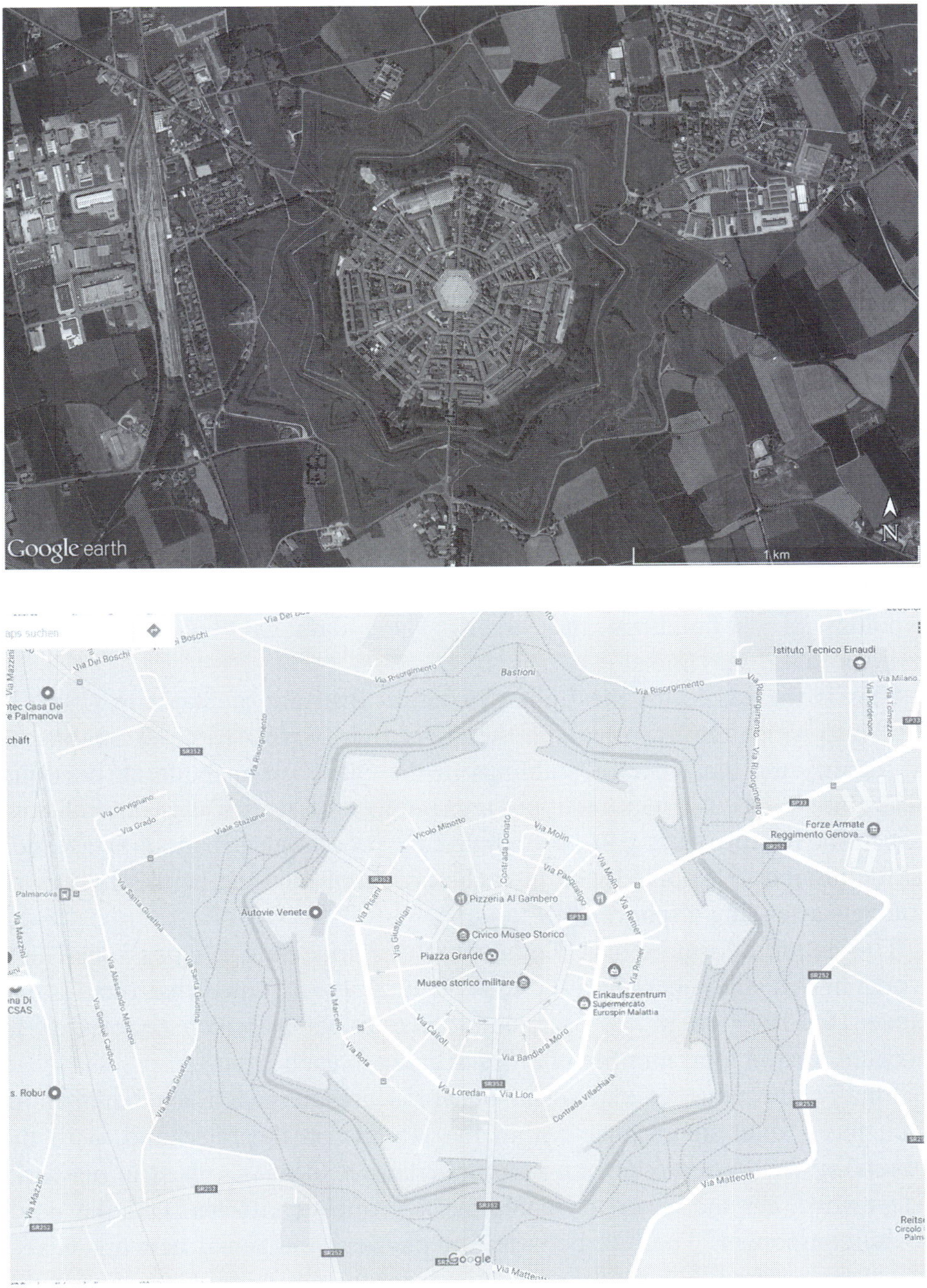


Fig. 8.3. Picture of the city of Palma Nuova and its typically star shaped planimetry. Google Maps (2017c)

of a discrete set of nodes, $\mathcal{N} = \{1, 2, \ldots, n\}$, and a discrete set of links $L \subseteq \mathcal{N} \times \mathcal{N}$.

In terms of modeling the HSPN, two sets of links and two sets of nodes are modeled to create the social and physical network. The former is produced by the set of nodes representing the residential buildings, $\mathcal{N}_b$, and the set of door links, $L_b$, which connect each building to street junction's nodes. Meanwhile, the latter is constituted by the set of nodes, $\mathcal{N}_s$, which represents the street junctions, and the set of links, $L_s$, which represents urban street patterns, where the number and length of the links representing the streets are also taken into account.

Finally, the city's HSPN is obtained and denoted as $G\,(\mathcal{N}_b \cup \mathcal{N}_s \cup L_b \cup L_s)$. To avoid traffic modeling issues, a further simplification is achieved regarding the vehicular and inhabitant flow, which is assumed to be bidirectional in each street. As a consequence, the HSPN is defined as an undirected graph, implying that the converse arch also exists for each arch linking the generic nodes $i$ and $j$. The proposed approach clearly enables us to model any kind of city, provided information is available about the location, number, and typology of its buildings and streets. These data can be acquired from national databases and surveys.

This is the case with the Italian National Institute of Statistics (ISTAT, 2011), to which the hypothesis in the present study refers. ISTAT enables us to identify the incidence of buildings in the entire urban building portfolio according to their structural typology, number of stories, and age of construction. Further data can be obtained by processing urban statistics such as the number of people living in each residential building or the number of students attending a school.

This latter information is fundamental for modeling a city's social network. In this study, based on the ISTAT data, it is assumed that there is one citizen for each 30 $m^2$ of each considered building, which enables us to account for the total number of citizens living in each city.

Finally, the physical and social networks are merged. Each building, in fact, has a dual significance within the HSPN: on the one hand, with the streets' system, it represents an essential component of the physical network. Meanwhile, on the other, it represents a group of citizens, i.e., the city's social component. The link between the physical and social network is also characterized and then modeled starting from the city's buildings. The building network is actually connected to the street network through the outgoing door links. It is also connected to the other buildings, which are social network nodes, through the outgoing street links that represent interactions that exist between groups of citizens.

### 8.2.2 Scenario Simulation: the Citizen-Citizen and Citizen-School Case Studies

Two different case analyses are presented and seismic scenarios are run for both of them. Once the HSPN is modeled, its efficiency is evaluated on the basis that no event has occurred. Consequently, an earthquake is simulated that causes disruption to the entire city model. In particular, this simulation damages and/or collapses buildings, while adjacent streets are also obstructed due to the debris falling from the buildings or safety measures that have been put in place.

The proposed study does not concentrate on the buildings' vulnerability from only a strictly engineering perspective; it also focuses on how this vulnerability affects the entire city's operation conditions whenever a catastrophic event occurs. So, damage to the city is simulated by assuming that structural vulnerability is uniformly distributed over the territory, and imposing diverse extreme seismic scenarios.

In this way, the damage suffered by the city's physical network is computed in a deterministic fashion. Two diverse earthquake severity levels are simulated for the HSPN by assuming, respectively, that between 15 percent and 30 percent of buildings collapse.

Furthermore, in each scenario, a particular assumption is made depending on the type of efficiency that has to be assessed. In evaluating citizen-citizen efficiency, a certain percentage (15 or 30 percent) of the residential buildings is assumed to collapse based on their identification number (ID). On the other hand, in the case study in which citizen-school efficiency has to be evaluated, the percentage of the buildings to be regarded as severely damaged or collapsed is evaluated for both the set of residential buildings and the set of school buildings. However, in this case, the same percentage is used to impose the buildings' collapse according to the same approach adopted for the citizen-citizen efficiency evaluation. Accordingly, the connectivity features of the network are evaluated with reference to both the links between couples of residential buildings and those between each residential building and school.

To achieve this in both the case analyses, a fully random methodology is implemented that generates a random permutation of integers in $1{:}N$. As a result, the numbers extracted from the permutation will determine which building will not survive by recognizing its ID.

Furthermore, the streets adjacent to such buildings are considered to be unusable according to a probability-based approach. Based on the ratio between each damaged building's height, $h$, and the width, $l$, of the street it is located on, the probability of street interruption is accounted for as shown in Eq. (8.1) (Cavallaro et al., 2014):

$$P_r(h,l) = \begin{cases} 1 & \text{if} \quad h \geq l \\ \dfrac{h}{l} & \text{otherwise} \end{cases} \tag{8.1}$$

A stream of uniform pseudorandom numbers is generated and values from the standard uniform distribution are selected on the open interval (0,1) and compared to the assessed values of the probability, $P_r$, that the streets are interrupted. As such values are larger or smaller than those obtained from the stream simulation, it will be decided respectively whether the street will or will not be obstructed by the adjacent buildings.

Note that a higher ratio between a building's height and a street's width has a cascade effect on the city's functionality level, meaning that there is a higher probability of street interruption. In fact, the higher an adjacent building, the greater the chance that debris will fall onto the street or that the street will be closed for safety reasons.

This has a further effect on the behavior of the HSPN itself: due to the street eventually becoming inaccessible, the link that represents it will not have value for network connectivity purposes. Bearing this in mind, and to evaluate efficiency in the aftermath of a catastrophic event, it has to be considered that whenever a seismic event occurs (particularly a seismic one), damage to and/or the collapse of buildings should be expected, as should street interruptions due to falling debris or other civil protection issues. As a consequence, nodes representing damaged buildings and links representing streets subjected to usage restrictions are considered to be inactive. Accordingly, they are removed from the network topology model, and efficiency in the aftermath of the earthquake is evaluated by only accounting for the city components that survive.

Moreover, damage to buildings means that for each simulated earthquake and damaged city configuration, a certain number of citizens have to be relocated. Accordingly, the same approach is also used to calculate the number of users being fed by the school service.

As a consequence, in both case analyses, a zero stage is recognized where a certain percentage of buildings (only residential or residential and scholastic) and streets are unusable, and so removed from the HSPN model, causing some parts of it to be disconnected.

As a case in point, at this stage, a suitable recovery strategy has to be selected and simulated to monitor progress in the HSPN restoration. In this study, a status quo down-up strategy is implemented (Cavallaro et al., 2014). This is directed to recover the urban HSPN to its initial configuration, with buildings being progressively put back in place, citizens being returned to their residences, and damaged streets being restored.

The recovery process is simulated though $n$ discrete stages, both for the citizen-citizen and the citizen-school case analyses. Each stage provides for a fraction $1/n$ of the displaced citizens to be relocated, starting from the smallest buildings, which are also the cheapest, and progressing step by step to the largest ones. In parallel, street links that were interrupted are also reactivated within the HSPN once the buildings that caused their interruption are reconstructed. So, street nodes and links and building nodes and door links are gradually reactivated, causing a certain quantity of buildings and streets to be restored and a certain number of inhabitants to be relocated in each stage of the selected recovery process. As a consequence, in each of the recovery stages, a different efficiency value is assessed for both the citizen-school and citizen-citizen case studies.

Three sets of measures are evaluated for each HSPN and each scenario and case study: (1) the number of damaged buildings and streets; (2) the values of the citizen-citizen efficiency, $E_{cc}$, and the citizen-school efficiency, $E_{cs}$, respectively, for the residential and the school HSPNs, with both being evaluated before the earthquake occurs, soon after it occurs and for each recovery stage; and (3) the systemic damage, $D$, the damage-dependent, $R^D$, and the damage-independent resilience, $R^E$, to finally quantify the recovery capacity.

### *8.2.3 A Novel Understanding of Complex Network Metrics: Assessing the Urban Systemic Damage*

In this chapter, urban efficiency is understood as the city network's capacity to feed urban services to its citizens, depending on their geographical location and the spatial configuration of the buildings. Efficiency is evaluated in the preevent network's configuration by accounting for all existing nodes and links as the global connectivity level of the studied urban environment. So, efficiency is a measure of the usability of services, and enables us to assess the damage to the quality of the urban services and entire city system, as perceived by its inhabitants.

Once the damaged configuration of the city's HSPN is known and efficiency has been evaluated, a recovery strategy has to be hypothesized and planned. According to the chosen strategy, recovery actions are then simulated within a discrete steps procedure, through which the streets and buildings are gradually restored. Accordingly, efficiency can be reassessed in each recovery stage by considering streets and buildings that have been repaired.

On the whole, to assess city efficiency according to graph theory, the quality of the connections between the pair of nodes $i$ and $j$ has to be evaluated. To do this, a cost is associated with each walk or path by summing up all the involved edges. Meanwhile, a walk from $i$ to $j$ is defined as an alternating sequence of nodes and edges. A walk is called a path when each node is crossed only once. So, a relationship on the distances between nodes is recognized in Eq. (8.2):

$$d = \frac{d_{ij}^{eucl}}{d_{ij}} \tag{8.2}$$

where $d_{ij}^{eucl}$ is the Euclidean distance between node $i$ and node $j$ and $d_{ij}$ is the length of the shortest path, namely the one between $i$ and $j$, which has a minimal length.

Typically, a 1-dimensional graph, $G$, can be defined through two measures: the characteristic path length, $L$, and the clustering coefficient, $C_c$ (Cardillo et al., 2006). The former is a global feature of the network, representing the mean graph distance over all the couples of vertices and is evaluated as shown in Eq. (8.3):

$$L = \frac{1}{N \cdot (N-1)} \cdot \sum_{i,j \in N,\, i \neq j} d_{ij} \tag{8.3}$$

where $N$ is the number of network nodes and $d_{ij}$ is the shortest path between each couple of nodes, and so the graph distance. $L$ can be defined if and only if the graph is connected, otherwise it cannot be a finite quantity, with $d_{ij}$ tending towards infinity. On the other hand, the clustering coefficient, $C_c$, is a local feature. When considering the generic node $i$, $C_c$ represents the subgraph of the neighbors of $i$, divided by the maximum possible number $k_i \cdot (k_i - 1)/2$, according to Latora and Marchiori (2001). This can be evaluated as shown in Eq. (8.4):

$$C_c = \frac{1}{N} \cdot \sum_i C_i \tag{8.4}$$

with $C_i$ being the number of edges in the graph, $G_i$, which is the subgraph of the given graph, $G$, which is induced by the first neighbors of $i$. In essence, the clustering coefficient enables us to evaluate the number of triangles in a real system.

According to Watts and Strogatz (1998), it is possible to independently and continuously rewire each edge of $G$ randomly with probability $p$. It can also be observed that this can be transformed from a regular lattice, when $p = 0$, into a random graph, when $p = 1$. In this transition, an intermediate state can be observed where, at small $p$, the system shows high clustering, like regular lattices, while still presenting small characteristic path lengths, like random graphs. This is a typical feature exhibited by real networks, such as social, informatics and biological networks, which are usually scale-free. This is described as small world behavior (Cardillo et al., 2006; Milgram, 1970; Watts and Strogatz, 1998). This means that such networks have a connection topology that is neither typically regular nor typically random. Nevertheless, according to Latora and Marchiori (2001), man-made urban networks and neural networks demonstrate small-world behavior, and so are efficient systems at both a local and global level. In this way, a single-variable definition is produced based on the general concept of efficiency, $E$, which enables us to withdraw all the constraints that are related to the system's unweightedness, connectedness, and sparseness.

Efficiency has a physical meaning that embraces the system's features at both the local and global levels and enables us to measure its functionalities in any condition. Of course, if we consider all the possible paths in the graph, $G$, from $i$ to $j$, $d_{ij}$ is the smallest sum of the physical distances. So, by supposing that the system is parallel, i.e., that material, information and/or people flows progress concurrently along the network through its arches from each node, it can be assumed that the global efficiency of a real network is inversely proportional to its shortest paths. As a consequence, this can be calculated as shown in Eq. (8.5) (Cavallaro et al., 2014; Latora and Marchiori, 2001; Watts and Strogatz, 1998):

$$E = \frac{1}{N \cdot (N-1)} \cdot \sum_{i,\, j \in N,\, i \neq j} \frac{1}{d_{ij}} \tag{8.5}$$

where $d_{ij}$ tends towards infinity and efficiency is zero when there is no path between the generic nodes $i$ and $j$. Furthermore, efficiency can be normalized in [0,1] by dividing the shortest path length between $i$ and $j$ by the Euclidean distance, $d_{ij}^{eucl}$, which is the geographical distance between $i$ to $j$ as the crow flies. Subsequently, the normalized pairwise efficiency can be calculated and averaged on each couple of nodes, thus representing global network efficiency according to Eq. (8.6) (Latora and Marchiori, 2001; Cavallaro et al., 2014):

$$E = \frac{1}{N \cdot (N-1)} \cdot \sum_{i,\, j \in N,\, i \neq j} \frac{d_{ij}^{eucl}}{d_{ij}} \tag{8.6}$$

In the present study, global efficiency has to be evaluated by accounting for the distance between the network's nodes that are feeding the city's inhabitants, but also for the number of citizens living in each building. As a consequence, a modified version of the proposed relationship is used herein that is consistent with the HSPN approach. This enables us to evaluate the connectivity level between groups of inhabitants, which is the citizen-citizen efficiency case, or between groups of inhabitants and urban services, for instance in the case of citizen-school efficiency. Equations (8.7) and (8.8) show the relationships (Cavallaro et al., 2014), with $E_{cc}$ representing citizen-citizen efficiency and $E_{cs}$ citizen-service efficiency (in the example, citizen-school efficiency):

$$E_{cc} = \frac{1}{H_{tot} \cdot (H_{tot} - 1)} \cdot \sum_{i \in B} H_i \cdot \left( (h_i - 1) + \sum_{j \in (B \setminus I)} H_j \cdot \frac{d_{ij}^{eucl}}{d_{ij}} \right) \tag{8.7}$$

Here, $i$ and $j$ are the IDs of the building nodes, $H_{tot}$ is the total number of the city's inhabitants, and $H_i$ and $H_j$ are, respectively, the number of citizens living in buildings $i$ and $j$. $B$ is the set of building nodes, $d_{ij}$ is the shortest path's length, $d_{ij}^{eucl}$ is the Euclidean distance between nodes $i$ and $j$, and $h_i$ is the number of inhabitants living in buildings with a zero distance from building $i$, which belongs to set $I$.

Accordingly, the efficiency of service HSPNs can also be assessed if the outer summation in Eq. (8.7) is replaced with a summation over the set $S$ of the buildings representing facilities such as schools. Furthermore, the term $H_{tot}$ is also replaced with the term $S_{tot}$, which is the summation of the total number of citizens using the buildings that supply the urban service being considered and represents their importance in the HSPN. Meanwhile, $S_i$ is the number of citizens who benefit from the service supplied by the facility building $i \in S$.

$$E_{cs} = \frac{1}{S_{tot} \cdot H_{tot}} \cdot \sum_{i \in S} S_i \cdot \left( h_i + \sum_{j \in (B \setminus I)} H_j \cdot \frac{d_{ij}^{eucl}}{d_{ij}} \right) \tag{8.8}$$

In essence, the difference in terms of the efficiency typology that can be evaluated depends on the distances used to compute it. As an example, in the case of citizen-citizen efficiency, the shortest path distances and the Euclidean distances are both evaluated between the couples of buildings that represent residences. Conversely, if citizen-school efficiency has to be computed, both distances must be evaluated between each city's physical component, representing a school, and each city's physical component, representing a residential building.

The global efficiency can therefore be evaluated for each city's HSPN, and this approach is employed to assess the damage to the city in a systemic fashion. Despite the traditional perspectives in civil engineering that focus on the single structure, it is possible in this way to obtain a global overview of the urban system by focusing on how urban damage affects the city's functionalities. In this way, once efficiency is evaluated before, $E_{pre}$, and soon after the event occurs, $E(t = 0) = E(0)$, or in any recovery stage, $E(t > 0)$, it is possible to define a function, called the recovery function, $Y(t)$, that returns the residual city system's capacity to feed services to its citizens. Equation (8.9) shows the formula:

$$Y(t) = \frac{E(t)}{E_{pre}} \tag{8.9}$$

where $t = 0$ is the time at which the seismic event occurs and the city's HSPN presents its "worst", i.e., damaged, configuration.

The proposed relationship can then be evaluated at each time stage of the recovery process, $Y(t)$, once efficiency at that time has also been assessed.

In keeping with this, it is possible to quantify the systemic damage to the entire urban network by simply observing the drop in the HSPN efficiency, $E(t)$, in terms of the recovery function, $Y(t)$.

This is the novelty of the study presented herein, which makes it possible to assess the state of service of the urban environment after the occurrence of a catastrophic event by merging civil engineering and complex network methodologies. Such an approach allows the after-event level of performance of the city, which is systemic and integral, to be measured. The systemic damage measure, which is normalized with respect to the city's preevent efficiency, $E_{pre}$, can simply be evaluated as shown in Eq. (8.10):

$$D(t) = \frac{E_{pre} - E(t)}{E_{pre}} = 1 - Y(t) \tag{8.10}$$

and is defined in the close interval [0,1]. The observation of such an indicator becomes critical when examining the city's HSPN soon after the event occurs (at the zero stage), $D(0)$.

Obviously, the more that the value of $D$ tends to unity, the more the observed systemic damage is severe. As a consequence, the two limit cases can be defined as $D(0) = 1$ "total damage", while $D(0) = 0$ means "no damage".

### 8.2.4 The Quantification of Urban Resilience

Urban resilience is understood as a fundamental component of sustainability, in particular as the capacity of a city ecosystem to be sustainable during the hazardous event occurrence phase. In essence, a city has to show readiness and promptness in its disaster response and has to effectively bounce back to an equilibrium condition, which can be new or the same as before the event occurred. For resilience quantification purposes, this study proposes a novel approach that enables us to evaluate urban life quality, disaster resilience and city robustness to structural damage according to a humanitarian approach. Damage suffered in an urban context is, in fact, evaluated here in an integral fashion as the decaying of the city's service state after the occurrence of an adverse event. This is an approach that does not look at the city as a global system, but analyzes it by accounting for each single city's components, both physical and human, and their interrelationships. As a consequence, the city model is built through the gradual annexation of such components according to modern multiscale approaches, from the lowest to the highest degree of network complexity.

Resilience can be assessed by integrating in time the recovery function, $Y(t)$, at all the recovery stages. Physically, this can be interpreted by observing the trend of the recovery function in the $t - Y(t)$ plane, which describes the recovery curve (Figure 8.4).

Resilience is the area under the curve, divided by the time needed to implement the selected strategy, which is the time passed by from $t_s$, when the recovery process has started, to $t_c$, when it is completed. In this way, the resilience quantification can be performed through the calculation of Eq. (8.11) (Bruneau et al., 2003; Cavallaro et al., 2014; Reed et al., 2009):

$$R = \frac{\int_{t_s}^{t_c} Y(t)dt}{t_c - t_s} \tag{8.11}$$

In fact, the city's likelihood of recovering efficiently from a disaster is strictly related to a huge quantity of complex and often uncontrollable variables. It is worth noting that decision making in such a context is ascribable to disaster managers, who, depending on the time, money and human and material resources available, choose which strategy should be undertaken for the recovery. Issues related to this process affect the city's recovery in different ways, with almost all of them related to time, $t$, meaning that they cannot be considered in detail. As a consequence, a good approach should totally remove the dependence of resilience on time in order to avoid embedding further uncertainty in the evaluation process. To do this, both HSPN

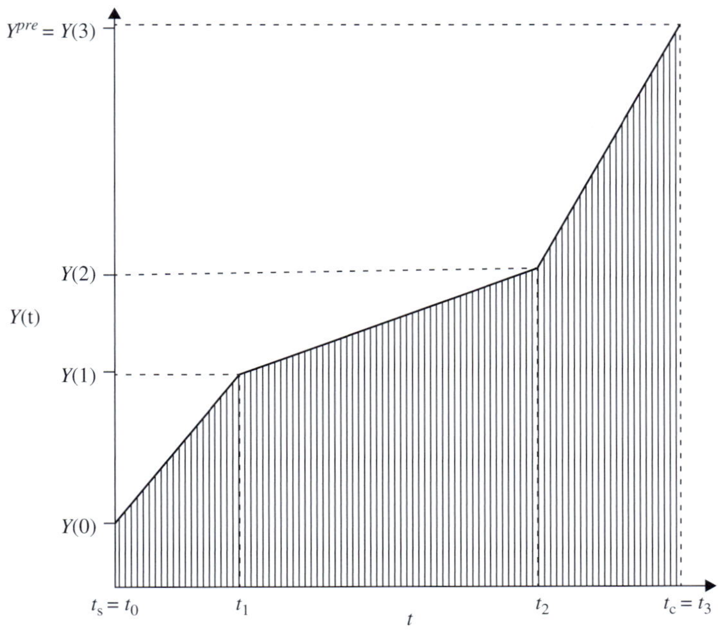


Fig. 8.4. Trend of the recovery function, $Y(t)$, across the recovery strategy against time, $t$

efficiency and the recovery function are defined as being dependent on the number of inhabitants being relocated in each recovery stage, as highlighted in Eq. (8.12) (Cavallaro et al., 2014):

$$Y(C) = \frac{E(C)}{E_{pre}} \tag{8.12}$$

This is the ratio between the city's efficiency level when $C$ inhabitants have been relocated and its efficiency level when no inhabitants need to be moved, i.e., when the seismic event has not yet occurred.

Furthermore, the dependence on the total state of damage is also removed, enabling us to evaluate a normalized recovery function, as highlighted in Eq. (8.13):

$$y(C) = \frac{Y(C) - Y(0)}{1 - Y(0)} \tag{8.13}$$

where $Y(0)$ indicates the residual HSPN efficiency soon after the event occurs (relocated citizens $C = 0$), and $Y(C)$ indicates the residual HSPN efficiency in each generic recovery stage ($C$ citizens relocated).

According to Cavallaro et al. (2014), with this, resilience can be finally quantified according to Eq. (8.14), as defined in [0, 1]:

$$R = \frac{\int_0^{C_{max}} y(C)dC}{C_{max}} \cong \frac{\sum_{c=0}^{C_{max}} y(C) \cdot \Delta C}{C_{max}} \tag{8.14}$$

where $C_{max}$ is the total number of citizens whose homes have been damaged and so need to relocate after the seismic event. The integral is simplified with a summation, which is the strategy implemented in a discrete number of steps.

In this context, a further issue is related to the dependence of resilience on the city's state of damage, given that it directly affects the quantification of the city's capacity to recover, according to the damage suffered soon after a certain event occurs.

In keeping with this, two alternative approaches are proposed to evaluate resilience, which are directly related to the physical meaning of resilience based on the observation of the recovery curve. In the former, resilience is evaluated as being independent of the initial state of damage, as shown in Eq. (8.13). Meanwhile, in the latter, it is evaluated as being dependent on this initial damage state. This last metric is based on the definition of the systemic damage, $D$, as set out in Eq. (8.10), being dependent this time on the number of relocated citizens, as shown in Eq. (8.15):

$$D(C) = \frac{E_{pre} - E(C)}{E_{pre}} = 1 - Y(C) \tag{8.15}$$

### *Quantification of Damage-Independent Resilience*

The recovery curve has to be observed to show the meaning of the proposed damage-independent resilience metric.

For instance, when considering that a city has been struck by an earthquake whose intensity is $I$, let us now suppose that local authorities undertake recovery actions that are completed in three stages; that is, the city's global efficiency bounces back to its preevent level.

The above-mentioned recovery path in the $E$-$C$ plane can then be described graphically, with $E$ being the normalized global efficiency and $C$ the number of relocated citizens (Figure 8.5).

Essentially, the normalized efficiency is evaluated by accounting for the drop in the efficiency of the city's HSPN in each recovery stage and also the efficiency drop between the HSPN condition before and soon after the event occurs.

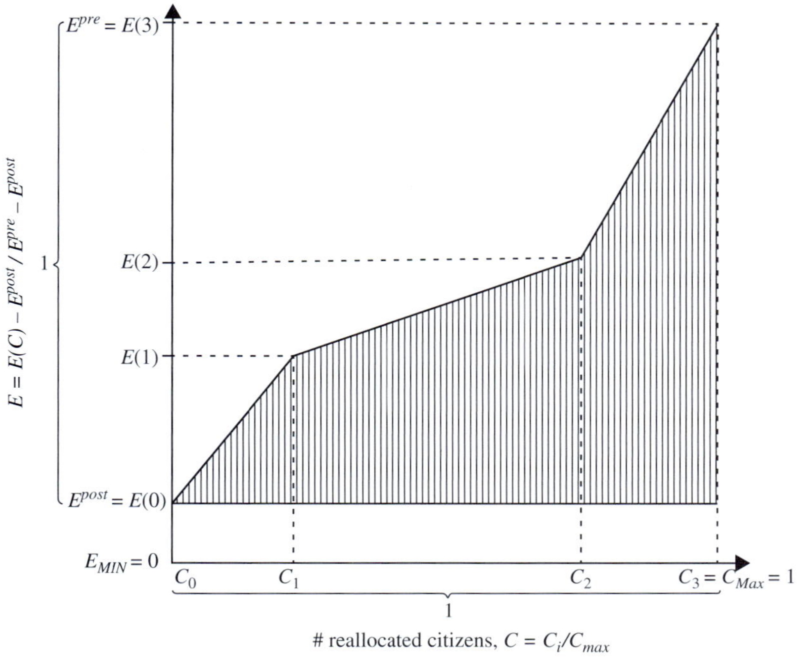


Fig. 8.5. Recovery curve in terms of global efficiency normalized to the preevent efficiency value as a function of relocated citizens in each stage of the recovery process. This is in turn normalized with respect to the maximum number of evacuated citizens

So, by considering that $C$ citizens are relocated at the *ith* reconstruction stage, normalized to $C_{max}$; examining the maximum number of citizens to be relocated; and by normalizing efficiency with respect to the efficiency drop soon after the earthquake occurs ($E^{pre} - E^{post}$), the normalized global efficiency, $E$, can be evaluated as shown in Eq. (8.16):

$$E(C) = \frac{E(C) - E^{post}}{E^{pre} - E^{post}} \tag{8.16}$$

This result is the same as the finding that can be obtained analogously if we consider Eq. (8.13) and replace the recovery function formula, as set out in Eq. (8.12), in terms of efficiency.

So, the proposed relationship for the normalized efficiency is substantially the same as the recovery function defined in the previous section, resulting in $y(C) = E(C)$. In parallel, the recovery function is now explicitly defined as the ratio between the efficiency drop when $C$ citizens have been relocated with respect to the efficiency soon after the event occurs, and the efficiency

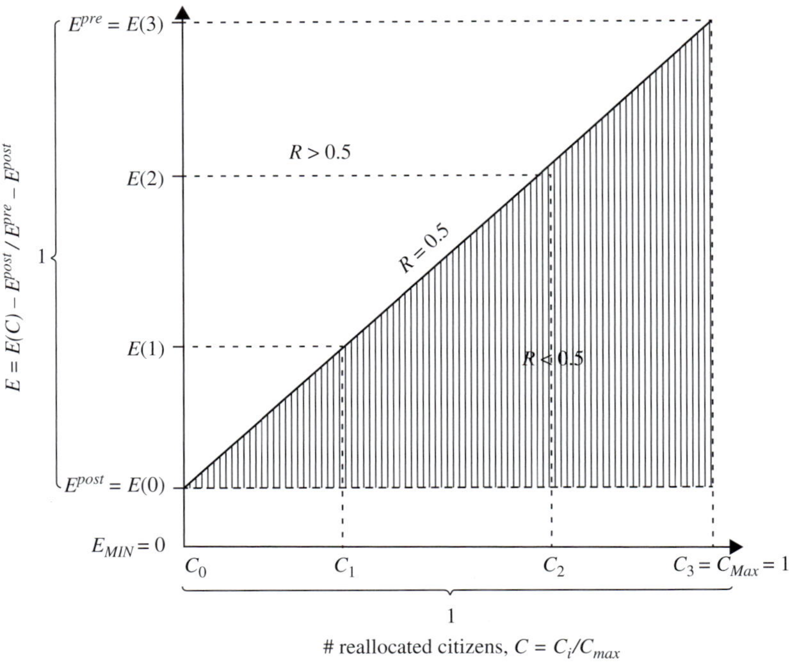


Fig. 8.6. Perfectly linear recovery curve

drop between the pre and postevent stages. Equation (8.17) shows the formula:

$$y(C) = \frac{E(C) - E^{post}}{E^{pre} - E^{post}} \tag{8.17}$$

Accordingly, resilience can be evaluated as the area underneath the recovery curve, i.e., as the integral of the recovery function across all the recovery stages. The proposed relationship is the following, Eq. (8.18):

$$R^E = \frac{\int_0^{Cmax} y(C) \cdot dC}{C_{max}} \cong \sum_i \frac{[y_i(C_i) + y_{i+1}(C_{i+1})]}{2} \cdot \Delta C_{i,\, i+1} \tag{8.18}$$

where $\Delta C_{i,i+1} = \frac{C_{i+1} - C_i}{C_{max}}$, namely the relocated citizen share normalized to $C_{max}$. As a consequence, resilience is defined in the [0,1] interval, where a particular condition is recognized for the 0.5 value. This is a crossing point between the recovery of the city's functionalities according to a sub- or super-linear trend, respectively. Such limit conditions are shown in Figures 8.6 and 8.7.

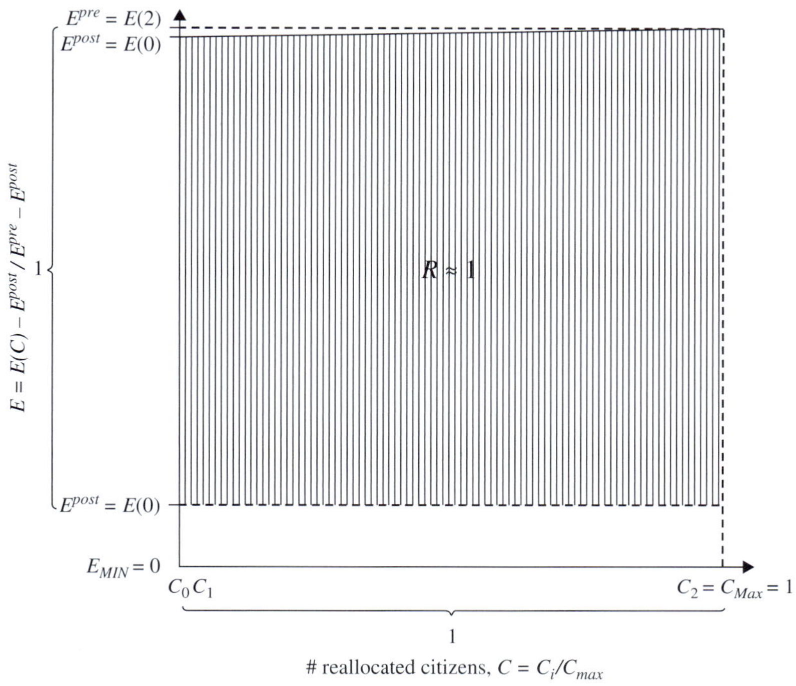


Fig. 8.7. Limit case in which the recovery is almost instantaneous and resilience attains its maximum value

It is clearly the case that if the curve is linear, resilience would attain the value $R = 0.5$. Conversely, the curve trend is super-linear when $R > 0.5$. Accordingly, a limit case can be recognized in which the recovery is almost instantaneous and resilience attains its maximum value. This is the case in which the area under the curve would tend to a unit-side polygon, being normalized in terms of both efficiency and the number of relocated citizens.

### *Quantification of Damage-Dependent Resilience*

Depending on the kind of issue being dealt with, it may be necessary to evaluate resilience without removing its dependence on the total state of damage soon after the event has occurred, instead specifically accounting for it. This is the case when a damage-dependent resilience metric is required. The proposed approach is basically the same as the previous one used for the quantification of the damage-independent resilience. The only difference is that resilience is evaluated as being dependent on the systemic damage, i.e., the global damage to the city's HSPN functionalities, $D$, as evaluated in Eq. (8.15). So, resilience is evaluated by accounting for the global city's

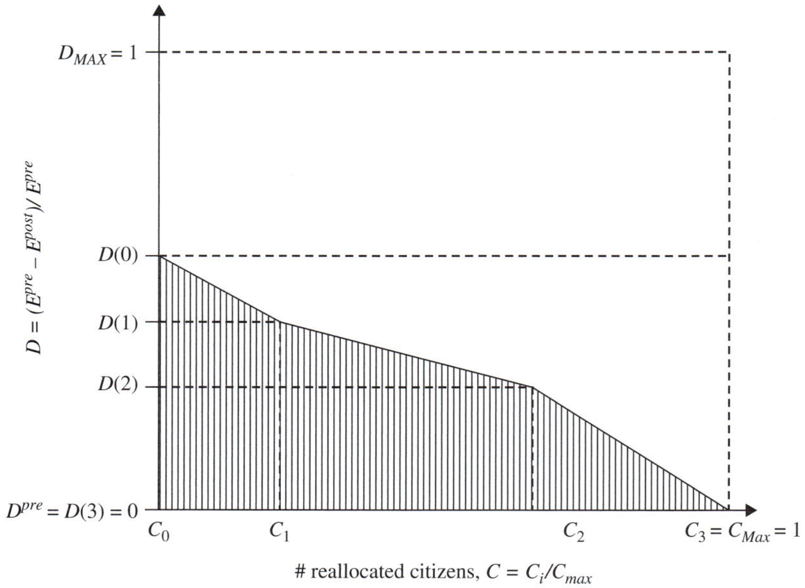


Fig. 8.8. Graphical representation of the recovery path in the *C-D* plane

efficiency, which is this time not normalized with respect to the preevent performance level in Eq. (8.19).

In parallel with this, by representing the recovery curve in the *C-D* plane (Figure 8.8), with *C* being the number of relocated citizens and *D* the systemic damage level in each recovery stage, resilience is also this time clearly represented by the area under the curve.

$$R^D = \frac{\int_0^{Cmax} D(C) \cdot dC}{C_{max}} = \sum_i \frac{[D_i(C_i) + D_{i+1}(C_{i+1})]}{2} \cdot \Delta C_{i,\ i+1} \qquad (8.19)$$

This is because using a synthetic indicator to quantify resilience may be misleading if one does not consider efforts made to bounce back to an equilibrium condition after an event. With this, one has to consider the damage condition that the city starts from with respect to its initial performance level, $E_{pre}$.

## 8.3 Case study: Measuring Physical Resilience in Naples (Italy)

### *8.3.1 Modeling Quartieri Spagnoli as a Complex Network*

Here, a real case study is developed to validate the approach and verify the robustness of the proposed metrics. The historical center of Naples, i.e., the Quartieri Spagnoli area, is modeled as an HSPN and earthquake scenarios

Fig. 8.9. Map of Naples' historical center (grey markers represent school buildings). After Google Maps (2017d).

are simulated to assess its resilience level according to the recovery strategy highlighted in Section 8.2.2 (status quo down-up strategy).

The Quartieri Spagnoli area is located in the inner city of Naples (Figure 8.9), and is composed of the Avvocata, San Ferdinando, and Montecalvario neighborhoods. Quartieri Spagnoli dates back to the sixteenth century, when it was built to host the Spanish military garrisons, which were in Naples to repress insurrections from the Neapolitan population.

Despite the poor conditions and reputation of this area, it represents the core of the historical and cultural local tradition. It comprises mostly masonry buildings that accommodate small artisan shops, places of worship, and typical local residences.

The selected area has a perimeter of 3.57 km and a wide in-plane geometry of 0.569 $km^2$. There are 614 residential buildings made of masonry and the local population is estimated to be 30,007.

Initially, only residential buildings are considered. The selected area has been modeled as an HSPN (Figure 8.10) and two diverse seismic scenarios are simulated.

In the first of these, collapse or severe damage is assumed for 15 percent of the buildings, while in the second this figure is taken to be 30 percent. As already outlined in Section 8.2.2, the damaged buildings are selected according to a fully randomized procedure.

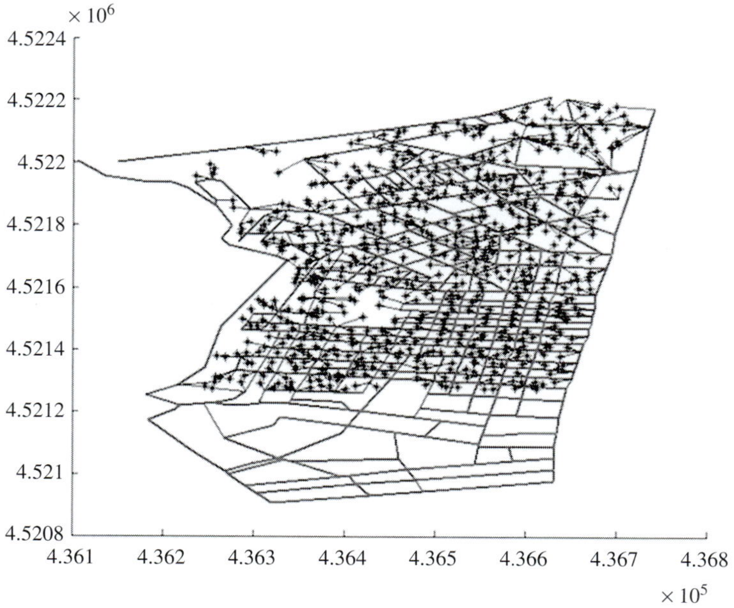


Fig. 8.10. HSPN of the historical center of the city of Naples (the Quartieri Spagnoli area) where only residential buildings have been modeled. (A black and white version of this figure will appear in some formats. For the color version, please refer to the plate section.)

As also shown in Section 8.2.2, street usability after the earthquake is evaluated according to a probability-based approach. Accordingly, the probability of street links becoming inaccessible is evaluated as a function of both their width and the buildings located along them.

Finally, the connectivity between couples of residential buildings is quantified across the simulated recovery strategy as the citizen-citizen efficiency.

A further HSPN of the Quartieri Spagnoli area is also modeled (Figure 8.11), and this also accounts for school buildings (17 buildings).

When added to the residential HSPN, these 17 buildings account for about 3,000 users.

According to the case analysis shown previously, two earthquake scenarios are simulated (15 and 30 percent of the buildings collapse), as is the status quo down-up strategy for the recovery. The only difference is in the assessed city efficiency, which this time refers to the citizen-school connectivity.

In both case analyses, each simulation is iterated ten times to produce different scenarios and to eventually enable substantial gaps to be evaluated. The following section shows and discusses the results of the analyses.

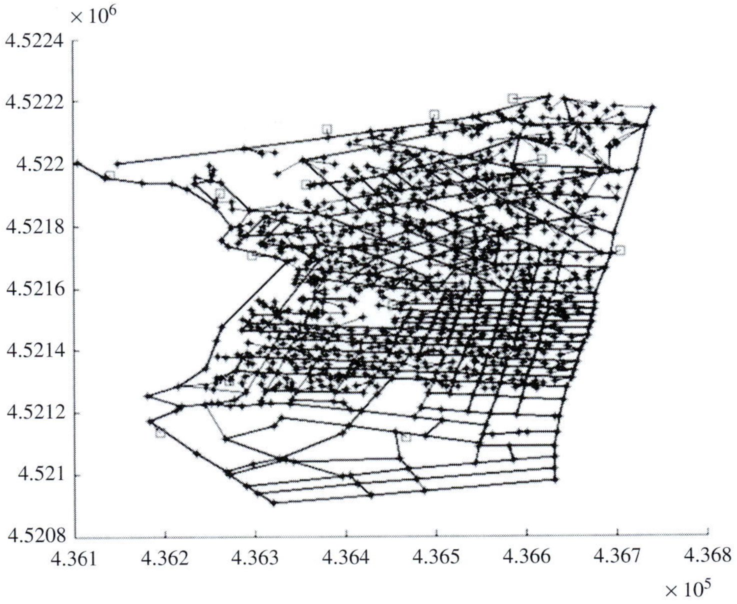


Fig. 8.11. HSPN of the historical center of the city of Naples (the Quartieri Spagnoli area) where both school (grey-squared markers) and residential buildings (black-starred dots) have been modeled. (A black and white version of this figure will appear in some formats. For the color version, please refer to the plate section.)

### *8.3.2 Discussion of the Results*

Once a city has been modeled, the methodology allows for a damage assessment to be conducted immediately after a seismic event occurs. Such an assessment considers damaged street patterns and buildings, as well as citizens who cannot remain at home and need to be relocated. As the scenario analysis is iterated ten times, drawing values from the pseudo-random number simulation, median values can ultimately be computed.

The damage suffered by the city is measured starting from the single building perspective. This is conceived as a physical structure, but also as an ideal reference point in which citizens live and from which they are served. Urban services, such as gas and water pipelines, electric grids, and road infrastructures, are linked to such buildings. So, once the links between buildings and all urban services are modeled, one can simply assume that when the building becomes unavailable, all the services that are linked to it are useless. In particular, in this work, the link between the couples of nodes representing buildings is modeled based on the street patterns of the studied city. This

Table 8.1. *Results of the analysis in the various case analyses and damage levels simulated*

| Case analysis | Damage level (in percentage) | $E^{pre}$ | $E^{post*}$ | $Y(0)^*$ | $D(0)^*$ | $R^{D*}$ | $R^{E*}$ |
|---|---|---|---|---|---|---|---|
| Citizen-citizen | 15 | 0.62 | 0.09 | 0.15 | 0.85 | 0.33 | 0.60 |
| | 30 | 0.62 | 0.01 | 0.02 | 0.98 | 0.46 | 0.53 |
| Citizen-school | 15 | 0.12 | 0.02 | 0.13 | 0.87 | 0.41 | 0.55 |
| | 30 | 0.12 | 0.004 | 0.03 | 0.97 | 0.52 | 0.47 |

* Median values

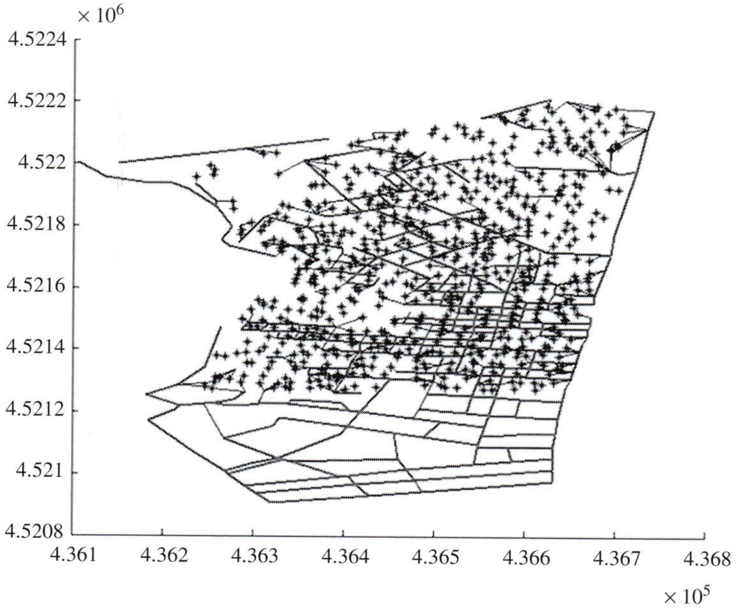


Fig. 8.12. Residential HSPN of the historical center of the city of Naples after the occurrence of a seismic event that caused 15 percent of buildings to collapse. (A black and white version of this figure will appear in some formats. For the color version, please refer to the plate section.)

assumption is justified by the fact that, in urban centers, the infrastructure for urban services (pipelines for instance) is usually located on the streets.

Table 8.1 and Figures 8.12–8.15 show the results of the 15 and 30 percent case analyses in terms of: the HSPN citizen-citizen efficiency, $E$; the recovery function and systemic damage values in the aftermath of the event, $Y(0)$ and $D(0)$; and the urban resilience, which is assessed as both damage-dependent, $R^D$, and damage-independent, $R^E$. Note that, for the sake of simplicity, both $E_{cc}$ and $E_{cs}$ are referred to as $E$ from this point onwards.

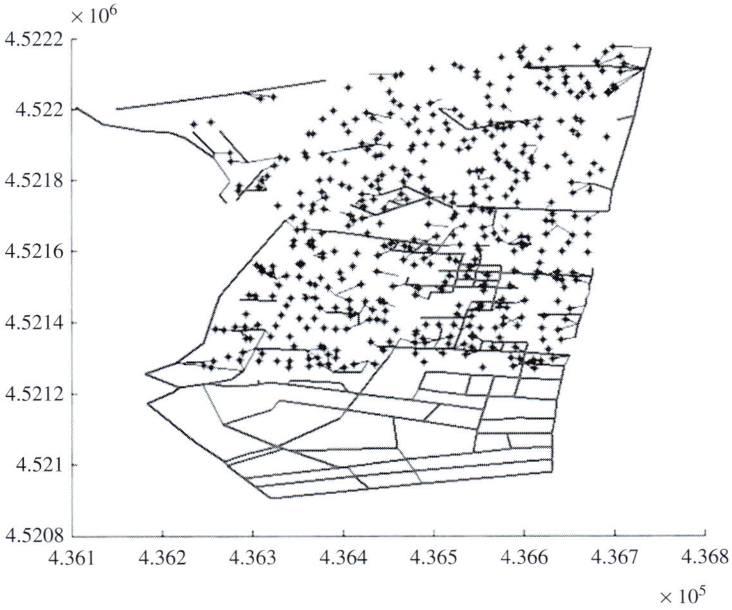


Fig. 8.13. Residential HSPN of the historical center of the city of Naples after the occurrence of a seismic event that caused 30 percent of buildings to collapse. (A black and white version of this figure will appear in some formats. For the color version, please refer to the plate section.)

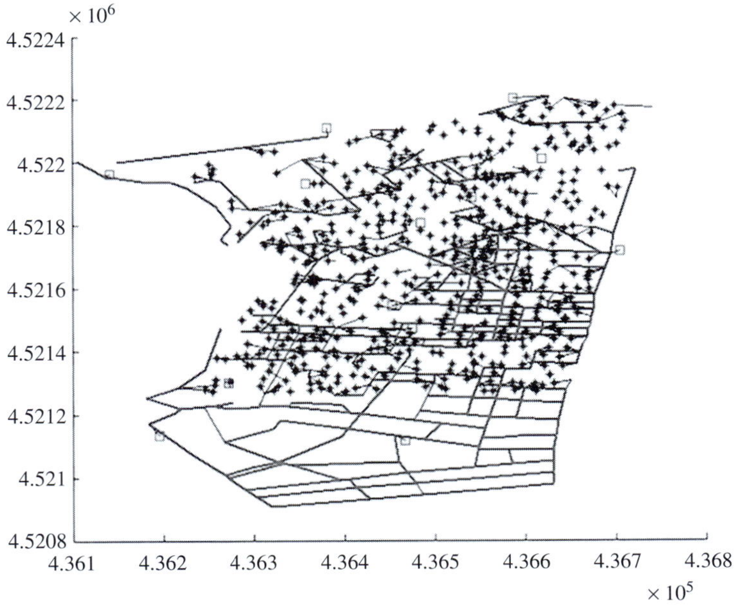


Fig. 8.14. School HSPN of the historical center of the city of Naples after the occurrence of a seismic event that caused 15 percent of buildings to collapse. (A black and white version of this figure will appear in some formats. For the color version, please refer to the plate section.)

Fig. 8.15. School HSPN of the historical center of the city of Naples after the occurrence of a seismic event that caused 30 percent of buildings to collapse. (A black and white version of this figure will appear in some formats. For the color version, please refer to the plate section.)

The HSPN configuration in the aftermath of the event can also be observed in Figures 8.12–8.15, where the black starred points represent residential buildings and black lines represent the door and street links, respectively.

As can be seen, the preevent efficiency is very different in the two case studies: 0.62 in the citizen-citizen example and 0.12 in the citizen-school version. This can be easily explained if we consider that efficiency is evaluated as being inversely proportional to the shortest path distances. In the citizen-school case study, such efficiency is the measure of the urban connectivity between each school building and each residential building, which is then averaged for the complete HSPN. So, when there are fewer schools than residential buildings, the shortest path distances are obviously higher. As a consequence, the resulting efficiency is lower in the case of the citizen-school case study than the citizen-citizen version.

Similar results for both case analyses are underlined in terms of the postevent efficiency, $E^{post} = E(0)$, the systemic damage, $D(0)$, and the recovery function, $Y(0)$. In both the citizen-citizen and citizen-school case analyses, the preevent efficiency drop is about 83 percent in the case where it is assumed that 15 percent of the buildings collapse, and about 97 percent in the 30 percent case. So, when it is assumed that the damage to the buildings increases by 15 percent (from 15 percent to 30 percent), this results in a proportional difference in the efficiency drop (about 13 percent).

The same trend is observed for the systemic damage, which is obviously directly related to the efficiency of the HSPN, when comparing the difference in terms of $D(0)$ between the 15 percent and 30 percent scenarios. On the other hand, in terms of the order of magnitude, the systemic damage is different when related to the efficiency values. In the citizen-citizen case analysis, for example, the damage is 0.85 in the 15 percent scenario and 0.98 in the 30 percent version. As the postevent efficiency is 0.09 in the 15 percent scenario and 0.01 in the 30 percent scenario, the systemic damage and the HSPN efficiency are inversely correlated. It should be noted that when the 30 percent case analysis is being considered, $D(0) = 0.98$ means that the HSPN is almost totally destroyed and its residual efficiency is therefore minimal ($E^{post} = 0.01$).

It is instead clearly evident that the recovery function is complementary to the systemic damage, being $Y(0) = 0.02$ in this case and so equal to $(1-D(0))$.

Finally, the HSPN's resilience can also be observed; both the damage-dependent and damage-independent resilience indicators are defined in [0,1], and so their order of magnitude is comparable. On the other hand, they have different meanings.

When it comes to the damage-dependent resilience, this is $R^D = 0.33$ in the 15 percent case analysis and $R^D = 0.46$ in the 30 percent version, which is a 39 percent increase. This resilience increases with the damage level, because the HSPN demonstrates a major capacity to recover. Indeed, it bounces back to the preevent equilibrium in the same number of steps, even though it starts from a more severe damage condition, which means that many more citizens need to be relocated and many more buildings have to be restored. This means that the most damaged HSPN is faster and more efficient in terms of resource use than the least damaged version.

Conversely, when considering the damage-independent resilience values, a 12 percent decrease is observed from the 15 percent to the 30 percent case analysis. This is because this metric is directly related to the attained efficiency values, as well as to the drop suffered from the pre- to the postevent condition and across all the recovery stages.

It can be asserted that the two proposed metrics are not mutually exclusive. Indeed, they can be used complementarily, as they cover diverse aspects of urban resilience.

$R^D$ is useful for comparing urban contexts struck by the same catastrophic event, enabling the systemic damage and the bouncing back capability at the local level to be evaluated contextually. In parallel with this, $R^E$ can be used to compare urban contexts that are very different or have been affected by different types of event. So, this can be used effectively to collect and compare best practices according to the event typology, even though the events occur in different geographical and urban contexts. This enables observations to be made and understandings to be achieved related to resilience issues on the global scale.

## 8.4 City Resilience Against Urban Size and Shape

### *8.4.1 Numerical Simulation and Graph Modeling*

Four different city shapes are artificially built as HSPNs, referring to the planimetry of real urban centers such as Barcelona, Paris and Los Angeles. In particular, rectangular, circular, hexagonal, and star-shaped HSPNs are modeled, with their size being increased according to their geographical extent and number of buildings, and so the number of citizens living in them. In particular, HSPNs with 50 (about 2,000 inhabitants), 200 (about 9,000 inhabitants), 1,250 (about 55,000 inhabitants) and 5,000 buildings (about 225,000 inhabitants) are modeled.

Primarily, only residential buildings are considered, and the HSPN efficiency is computed as the citizen-citizen version. Then, in each modeled HSPN, two percent of the buildings are assumed to be school buildings in order to evaluate the efficiency of the city's connectedness between residential and school buildings, and so between pairs of citizens being fed from the urban school service. Urban system modeling is performed through the use of a GIS, which enables us to integrate a wide range of data and identify more specific information through acquisitions, georeferencing, and documenting data. In particular, data about the population, the geographical extent of an area, and the buildings' spatial distribution, number and structural typology are also embedded in the GIS-based software. Some assumptions are made and hypotheses are produced in order to characterize the HSPNs:

- The structural typology is assumed to be frame buildings made of reinforced concrete, with all the buildings designed for gravity loads that are regular in terms of both their plane and height.
- The buildings considered for the city scenario simulations are assumed to be typical European 1970s/1980s constructions, with between two and five stories.
- The citizens living in each city are accounted for depending on the total floor area of each structural typology and assuming about one citizen for each 30 m$^2$, as suggested by ISTAT (2011).
- The percentage of buildings with reference to their number of stories is taken to be fixed: residential buildings are modeled as 10 percent being 2-story, 40 percent 3-story, 30 percent 4-story and 20 percent 5-story.
- The school buildings are all assumed to have five stories.

- Each urban geometry is modeled with an increasing number of buildings and its territorial extent is scaled accordingly.

Figure 8.16 shows the four different city shapes for the case in which 200 buildings are modeled and the citizen-citizen efficiency is evaluated:

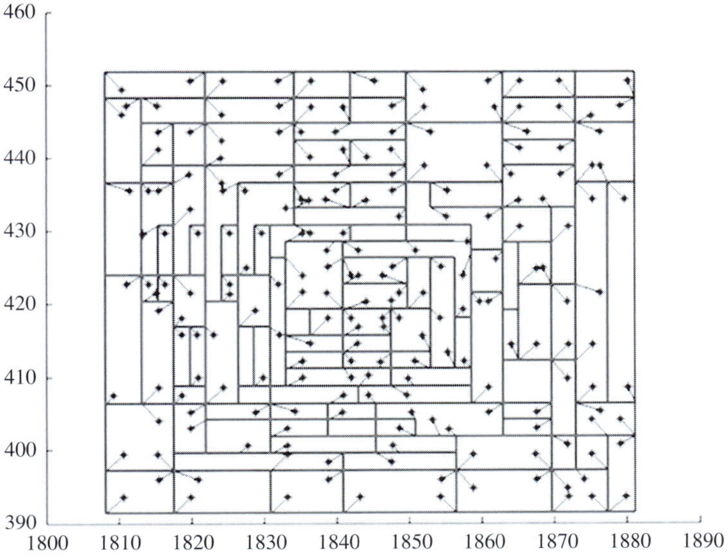


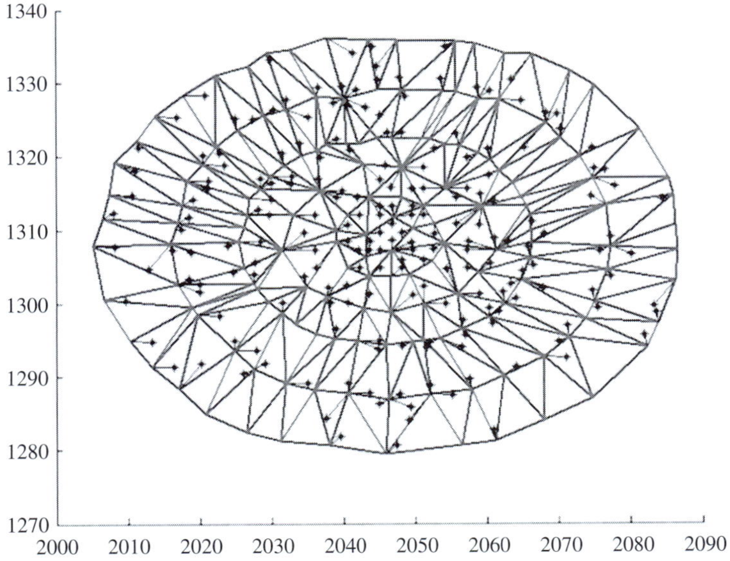


Fig. 8.16. City shapes: rectangular (A), circular (B), hexagonal (C), and star (D) modeled for the case analysis of 200 buildings. (A black and white version of this figure will appear in some formats. For the color version, please refer to the plate section.)

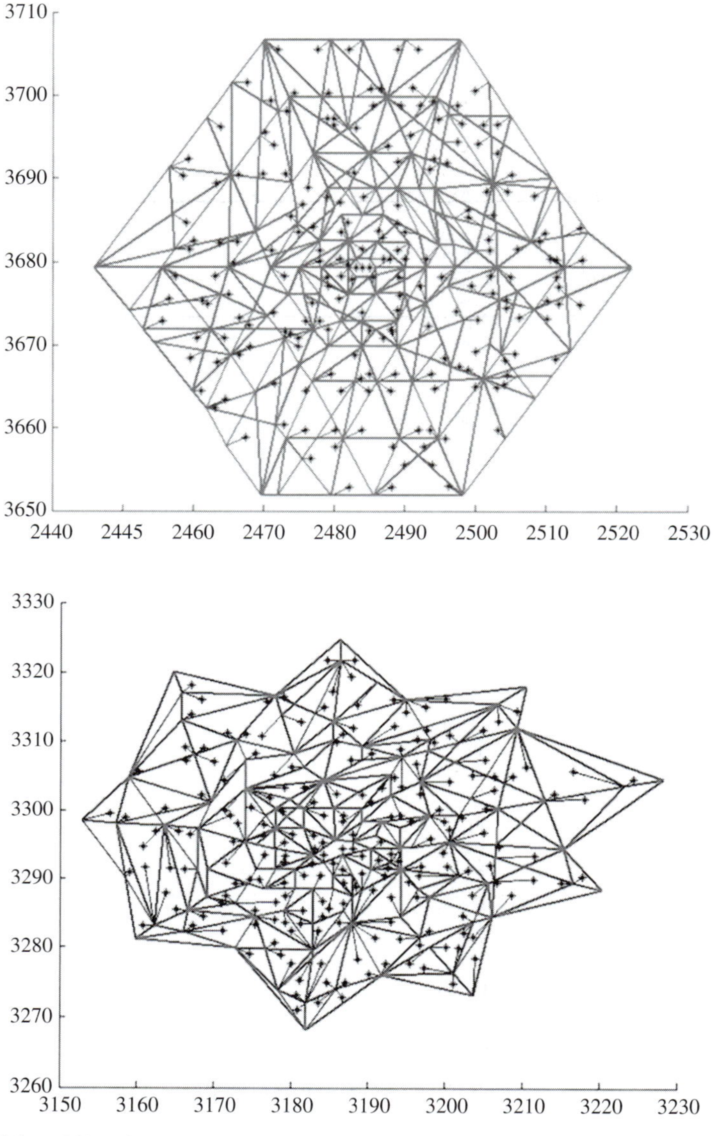


Fig. 8.16. (*Cont.*)

The grey lines represent street patterns, whose intersections are street junctions; the black lines represent the door links between the building and street nodes; and the black starred points represent residential building nodes.

On the other hand, to compute the citizen-school efficiency, building nodes representing schools are also modeled and spatially distributed in a uniform fashion, as can be seen in Figure 8.17 in the case of the 1,250 buildings' case analysis.

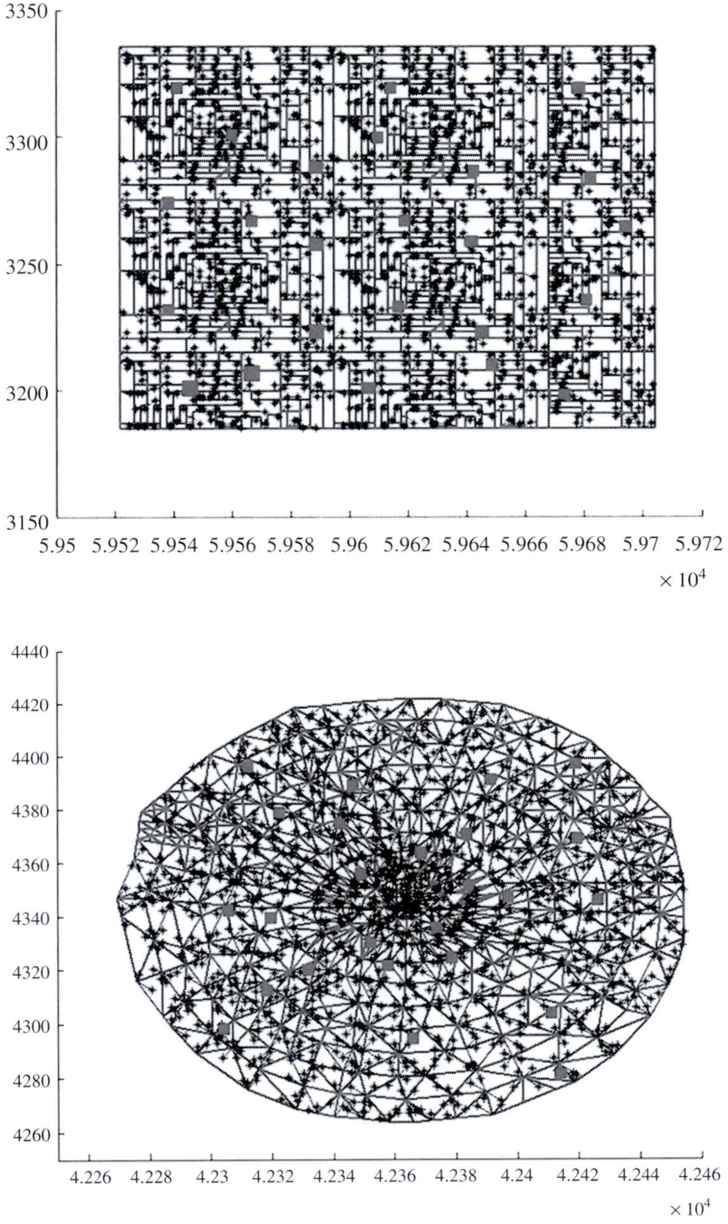


Fig. 8.17. City shapes: rectangular (A), circular (B), hexagonal (C), and star (D) modeled for the case analysis of 1,250 buildings (25 of which are considered to be schools). (A black and white version of this figure will appear in some formats. For the color version, please refer to the plate section.)

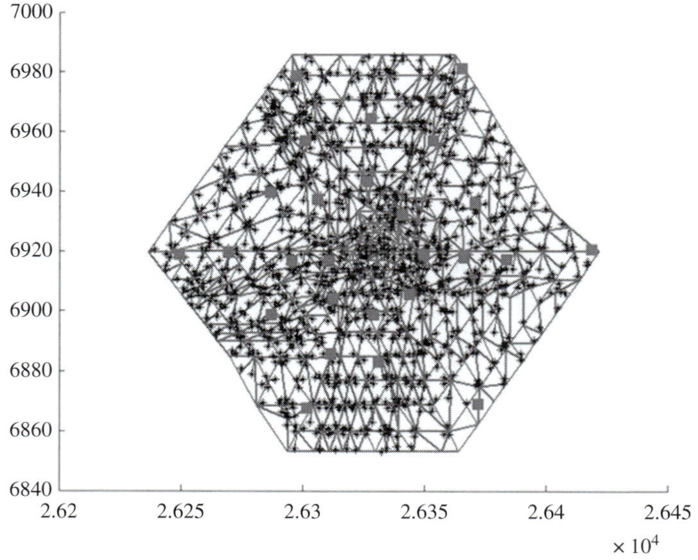


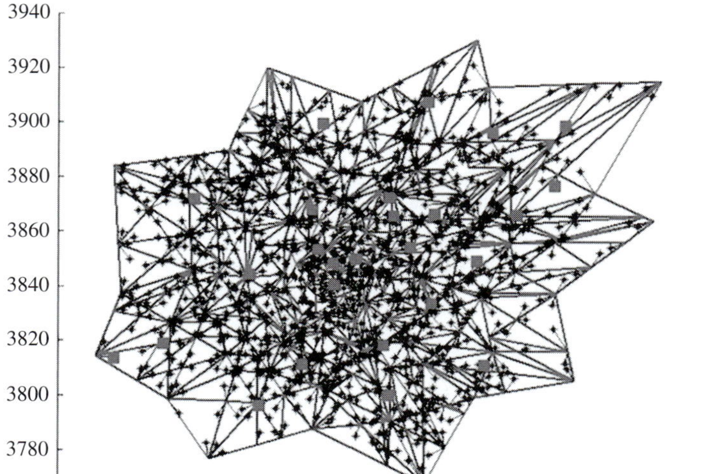


Fig. 8.17. (*Cont.*)

Here, the grey squared nodes represent the city's schools and the black starred nodes the residential buildings, which number 25. In the case in which 200 buildings are modeled, five of them are considered to be schools. Meanwhile, in the 5,000 building case, 100 are considered to be schools. Finally, the case of 50 buildings is not run for the citizen-school efficiency evaluation because, if a 50 building city exists at all, it will have only one or two schools at most, and it would therefore make no sense to simulate a certain percentage of schools collapsing according to the aims of the present study.

Tables 8.2 and 8.3 show the modeled HSPN shapes on which the scenario analyses are run (with their related features) in the case of the citizen-citizen and citizen-school efficiency assessments:

Table 8.2. *Modeled HSPN shapes and related features in the case of the citizen-citizen efficiency assessment*

| Shape | Size [CAD units] | Number of residential buildings | Number of inhabitants |
|---|---|---|---|
| Rectangular | 918 | 50 | 1,945 |
| | 4,380 | 200 | 9,354 |
| | 27,482 | 1,250 | 57,547 |
| | 110,230 | 5,000 | 230,246 |
| Circular | 875 | 50 | 1,945 |
| | 4,536 | 200 | 8,583 |
| | 29,230 | 1,250 | 55,931 |
| | 112,167 | 5,000 | 226,355 |
| Hexagonal | 972 | 50 | 1,945 |
| | 4,180 | 200 | 8,369 |
| | 24,605 | 1,250 | 57,368 |
| | 129,944 | 5,000 | 222,682 |
| Star | 1,122 | 50 | 2,029 |
| | 4,256 | 200 | 8,506 |
| | 30,176 | 1,250 | 53,867 |
| | 107,146 | 5,000 | 225,593 |

Table 8.3. *Modeled HSPN shapes and related features in the case of the citizen-school efficiency assessment*

| Shape | Size [CAD units] | Number of residential buildings | Number of inhabitants | Number of schools |
|---|---|---|---|---|
| Rectangular | 4,380 | 195 | 9,354 | 5 |
| | 27,482 | 1,225 | 57,547 | 25 |
| | 110,230 | 4,900 | 230,246 | 100 |
| Circular | 4,536 | 195 | 8,583 | 5 |
| | 29,230 | 1,225 | 55,931 | 25 |
| | 112,167 | 4,900 | 226,355 | 100 |
| Hexagonal | 4,180 | 195 | 8,369 | 5 |
| | 24,605 | 1,225 | 57,368 | 25 |
| | 129,944 | 4,900 | 222,682 | 100 |
| Star | 4,256 | 195 | 8,506 | 5 |
| | 30,176 | 1,225 | 53,867 | 25 |
| | 107,146 | 4,900 | 225,593 | 100 |

As a further example, Figure 8.18 shows the scaling for the star-shaped HSPN.

When performing the scenario analyses using the randomized procedure highlighted previously, two earthquake intensities are considered; damage to

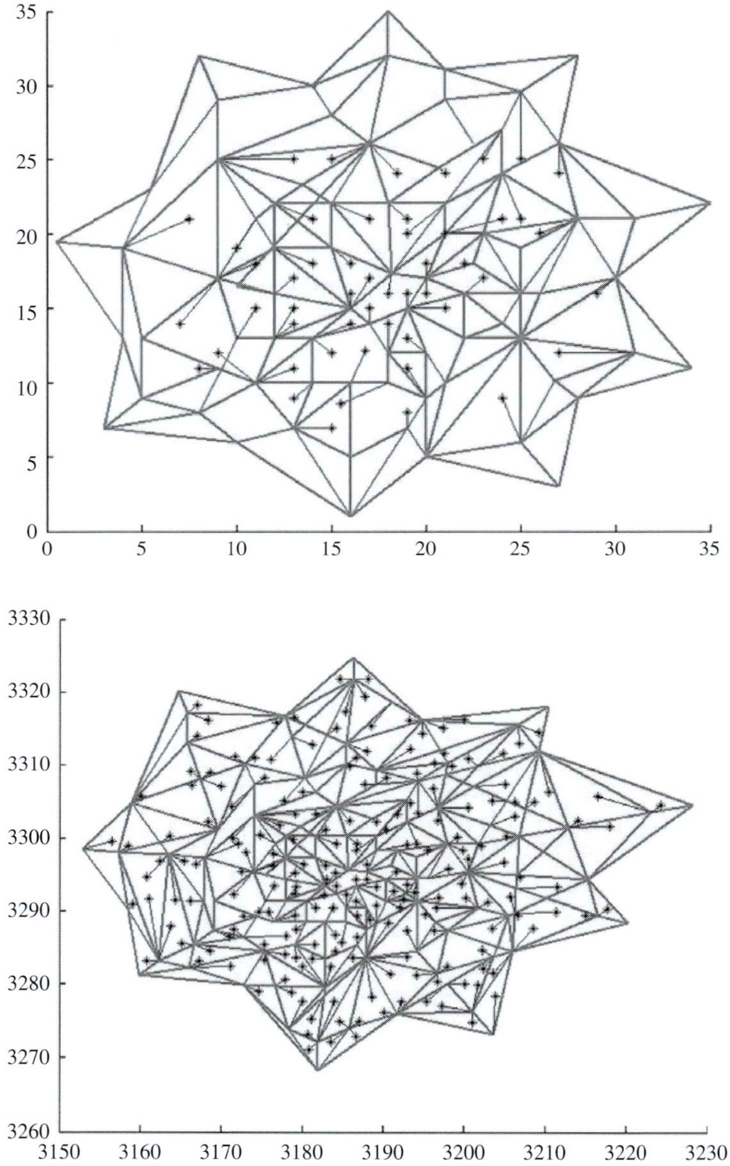


Fig. 8.18. Example of scaling in the case of a star-shaped city with 50 (A), 200 (B), 1,250 (C) and 5,000 (D) residential buildings. (A black and white version of this figure will appear in some formats. For the color version, please refer to the plate section.)

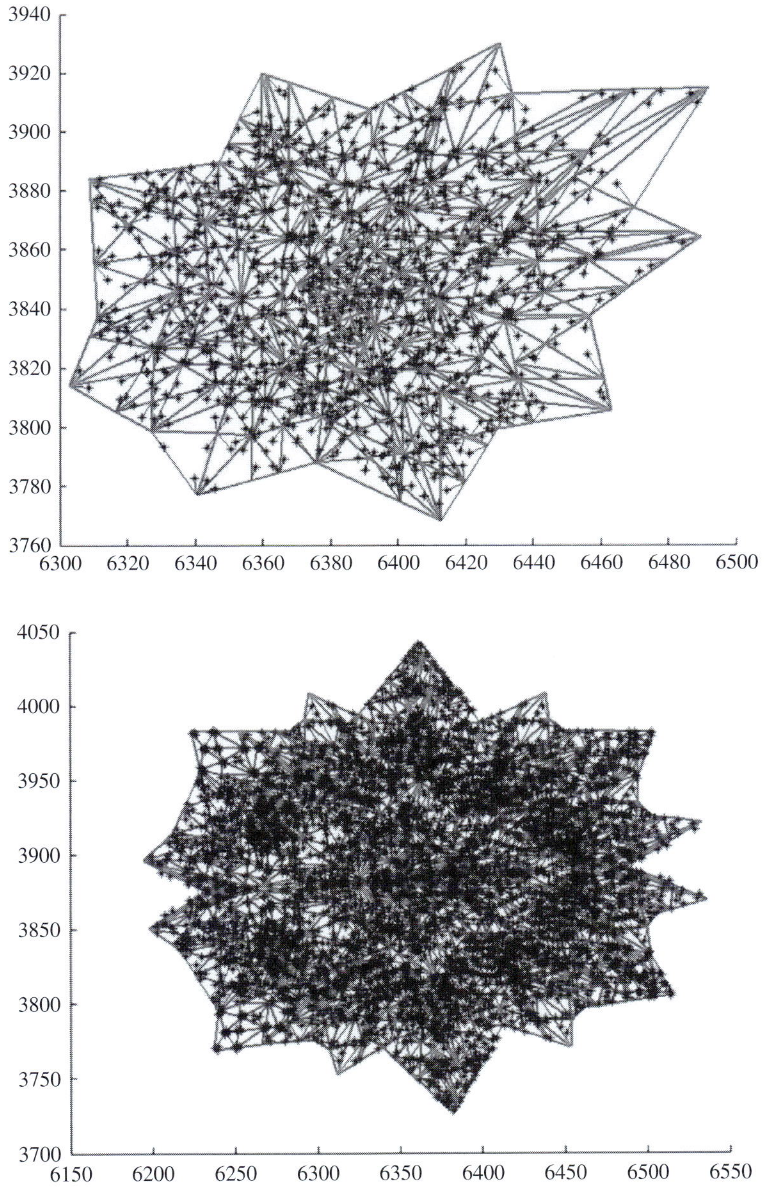


Fig. 8.18. (*Cont.*)

each HSPN is assumed to cause the collapse of 15 percent and 30 percent of, at first, only residential buildings, and then both residential and school buildings. A strategy is designed which considers the HSPN structure that has to be restored to the preevent configuration, and so has the global efficiency to bounce back to the preevent value (status quo down-up).

Table 8.4. *Results of the analyses for each HSPN shape and size for the 15 percent citizen-citizen case*

| Strategy citizen-citizen 15 percent | | | | | | | | | | | | |
|---|---|---|---|---|---|---|---|---|---|---|---|---|
| Shape | Rectangular | | | | | | Circular | | | | | |
| #bldgs | $E_{cc,pre}$ | $E_{cc,post}$ | $Y(0)$ | $D(0)$ | $R^D$ | $R^E$ | $E_{cc,pre}$ | $E_{cc,post}$ | $Y(0)$ | $D(0)$ | $R^D$ | $R^E$ |
| 50 | 0.60 | 0.25 | 0.426 | 0.574 | 0.24 | 0.57 | 0.94 | 0.51 | 0.540 | 0.460 | 0.22 | 0.55 |
| 200 | 0.75 | 0.38 | 0.500 | 0.500 | 0.25 | 0.51 | 0.93 | 0.43 | 0.458 | 0.542 | 0.27 | 0.51 |
| 1,250 | 0.75 | 0.39 | 0.521 | 0.479 | 0.22 | 0.53 | 0.95 | 0.23 | 0.243 | 0.757 | 0.40 | 0.46 |
| 5,000 | 0.78 | 0.40 | 0.510 | 0.490 | 0.23 | 0.53 | 0.94 | 0.18 | 0.193 | 0.807 | 0.37 | 0.55 |
| Shape | Hexagonal | | | | | | Star | | | | | |
| #bldgs | $E_{cc,pre}$ | $E_{cc,post}$ | $Y(0)$ | $D(0)$ | $R^D$ | $R^E$ | $E_{cc,pre}$ | $E_{cc,post}$ | $Y(0)$ | $D(0)$ | $R^D$ | $R^E$ |
| 50 | 0.86 | 0.51 | 0.588 | 0.412 | 0.19 | 0.53 | 0.91 | 0.54 | 0.598 | 0.402 | 0.17 | 0.57 |
| 200 | 0.93 | 0.45 | 0.487 | 0.513 | 0.28 | 0.48 | 0.90 | 0.37 | 0.414 | 0.586 | 0.29 | 0.50 |
| 1,250 | 0.93 | 0.20 | 0.216 | 0.784 | 0.36 | 0.54 | 0.93 | 0.20 | 0.212 | 0.788 | 0.39 | 0.50 |
| 5,000 | 0.92 | 0.20 | 0.213 | 0.787 | 0.37 | 0.53 | 0.92 | 0.14 | 0.152 | 0.848 | 0.36 | 0.56 |

The recovery strategy in this case is also implemented in a discrete number of steps, $n$. Finally, efficiency, systemic damage and resilience values are calculated.

### 8.4.2 Discussion of the Results of the Analyses

In terms of the Quartieri Spagnoli case study, in Tables 8.4 and 8.5 the results of the 15 percent and 30 percent case analyses are presented with respect to: the HSPN citizen-citizen efficiency, $E$; the recovery function and systemic damage values in the aftermath of the event, $Y(0)$ and $D(0)$; and the urban resilience, which is assessed as both damage-dependent, $R^D$, and damage-independent, $R^E$.

The citizen-school efficiency, the systemic damage and the two proposed resilience metrics are also assessed for each HSPN size and shape when modeling the HSPN by also accounting for the school buildings. The analyses of the results are shown in Tables 8.6 and 8.7.

In the main, it can be seen that the higher the number of buildings that collapse, the greater the systemic damage. Subsequently, the resilience of the HSPNs is also greater, both in the damage-dependent and the damage-independent assessments. In parallel with this, a major drop is observed in the efficiency level, $E$, as greater systemic damage, $D(0)$, is evaluated in the after-event assessment.

Table 8.5. *Results of the analyses for each HSPN shape and size for the 30 percent citizen-citizen case*

| Strategy citizen-citizen 30 percent | | | | | | | | | | | | |
|---|---|---|---|---|---|---|---|---|---|---|---|---|
| Shape | Rectangular | | | | | | Circular | | | | | |
| #bldgs | $E_{cc,pre}$ | $E_{cc,post}$ | $Y(0)$ | $D(0)$ | $R^D$ | $R^E$ | $E_{cc,pre}$ | $E_{cc,post}$ | $Y(0)$ | $D(0)$ | $R^D$ | $R^E$ |
| 50 | 0.60 | 0.13 | 0.224 | 0.776 | 0.35 | 0.56 | 0.94 | 0.20 | 0.208 | 0.792 | 0.38 | 0.53 |
| 200 | 0.75 | 0.13 | 0.170 | 0.830 | 0.43 | 0.48 | 0.93 | 0.16 | 0.171 | 0.829 | 0.43 | 0.48 |
| 1,250 | 0.75 | 0.07 | 0.095 | 0.905 | 0.41 | 0.52 | 0.94 | 0.01 | 0.013 | 0.987 | 0.58 | 0.41 |
| 5,000 | 0.78 | 0.08 | 0.100 | 0.900 | 0.41 | 0.55 | 0.94 | 0.00 | 0.003 | 0.997 | 0.51 | 0.49 |
| Shape | Hexagonal | | | | | | Star | | | | | |
| #bldgs | $E_{cc,pre}$ | $E_{cc,post}$ | $Y(0)$ | $D(0)$ | $R^D$ | $R^E$ | $E_{cc,pre}$ | $E_{cc,post}$ | $Y(0)$ | $D(0)$ | $R^D$ | $R^E$ |
| 50 | 0.86 | 0.32 | 0.368 | 0.632 | 0.27 | 0.57 | 0.91 | 0.32 | 0.349 | 0.651 | 0.28 | 0.57 |
| 200 | 0.93 | 0.22 | 0.237 | 0.763 | 0.41 | 0.47 | 0.90 | 0.07 | 0.072 | 0.928 | 0.49 | 0.46 |
| 1,250 | 0.93 | 0.01 | 0.009 | 0.991 | 0.51 | 0.48 | 0.93 | 0.01 | 0.009 | 0.991 | 0.57 | 0.43 |
| 5,000 | 0.92 | 0.00 | 0.005 | 0.995 | 0.52 | 0.49 | 0.92 | 0.00 | 0.002 | 0.998 | 0.52 | 0.48 |

Table 8.6. *Results of the analyses for each HSPN shape and size for the 15 percent citizen-school case*

| Strategy citizen-school 15 percent | | | | | | | | | | | | |
|---|---|---|---|---|---|---|---|---|---|---|---|---|
| Shape | Rectangular | | | | | | Circular | | | | | |
| #bldgs | $E_{cc,pre}$ | $E_{cc,post}$ | $Y(0)$ | $D(0)$ | $R^D$ | $R^E$ | $E_{cc,pre}$ | $E_{cc,post}$ | $Y(0)$ | $D(0)$ | $R^D$ | $R^E$ |
| 200 | 0.16 | 0.05 | 0.343 | 0.657 | 0,41 | 0.30 | 0.185 | 0.08 | 0.435 | 0.565 | 0.36 | 0.34 |
| 1,250 | 0.15 | 0.06 | 0.393 | 0.607 | 0,40 | 0.35 | 0.189 | 0.03 | 0.180 | 0.820 | 0.47 | 0,.43 |
| 5,000 | 0.16 | 0.07 | 0.439 | 0.561 | 0,35 | 0.38 | 0.187 | 0.03 | 0.179 | 0.821 | 0.46 | 0.44 |
| Shape | Hexagonal | | | | | | Star | | | | | |
| #bldgs | $E_{cc,pre}$ | $E_{cc,post}$ | $Y(0)$ | $D(0)$ | $R^D$ | $R^E$ | $E_{cc,pre}$ | $E_{cc,post}$ | $Y(0)$ | $D(0)$ | $R^D$ | $R^E$ |
| 200 | 0.183 | 0.07 | 0.379 | 0.621 | 0.37 | 0.33 | 0.177 | 0.05 | 0.262 | 0.738 | 0.50 | 0.34 |
| 1,250 | 0.186 | 0.03 | 0.182 | 0.819 | 0.46 | 0.43 | 0.187 | 0.03 | 0.150 | 0.850 | 0.48 | 0.45 |
| 5,000 | 0.185 | 0.03 | 0.167 | 0.833 | 0.45 | 0.45 | 0.184 | 0.02 | 0.119 | 0.881 | 0.47 | 0.46 |

It is evident that the results in terms of the resilience of the HSPNs, $R^D$ and $R^E$, are not always in agreement. In fact, the damage-independent resilience provides information about the capability of the studied HSPN to react to a seismic event in terms of responsiveness, quickness, resourcefulness, and robustness. The latter is a very important feature, because when a city's physical system is robust enough to suffer less damage when an adverse event

Table 8.7. *Results of the analyses for each HSPN shape and size for the 30 percent citizen-school case*

| Strategy citizen-school 30 percent | | | | | | | | | | | | |
|---|---|---|---|---|---|---|---|---|---|---|---|---|
| Shape | Rectangular | | | | | | Circular | | | | | |
| #bldgs | $E_{cc,pre}$ | $E_{cc,post}$ | $Y(0)$ | $D(0)$ | $R^D$ | $R^E$ | $E_{cc,pre}$ | $E_{cc,post}$ | $Y(0)$ | $D(0)$ | $R^D$ | $R^E$ |
| 200 | 0.16 | 0.02 | 0.119 | 0.881 | 0.55 | 0.36 | 0.19 | 0.02 | 0.116 | 0.884 | 0.61 | 0.34 |
| 1,250 | 0.15 | 0.01 | 0.064 | 0.936 | 0.65 | 0.30 | 0.19 | 0.00 | 0.008 | 0.992 | 0.68 | 0.32 |
| 5,000 | 0.16 | 0.01 | 0.043 | 0.957 | 0.61 | 0.36 | 0.19 | 0.00 | 0.002 | 0.998 | 0.66 | 0.34 |
| Shape | Hexagonal | | | | | | Star | | | | | |
| #bldgs | $E_{cc,pre}$ | $E_{cc,post}$ | $Y(0)$ | $D(0)$ | $R^D$ | $R^E$ | $E_{cc,pre}$ | $E_{cc,post}$ | $Y(0)$ | $D(0)$ | $R^D$ | $R^E$ |
| 200 | 0.18 | 0.04 | 0.218 | 0.783 | 0.55 | 0.35 | 0.18 | 0.01 | 0.071 | 0.929 | 0.71 | 0.28 |
| 1,250 | 0.19 | 0.00 | 0.006 | 0.994 | 0.66 | 0.33 | 0.19 | 0.00 | 0.005 | 0.995 | 0.69 | 0.31 |
| 5,000 | 0.19 | 0.00 | 0.004 | 0.996 | 0.66 | 0.34 | 0.18 | 0.00 | 0.001 | 0.999 | 0.67 | 0.33 |

occurs, it is consequently also more resilient. Conversely, when it comes to damage-dependent resilience, which is directly related to damage suffered by the HSPN, this highlights the capability of the HSPN to react to the event in terms of responsiveness, quickness, and resourcefulness.

The seismic performance of the modeled HSPNs can also be investigated from a strictly civil engineering perspective in terms of the systemic damage after the event, $D(0)$.

In both case analyses and both seismic scenarios, the rectangular HSPN is the one that sustains less damage in almost all the cases, followed by the hexagonal HSPN. The major robustness of such HSPN shapes can thus be asserted. In fact, in terms of the systemic damage, both the rectangular and hexagonal shapes had the lowest values. In parallel with this, they are also the most resilient geometries in terms of the damage-independent resilience, $R^E$.

On the other hand, when considering the damage extent, the star-shaped HSPN is the most resilient according to the damage-dependent resilience metric, $R^D$. This can be understood as the star HSPN bouncing back to its preevent efficiency within the same number of stages as the other HSPNs, but starting from a more major level of damage.

Accordingly, the results also have to be understood in light of the preevent efficiency level, which each HSPN exhibits. This is because, in order to quantify resilience effectively, consideration has to be given to the damage level from which the recovery starts, the initial efficiency level, and the postevent, i.e., the residual efficiency level.

As an example, if we consider the citizen-citizen case analysis, when it is assumed that 30 percent of buildings collapse, the lowest systemic damage value is attained for the rectangular and hexagonal HSPN shapes, regardless of their size. On the other hand, an important difference is observed in terms of the preevent efficiency level, which is lower in the case of the rectangular-shaped HSPN than the hexagonal-shaped HSPN. As a consequence, this last geometry obviously suffers less damage, $D(0)$. Meanwhile, in the rectangular HSPN case, the lower level of damage is effectively understood as a higher urban network's robustness, which also has its equivalent in a lower preevent efficiency level.

These are, however, circumstances whose consideration is embedded in both the systemic damage and resilience assessments, as they refer to the normalized efficiency with respect to the preevent efficiency. As a result, $D(0)$ makes all the HSPN sizes and shapes comparable, regardless of their higher or lower preevent efficiency compared to their postevent efficiency.

The assessed resilience is also observed with reference to the HSPN size. According to Bettencourt et al. (2007), processes governed by community-based dynamics usually exhibit a sublinear trend against the city size, while processes governed by economies of scale exhibit a superlinear trend. Nonetheless, the same cannot be asserted when observing the trend of the proposed resilience metrics in terms of the city size, as fluctuations are observed in their values in both the case analyses when compared with the HSPN scaling.

## 8.5 Conclusion

Contemporary dynamics of urbanization led cities to be spaces where a huge quantity of physical assets and people concentrate. As a consequence, urban environments result to be ever more exposed and mitigating potential impacts of external stresses is paramount. This is ever more important when dealing with natural events, particularly earthquakes, that typically cause widespread and severe damage. Accordingly, enhancing physical resilience of cities according to a human-centric perspective can help to limit urban business disruptions and ensure higher people safety and life quality.

This chapter has outlined the theoretical basis of the concept of resilience and its strict interrelation with the ecosystem theory and sustainability principles. Resilience is understood as the capability of an urban system to suffer limited damages in the face of an extreme event, and to restore its functionalities in an efficient fashion. According to this understanding, the recovered urban configuration can be the same as the preevent or a new one, provided that urban functioning is guaranteed to be at least equivalent to the preevent condition.

An engineering framework is described for modeling urban environments and quantifying resilience. Essentially, four main phases can be recognized within the methodology:

- Modeling the studied urban system as a HSPN, by overlapping planar graphs, that can represent different typologies of the urban physical nodes or residential buildings, where citizens reside.
- Characterizing structural vulnerability and the probability of street interruption, by integrating fragility curves and probability-based models within the framework.
- Generating seismic scenarios and hypothesizing a recovery strategy to be simulated to quantify resilience.
- Monitoring and assessing urban efficiency, in terms of its connectivity features, in the preevent stage, after the event, and in each recovery stage.
- Quantifying resilience according to the developed indices.

Results are presented and discussed with reference to two different damage scenarios and two alternative recovery strategies:

- Seismic scenarios causing 15 percent or 30 percent of buildings to collapse.
- Recovery strategy focusing on residential buildings, hence to restore the urban connectivity between couples of citizens, or recovery strategy that focuses on school buildings, namely aimed at restoring the connectivity between citizens and the school service.

Seismic scenarios and the recovery strategy have been simulated on the complex network modeling the real case study of the inner city Naples (Quartieri Spagnoli area), and the case study of synthetically designed urban shapes with different sizes.

Analysis results highlighted the direct correlation between the number of buildings located within an urban context and the systemic damage observed after a seismic event. Hence, urban contexts that exhibit the capability to recover in cases where severe damages are observed also show greater resilience.

In addition, findings in terms of the damage-dependent and the damage-independent indices are not always in agreement, because of the different urban features that each of them highlights.

In fact, the damage-independent resilience enables assessment of the capability of a city to react to an event in terms of responsiveness and robustness. On the other hand, the damage-dependent resilience highlights the capability of the HSPN to react to the event in terms of quickness and resourcefulness, being directly related to the damage suffered.

# 9

# Social Resilience

GÉRARD HUTTER AND DANIEL F. LORENZ

## 9.1 Introduction

Currently, we experience the spread of the word "resilience" in most diverse research areas and policy fields of modern society (a world society as sociologists tell us). To name only a few examples: psychologists ask under which conditions and through which processes those children that face adverse risk situations (e.g., poverty, child abuse) may still build and display "normal" development competencies in specific social and/or cultural settings (e.g., Masten, 2014; Zander, 2010). Social psychologists and management scholars like Karl Weick ask through which social processes teams may be able to deal with surprises and sudden dangers in the context of firefighting (e.g., see the seminal paper on the Mann Gulch Disaster by Weick, 1993; see also Weick and Sutcliffe, 2015). Organizational scholars are concerned about organizational performance, often of business organizations, in the face of ever-increasing levels of complexity, uncertainty, and ambiguity in the context of competition about scarce resources and claims for legitimacy (e.g., Kayes, 2015). Resilience is especially prominent in research on so-called "socio-ecological systems" (e.g., Boyd and Folke, 2012) and on the prospects of understanding, analyzing, and promoting disaster resilience in the context of natural hazards, climate change, and the possibility of cascading disastrous events (e.g., Coaffee and Lee, 2016; Tierney, 2014). Given this high diversity of research streams and policy contexts and some communication across streams and contexts, no wonder that there is also a high diversity of definitions, theories, and methods to use "resilience" in natural hazards research and management practice.

Fortunately, there are useful book-length publications that review resilience definitions in diverse scientific disciplines and related theoretical approaches as well as their methodological and practical implications (e.g., Comfort et al., 2010; Coaffee and Lee, 2016; Tierney, 2014). We do not seek, therefore, to provide another overview over social resilience research that discusses the advantages and disadvantages of diverse perspectives. We are more interested in promoting a specific sociological perspective on resilience. To accomplish this, we adopt the following preliminary wide definition of social resilience (based on Boin et al., 2010: 9):

> Social resilience is the capacity of a social entity (social order of actions) to proactively adapt to and recover from disturbances that are perceived within the social entity to fall outside the range of "normal" and "expected" disturbances.

This definition of resilience provides an anchor in the expanding "landscape" of resilience research in the natural sciences, civil engineering, and the social sciences. Therefore, some remarks on the definition are in order: First, resilience is understood as a capacity of social entities. Questions of how and why actors address "resilience" in ideological processes are certainly important (e.g., "resilience" as manifestation of the politics of neoliberalism, Coaffee and Lee, 2016). However, the following places such questions at the end of our argumentation, not at the beginning. Second, the definition highlights social entities understood as social order of actions. Without doubt, ecological systems and individual persons influence social action. However, from a conceptual viewpoint, they are placed in the context of social resilience. Third, the definition may be characterized as a wide definition because efforts to proactively and reactively manage disturbances are addressed (see also Weick and Sutcliffe, 2015, for an alternative of understanding resilience mainly as the capacity to react to the unexpected). Fourth, the definition explicitly refers to perceptions and interpretations. In line with principles of qualitative social research (e.g., Strübing, 2013), social resilience is seen as a capacity to manage the unusual and unexpected from the viewpoint of the focal social order of actions. We are interested in understanding social action that shapes perceptions, expectations, interpretations and learning processes based on context conditions that are characterized by power relations, networks, and institutions.

To illustrate our arguments, we often use examples of resilience (or the absence of it) in the context of dealing with flood hazards. Further types of hazards and disasters are also addressed (e.g., large-scale disasters that are characterized by cascades of disastrous events). Through promoting a sociological perspective on resilience, we encourage the reader to consider the following three learning aims:

- Learning to distinguish between multiple social levels of organizing for resilience, especially from a sociological perspective on "human agency";

- Learning to understand social resilience as a set of capacities of actors in modern society that is characterized by social processes of perception, interpretation, and meaning-making; and
- Learning to consider the challenges of analyzing and assessing social resilience with regard to multiple social levels and a complex set of capacities.

The following considers a wide range of conceptual and empirical publications on resilience in general, social resilience in the context of natural hazards in particular. We developed our own publications through conducting a series of research projects, often investigating collective efforts of dealing with floods in urban and rural regions. Our learning aims help to structure this contribution to the textbook on vulnerability and resilience into three main chapters and one concluding chapter: Section 9.2 specifies the meaning of the word "social entity" based on the notion of human agency (Emirbayer and Mische, 1998) and with regard to different types of social entities (small social groups, organizations, and network of organizations). Then, Section 9.3 explains that social resilience refers to a complex and dynamic set of capacities that may be attributed to "actors of modern society" (Meyer and Jepperson, 2000). Section 9.4 discusses some challenges of empirical analysis and of attempts to assess social resilience in the context of natural hazards. Section 9.5 concludes our contribution to social resilience research.

## 9.2 Organizing for Resilience on Multiple Social Levels: From the Perspective of Human Agency

Organizing for resilience in modern societies potentially involves and affects actors from various spheres and organizational fields. Researchers and practitioners alike often mention public leaders (elected politicians and public officials), market-oriented business organizations, organizations of the "third sector", and, last but not least, citizens and self-organized community forms based partially on solidarity between citizens (e.g., Hutter, 2016; Jessop, 2011; Tierney, 2014). This inclusive perspective on ways to organize for resilience is not surprising, because the call for resilience materializes especially with regard to nonroutine events, crisis, disasters, as well as catastrophes. In such situations, the limitations and shortcomings of state-centered approaches like anticipation-based "command and control" as well as plan-based disaster management are often salient, which paves the way for discussions about policy reform and social change (Boin and McDonnell, 2007; Coaffee and Lee, 2016; Weick and Sutcliffe, 2015; Wildavsky, 1988).

We do agree that the notion of social resilience encourages researchers and practitioners to follow an inclusive approach to natural hazard and disaster management (Goldstein 2012). In line with a pragmatic understanding of "human agency" (Emirbayer and Mische, 1998), Section 9.2 is somehow optimistic[1] that actors in modern societies are – to some extent – able to proactively adapt to unexpected disturbances in the future even in the face of diverse barriers to such resilience organizing[2]. Furthermore, we are interested in understanding people, tools, and tasks as well as organizational goals, technical systems, and relations between organizations, and how all this fits together to organize for resilience in the context of natural hazards in modern society[3]. Hence, organizing for resilience as manifestation of human agency requires both an inclusive approach and a multilevel approach to distinguish between social levels of organizing that encompass, for instance, collective resilience efforts of teams, organizations[4], and networks of organizations. Before we elaborate on this, the following briefly introduces our understanding of "human agency".

Nowadays, most social science researchers contend that both "human agency" (e.g., the "free will" or "being able to act otherwise") as well as social structures and systems are important to understand relations between actors, social action, and change. However, there are ongoing discussions about how to exactly conceptualize and study the relations of agency and structures in general, with regard to actions in the context of natural hazards in particular. Of course, the following does not claim to resolve these issues. But it assumes that it is plausible to ask whose resilience is highlighted and what the conditions and consequences of such a focus are (Coaffee and Lee, 2016; Hutter et al., 2014). This is so because the preliminary wide definition of social resilience given above focused on capacities of social action, but left open who the actors of action are. As with regard to "resilience", research efforts of social scientists have led to an abundance of conceptions of actors, social action, and human agency (e.g., see Joas and Knöbl, 2013, for an overview over social theory). In this contribution, we highlight human agency (instead of social structures) and actors (instead of abstract social systems). To be more specific about actors, agency, social action, and relations to structures and systems, the following adopts a time-oriented perspective of "human agency" to integrate agency efforts of iteration (oriented towards past accomplishments), projectivity (oriented towards an imagined future or envisioned futures), and practical evaluation (oriented towards an emerging present in the face of structurally shaped contingencies) (Emirbayer and Mische, 1998; for agency and resilience see Lorenz and Dittmer, 2016).

We propose that social resilience as a manifestation of human agency shows distinctive characteristics at the

- Intra-organizational level,
- Organizational level, and
- Inter-organizational level.

We consider all these three social levels to be important to promote community and urban resilience in the context of natural hazards (e.g., Coaffee and Lee, 2016; Tierney, 2014).

Proactively adapting to future disturbances that fall outside the perceived range of "normal" events requires that organizational members are able to imagine "strange futures" (projectivity) while simultaneously dealing with the demands of an evolving present (practical evaluation). Organizational scholars (e.g., March, 1991) have long pointed out that organizations have difficulties in maintaining both learning processes termed exploration to deal with possible, distant, and perhaps strange futures and exploitation to reap the benefits from knowledge accumulation based on experience (iteration). Organizations tend to prefer exploitation while neglecting to sustain exploration. However, some researchers also show that face-to-face communication between organizational members may facilitate intra-organizational processes of developing formal or informal social groups that are eager to prepare for rare and even "strange" events in the future[5]. Furthermore, resilience at this social level is also characterized by the collective capacity to discuss the specific implications of imagination for how an organization runs right now.

From the viewpoint of the whole organization, results of efforts to imagine possible, but strange disturbances are rather "useless," if they do not connect to organizational strategy, structure, incentives, and routines. In modern society, organizations responsible for natural hazard management are often to some extent bureaucratic organizations that display a significant degree of formalization (e.g., formalized responsibilities, authority, standardized processes of communication, and routines like "Standard Operating Procedures (SOPs)"). Social resilience at this level may require that people "at the top" of the hierarchy are willing and able to provide resources and intra-organizational arrangements to facilitate and exploit processes of team-based learning for dealing with future imagined disturbances and for handling current emergencies through mindful process management. The careers of people motivated to focus on their own failures and the failures of others may be an indicator of social resilience at the organizational level: social resilience may be high in organizations that promote such people, resilience may be low in organizations that negatively sanction such people (Weick and Sutcliffe, 2015).

Increasingly, there is research that focuses on inter-organizational relations to improve natural hazard management and to enhance resilience (e.g., Boyd and Folke, 2012; Comfort et al., 2010; Goldstein, 2012; Thaler et al., 2016). With regard to social resilience, we highlight the following points: A wide preliminary definition requires that actors interested in proactively adapting to futures disturbances are eager and able to combine diverse domains and types of knowledge in processes of inter-organizational collaboration that are often resource-demanding and far from risk-free in terms of realizing the intended collaborative advantage (e.g., Huxham and Vangen, 2005; Zimmermann, 2010). For instance, the flood disaster in Dresden (Germany) in August 2002 triggered collaboration of organizations from research and practice at the spatial level of the whole catchment of the river Weisseritz (a tributary of the river Elbe). This inter-organizational collaboration included state authorities, municipalities, and organizations from civil society. Collaborators were able to jointly produce some innovative products for dealing with future flood events (Wirth et al., 2010), but showed only rather limited capacity to facilitate collaborative pro-active adaptation to future disturbances among the participating organizations (Vulturius, 2013). Collaboration among organizations from diverse societal realms may be especially difficult when it comes to collective efforts that address future disturbances and even strange events that are difficult to interpret based on the diversity of the existing strategies, structures, and routines of the participating organizations. Table 9.1 summarizes this chapter and adds some details and references to the argumentation.

## 9.3 Social Resilience as a Set of Capacities: From a Sociological Perspective

When it comes to disaggregate the overall concept of resilience into components in order to understand how "a social entity [...] proactively adapt[s] to and recover[s] from disturbances that are perceived within the social entity to fall outside the range of 'normal' and 'expected' disturbances" (Boin et al., 2010: 9), discussions usually focus on separate, though interlinked capacities (e.g., Tierney, 2014 based on an extensive review of the literatures relevant for disaster resilience). As the discourse on resilience is nowadays situated in various scientific disciplines and research streams, not only a high diversity of definitions of social resilience is to be found, but also of the capacities that constitute resilience.

Table 9.1. *Social resilience at the intra-organizational, organizational, and inter-organizational level*

| | Characteristics of social entity | Social resilience as capacity (examples) |
|---|---|---|
| Intra-organizational level | Social entities are often characterized by regular direct interaction and face-to-face communication (e.g., project team)<br>Social relations of individual persons in a common organizational context (e.g., communication in the shadow of the organizational hierarchy)<br>To some extent formalized social relations (e.g., Top Management Team (TMT), project team) | Capacity to proactively imagine possible disturbances (e.g., scenario planning to consider low-probability/high-impact events)<br>Capacity to discuss the implications of imagination for the organization (e.g., through role playing in teams)<br>Capacity of recovery from disturbances that threaten team/group performance |
| Organizational level | Goal-directed and partly formalized social order of actions of individual persons; an organization encompasses multiple teams and groups<br>Organizations are often characterized by a formalized hierarchical authority structure, technical systems, incentives, and routines<br>Organizational membership may be due to working contracts and mandates | Capacity to proactively provide sufficient resources and organizational arrangements for team-based efforts of resilience building<br>Capacity to promote people that consider the liabilities of success (e.g., people that are preoccupied with failure, Weick and Sutcliffe, 2015)<br>Capacity to learn from the experience of disturbances at various scales (e.g., small to large) |
| Inter-organizational level | Set of nodes (= organizations) and social relations based on human communication and collaboration between organizations<br>Goal-directed networks (Provan and Kenis, 2008) are networks with high agency<br>Different forms of network governance (e.g., lead-organization network, shared governance) | Capacity to proactively combine resources and knowledge in the face of high heterogeneity between the participating organizations<br>Capacity to communicate the results of heterogeneous collaboration<br>Capacity to recover from disturbances that threaten the accomplishment of network goals |

Adapted from Emirbayer and Mische (1998) on "human agency" and Weick and Sutcliffe (2015) on agency in the context of resilience; see Baum (2002), Crossan et al. (1999), Provan and Kenis (2008), and Sutcliffe and Vogus (2003) to distinguish between social levels in organizational studies.

Section 9.3 will at first introduce and discuss the triad of coping capacity, adaptive capacity, and transformative capacity, before it subsequently will highlight certain conceptual arguments in order to provide an understanding of social resilience from a sociological perspective. The chapter underlines the contrast between an understanding of resilience that distinguishes between coping, adaptive, and transformative capacity on the one hand and a more sociological informed view on the other that argues to highlight coping and participative capacity as core components of social resilience because they shape how adaptive and transformative capacity may develop. Then, Section 9.3 discusses how this sociological understanding of resilience is related to the social levels identified in Section 9.2.

### *9.3.1 The Triad of Coping Capacity, Adaptive Capacity, and Transformative Capacity*

Some authors (e.g., Béné et al., 2016; Davies et al., 2013; Keck and Sakdadolrak, 2013) argue that social resilience encompasses coping capacity, adaptive capacity and transformative capacity. This influential threefold division is useful to understand various degrees of re- and pro-active moments of resilience. The distinction of adaptation and transformation originally goes back to ecological or social-ecological research on resilience. Folke (2006), for instance, uses the terms adaptability and transformability to describe different aspects of resilience: the former in the case of short-term reactive measures, and the later as the establishment of entirely new structures and pathways.

As we experience a still expanding "landscape" of resilience the boundaries of the concept often appear blurred. While some authors, for instance, Pelling (2011: 78), distinguish a "conservative" notion of resilience being the "contained" persistence of functions and practices from fundamentally different transformations meaning the "deepest form of adaptation indicated by reform in over-arching political-economy regimes and associated cultural discourses", successive authors incorporated the idea of transformation into their "new" notion of resilience. Thus, the triad of coping capacity, adaptive capacity and transformative capacity reflects the stepwise evolvement of the resilience concept starting with the "conservative" persistence of a system, over adaptations of systems ending so far with "evolutionary" transformations of systems in the face of global challenges.[6]

All three, coping capacity, adaptive capacity, and transformative capacity, can be differentiated in terms of the response to threat, the temporal scope and the degree of change (Keck and Sakdadolrak, 2013). While coping capacity, in this interpretation, is manifesting in short-term "reactive" (Obrist et al., 2010: 289) measures after shock events (ex-post), adaptive capacity and transformative capacity display a proactive (ex-ante) and long-term temporal scope.

Coping capacity is related to persistence and survival in times of disaster as well as quick recovery thereafter, while adaptive capacity encompasses informed choices about various incremental changes and adjustments to ensure functioning in the face of future threats and shocks (Béné et al., 2015). Transformative capacity is needed (Béné et al., 2016; Nelson et al., 2007), when incremental adaptive changes prove to be insufficient or "untenable" (Walker et al., 2004: 5) and a permanent and fundamental altering of structures and functions is required in order to cope successfully in the future.

Thus, both, adaptive and transformative capacity, are orientated to prevent disasters to happen or at least mitigate their damage, before they strike by enabling pro-active change in case social entities face natural hazards or other types of hazards. Besides being future-orientated, both show different degrees of change and distinguished outcomes. Adaptive capacity describes the "ability to learn from past experiences and adjust themselves to future challenges in their everyday lives" (Keck and Sakdadolrak, 2013: 5), while transformative capacity is characterized by "the ability to craft sets of institutions that foster individual welfare and sustainable societal robustness towards future crises" (Keck and Sakdadolrak, 2013: 5).

Even though Béné et al. (2015) distinguish all three capacities by the intensity of the shock event and the intensity of the responses, they still argue that all three capacities might occur simultaneously without being dependent on the natural hazard and its impact. Coping capacity, for them, includes various existing strategies of risk and disaster management, such as preventive measures, that are used by social entities to cope with shocks and impacts. Adaptive capacity demonstrates the "capacity to learn, combine experience and knowledge, adjust responses [...] to changing external drivers and internal processes, and continue operating" (Berkes et al., 2003). By contrast, transformative capacity describes a qualitative difference in terms of learning and change by referring to "the capacity to create an enabling environment through investment in good governance, infrastructure, formal and informal social protection mechanisms, basic service delivery, and policies/regulations that constitute the necessary conditions for systemic change" (Béné et al., 2015: 10). As such transformative resilience addresses not only present risks and threats, but also the underlying risk and vulnerability drivers.

In order to fully grasp social resilience and its distinct features, the following sections introduce an alternative set of capacities with a broader sociological informed understanding of the social matrix enabling resilience and human agency. The rationale behind this alternative set is not a critique of the aforementioned three capacities, but rather their limited scope, when it comes to the social sphere and its specific characteristics. This alternative set is rather a broadening from a sociological point of view due to the fact that most important social features of social resilience and human agency have been neglected (Voss, 2008); without their consideration, the specific social aspects of adaptation and recovery as mentioned by Boin et al. (2010) in our preliminary wide definition of social resilience in the introduction cannot be grasped and understood.

By distinguishing coping capacity and participative capacity as well as adaptive and transformative capacity (e.g., Lorenz, 2013; Voss, 2008) not only the aforementioned aspects of dealings with risks and natural hazards can be understood, but also the specific social aspects of resilience, being in particular

- Social processes of perception, interpretation and meaning (see learning aims) as well as
- Power relations and ideological processes that influence social resilience and human agency substantially.

In the following Section 9.3.2, first, we will introduce the concept of coping capacity and highlight the specific aspects of coping in the social sciences. Section 9.3.3 explains participative capacity and the dimension of power as well as the implications for social resilience. After the introduction of both coping in terms of interpretation and meaning as well as power relations the ground is prepared to reconsider the understanding of adaptive capacity and transformative capacity.

### *9.3.2 Coping Capacity Reconsidered*

Coping in the discourse on resilience is often understood in terms of physical or organizational reactive responses. UNISDR (2009: 8), for instance, defines coping capacity as the "ability of people, organizations and systems, using available skills and resources, to face and manage adverse conditions, emergencies or disasters."

In the given definition, coping relates to the persistence or restoration of a (former) status quo. Here, the prominent and influential metaphor of "bouncing back" for resilience can be found (e.g., Manyena et al., 2011). To achieve such bouncing back, for instance, existing resources and preplanned responses such as risk and disaster management plans and measures can be utilized to moderate shocks and to restore core societal functions that have been affected by natural hazards or cascades of disastrous events (Béné et al., 2015).

From a sociological perspective, such an understanding of coping needs to be revised and widened. Sociological as well as anthropological studies teach us that social actors do not just restore a former status quo after extreme and unexpected events, but show a much more complex and interpretative dealing with such events (Erikson, 1994). A symbolic dimension of meaning (Lorenz, 2013) and social processes of remembering, perception, interpretation, imagination and expectation building come into play and need to be considered to understand resilience in terms of human agency (Emirbayer and Mische, 1998).

The first studies on resilience relevant for the social sciences were placed in the discipline of psychology. Here, for instance, Werner (1971, 1993) observed that a significant number of children could actually cope with adverse social conditions without being seriously affected. Today, psychology discusses the psychological consequences after traumatic and extreme events such as natural or technological disasters and the psychological processing of those events, respectively. The concept of a "sense of coherence" (Kimhi, 2014), for instance, describes human agency after traumatic events in terms of comprehensibility, manageability and meaningfulness.

But coping may not only be found in case of individual persons, but may be relevant for all social entities ranging from intra-organizational teams to local communities and to whole societies (Tierney, 2014). Here, coping capacity is understood as the cultural and social "dealing" with collective stress (Norris et al., 2008; Voss, 2008). As a result, coping especially comes to bear in the midst of or after a disaster. While endowment with meaning, i.e., attribute meaning to otherwise meaningless disasters, proves to be a significant precondition for the overcoming of disastrous occurrences (Norris and Stevens, 2007), the "collapse of collective sensemaking" can be viewed as being an integral component of disaster.

Weick (1993: 638), for example, shows how the "steady erosion of sense and structure" in a small intra-organizational team of smokejumpers resulted in the Mann Gulch disaster. For Weick, improvization, virtual role systems, an attitude of wisdom and norms of respectful interaction are sources of resilience and forestall disastrous disintegration of social entities.

Looking at the community level, Dittmer et al. (2016), for instance, show how a local rural community coped with the disaster of the Elbe River flood in the year 2013, both in psychological and physical terms. By establishing their own patterns of interpretation and communal narratives about the disaster, the community displayed significant efforts of human agency in terms of organizing themselves for weeks autonomously in the inundated area – while still experiencing the situation as meaningful and a situation of community-building.

It should not be surprising that communities are the paramount focus of resilience research as they have a certain degree of social cohesion before the disaster strikes (Aldrich, 2012; Eyre, 2006). As shown by Erikson (1976) in his seminal study concerning the loss of community after the Buffalo Creek dam failure and subsequent flooding, if collective coping fails, then there are reciprocal effects on social cohesion. Coping in this socio-scientific sense is therefore indispensable for coping on the physical level.

On the societal or even cultural level, certain arrangements with "disasters" can exist so that events that would be a disaster in other socio-cultural contexts become "ordering elements" of the social sphere (e.g., Bankoff, 2007). Voss (2008) argues that coping is highly dependent on what is socially and culturally perceived as a natural hazard or disaster and what is not. The ways how people actually "cope in an emergency or a disaster are shaped by [...] a cultural narrative that creates a set of expectations and sensitizes people to some problems more than others" (Furedi, 2007: 485).

The existence of such different expectations and ways to cope is illustrated by Macamo (2003) who, through the example of the flooding in Mozambique in 2000, demonstrates that certain coping mechanisms might cause that events (that are treated as disasters by global media) are not necessarily perceived as disasters by the affected people – but that they experience other disasters that are not acknowledged by the global public at all. Beyond such cultural cosmologies, every culture harbors within itself forms of dealing with the loss and the collapse of collective sensemaking which can be entirely different from those of other cultures (Heesen et al., 2014).

In addition to this, anthropological research among others has come to learn of numerous social meaning creation practices such as mourning rituals, split narratives and social cohesion, which serve to aid and help individual and collective coping.

### *9.3.3 Participative Capacity*

Whether persons, teams, organizations, networks, communities, and societies can actually cope with threats and disasters depends not only on the coping capacities of actors, but quite often to a greater degree also on the societal factors that facilitate or constrain these actors to do so. Actual social action is the function of complex social context conditions and processes, for instance, power relations, networks of different types and size, as well as institutional conditions (including cognitive-cultural aspects of institutions, Scott, 2014). Especially the underlying dimension of power has been neglected in the discussion of social resilience for a long time. But more recently, a significant number of scholars (e.g., Bohle et al., 2009; Lorenz, 2013; Voss, 2008; Voss and Funk, 2015) have addressed this aspect of social resilience.

The concept of human agency, as introduced by Emirbayer and Mische (1998), highlights the embeddedness of actors in a social "fabric". Agency is therefore

> on the one hand, a process of active shaping of one's given circumstances. On the other hand, it is at the same time embedded in social structures which are in turn historically, culturally, and societally framed in a specific manner. These structures provide 'radii of agency' to the actors thus determining who is in the position to act and who is not, or how specific ways of acting/agency are each symbolically charged, interpreted, and societally acknowledged
>
> *(Lorenz and Dittmer, 2016: 36).*

The concept of participative capacity enters the discussion of social resilience to acknowledge these different "radii of agency" and to direct attention to the interpretive power and influential prospects of actors regarding those local, regional, and global processes that affect them (see also Healey, 2009 on interpretative power in the context of urban development and spatial planning).

> [P]articipative capacity becomes a key category in the circle of disasters: the lower the participative capacity, the lower the resonance for critical developments, the lower the prevention activities, the lower the capacity to respond and to adapt and so on
>
> *(Voss, 2008: 52).*

The contraction of participative capacity limits the various potentials to affect the conditions of life as well as the possibilities to deal with natural hazards, disasters, and future threats.

Arguing for an inclusive perspective on ways to organize for resilience the concept of participative capacity is most relevant across and on all social levels. It is to be found in intra-organizational processes, for instance, team efforts to balance exploration and exploitation as introduced earlier (Section 9.2). In the context of "High-Reliability Organizations (HRO)" (Weick et al., 1999), rigid structures of power and formal hierarchies are seen as amplifying errors while flat and more participative hierarchies may contribute to stronger collective mindfulness and this is then translated into more social resilience in the context of technological as well as natural hazards (Weick and Sutcliffe, 2015).

Participative capacity is also highly relevant for local communities and on the societal level. Even in democratic systems that lawfully ensure the rights of freedom and participation, there is nonetheless an uneven distribution of interpretive power and hazards between groups due to social factors, as well as occasional serious constrictions on the ability of groups and communities to self-organize. The uneven distribution of resources (Adger, 2000b), the differing strength and breadth of available networks (Blaikie et al., 1994), exclusion/inclusion (Cutter et al., 2003), mobility (Adger, 2000b), gender (Enarson and Chakrabarti, 2009), language (Voss and Funk, 2015), to name only a few factors, interfere and culminate in the unequal distribution of power and the ability to affect change to the conditions of life.

### *9.3.4 Adaptive and Transformative Capacity Revisited*

Adaptive capacity is – in the context of social-ecological systems – discussed as the ability "to learn, combine experience and knowledge, adjust [...] responses to changing external drivers and internal processes" (Folke et al., 2010: 20) by taking advantage of the available body of knowledge and collected experience (Westley et al., 2002; Young et al., 2006). Therefore, in order to increase resilience, recognizing and examining uncertainty as well as the exigent nature of certain adaptation processes are seen as a first step (Folke, 2006; Gunderson, 2003). As to what the form and design of these adaptation processes should be, the authors point out the necessity to draw upon various differing sources of knowledge in dealing with natural hazards and disastrous events (Berkes, 2007; Voss, 2009). Furthermore, many authors (Adger, 2000a; Hagan and Maguire, 2007; Norris and Stevens, 2007; Voss, 2009) argue that the level of dependence on specific resources limits adaptive capacity. According to Folke et al. (2003), resource diversity and variability are other core elements of resilience.

To consider adaptive and transformative capacity from a sociological view, the revisions of coping and participative capacity need to be taken into account as they affect the conceptualizations of both capacities substantially. Adaptive capacity is not just – as, for instance, in the ecological discourse on resilience (e.g., Folke, 2006) – the ability to adjust in the face of natural hazards or unknown future threats. If we take the contribution of the social sciences seriously, adaptation as well as transformation need to be seen as deeply entangled with the respective social processes of perception, interpretation and meaning (coping capacity) and existing structures of power (participative capacity). These capacities always interfere with adaptive and transformative measures. Adaptive capacity, therefore, describes the ability to adjust and adapt in the face of natural hazards or other known as well as unknown threats under the conditions of coping capacity and participative capacity.

Ecological systems adapt or transform through processes of trial-and-error. Even though adaptations and transformations of social orders may also be sometimes due to processes of trial-and-error (Bohle, 2008; Voss, 2009), such changes are based on processes of perception, interpretation, expectation and meaning. Adaptation and transformation in the case of social action and order are most commonly not only processes of trial-and-error; social actors rather implement adaptations and transformations in a goal-oriented and reflexive manner – or at least try to do so and fail from time to time. The perspective of human agency reminds us to consider all three constitutive elements of iteration, projectivity, and practical evaluation. As such, social resilience combines the selective perpetuation of certain patterns of action from the past (iteration) with the social imagination of different futures, that may reconfigure past patterns with respect to certain future expectations or imaginations of the actors (projectivity) and the "practical and normative judgements among alternative possible trajectories of action, in response to the emerging demands, dilemmas, and ambiguities of presently evolving situations" (practical evaluation) (Emirbayer and Mische, 1998: 971).

Seen from the perspective of human agency and its notion of the embeddedness of social actors in a social fabric, the presented aspects of adaptive capacity must be seen in the light of coping capacity and participative capacity as both either enable or impede adaptive capacity. Which natural hazards, threats and disaster prevention measures are perceived at all, which gain priority, and which adaptive measures reach the necessary social acceptance and are undertaken in the end is highly dependent on interpretative processes, social expectations and the influence of social actors (e.g., organizations, local communities, pressure groups, nations).

The dependence on and the available resources for adaptive capacity are in most cases not issues of choice, but rather distributed by all those factors that make up participative capacity. In the discussion on adaptive capacity the significance of social capital is discussed at length (Adger, 2000b; Hagan and Maguire, 2007; Tierney, 2014). What is often missing and what is brought to the fore by participative capacity is that the distribution of social capital is never natural, but always the result of social actors struggling for more social capital and that there is always a so-called "dark side of social capital" (Aldrich, 2012; van Deth and Zmerli, 2010) in terms of social exclusion, marginalization, etc.

Given the often much more normative actions and goals of transformation, for instance in terms of "individual welfare" (Keck and Sakdadolrak, 2013: 5) or "good governance" (Béné et al., 2015: 10), the entanglement of transformative capacity with socially build expectations, fundamental norms and values, normative judgments, ethical conceptions of the "good life", but also power structures becomes even more apparent.

### *9.3.5 Human Agency and the Capacities of Social Resilience*

It seems reasonable to expect that all capacities of social resilience may be found on all social levels (see Table 9.1 in Section 9.2). However, given specific context conditions, some selected capacities might be more salient on specific levels than others (Ansell and Gash, 2008). Therefore, any attempt to apply social resilience as "matrix" of levels and capacities needs careful contextual analysis to consider the case that matters for dealing with natural hazards (see Yin, 2014 on case study design). To illustrate this methodological argument, the following briefly describes a hypothetical "optimistic scenario" in which people based on collective hazard experience (e.g., a disastrous regional flood event) and concern about the current state of natural hazard management set out to develop *new* ideas for improving hazard management in an inter-organizational context:

- On the intra-organizational level, participative capacity may be key because the inclusion of new actors into organizational processes that tend to conservative problem solving facilitates social processes of imagining new possible disturbances and new ideas to improve natural hazard management. On the team level, the underlying dimension of power is often more "visible" compared to more abstract social relations and structures (e.g., institutionalized organizational practices in a specific field of society like energy policy, Tierney, 2014). In small teams, members are embedded in the team's social structure that distributes different "radii of agency". In an optimistic scenario, the combination of (1) common event experience, (2) the existence of teams that seek to develop new management solutions, (3) facilitative organizational arrangements, and (4) people in "high places" that act as sponsors of innovation increases the likelihood that social resilience for dealing with future hazards is enhanced (e.g., Hutter, 2007, 2016; Van de Ven et al., 1999).
- On the organizational level, coping capacity may be crucial because new ideas developed through team efforts need collective recognition and approval in a situation where political pressures to demonstrate "lessons

learned" from natural hazard experience are high and demands on the allocation of public resources are diverse. Explicit interpretations embedded in organizational strategies, structures, and routines shape foremost how organizations define vital organizational processes, "normal" and "unexpected" disturbances or potential disaster events, and prioritize their measures accordingly (Bazerman and Watkins, 2008).

- On the inter-organizational level, adaptive capacity may be key. Within a functional differentiated world (society), all kinds of tasks and processes (potentially) involve diverse organizations. In order to proactively adapt to futures disturbances, truly collaborative endeavors require "networked capacities" (Tierney, 2014: 182) that combine diverse knowledge domains and types of knowledge in processes of inter-organizational learning. On the one hand, achieving resilience on this level of organizations from diverse societal realms is especially difficult as it involves a diversity of existing strategies, structures, norms as well as coping and participative capacities of (possible) collaborators. On the other hand, if inter-organizational collaboration is successfully established, high diversity in abilities facilitates *unique* forms of social resilience in terms of adaptive (or transformative) processes.

This is, as mentioned earlier, a hypothetical "optimistic scenario" to illustrate the idea that specific selected capacities may be placed in the foreground of argumentation for "good reasons". Deeper analysis may try to adopt the whole conceptualization of social levels (intra-organizational, organizational, and inter-organizational level) and capacities (coping, participative, adaptive, and transformative capacity) to understand how the prospects of promoting resilience in the face of natural hazards may be enhanced. However, social actions and social order that are constitutive for "human agency" to promote resilience are also constitutive of how potentials for risk, crises, and disaster were produced in the first place. Hence, there is a "pessimistic scenario" to every optimism.

## 9.4 Analyzing and Assessing Social Resilience in the Context of Natural Hazards

Sections 9.2 and 9.3 conceptualized social resilience from a sociological viewpoint as a "matrix" of social levels (intra-organizational, organizational, and inter-organizational level) and as a set of capacities (coping and participative capacity as well as adaptive and transformative capacity). We summarized some important social resilience research with regard to these levels and capacities. Overall, resilience research in general, social resilience research in particular, seems to be highly dynamic. Furthermore, policy makers in natural hazard and disaster management as well as in collective efforts for sustainable development increasingly use the notion of resilience to suggest ways to improve policies and practices in the face of the manifold pressing problems of our time (e.g., Tierney, 2014). For instance, Coaffee and Lee (2016) outline policies and practices to enhance resilience in the context of climate change, security concerns in urban regions, and large-scale disasters.

Given the dynamics of resilience research and the request of some policy makers for valid as well as user-friendly and context-specific results of scientific analysis, a widely acknowledged research paradigm about resilience could be helpful to facilitate joint qualitative as well as quantitative analysis and to provide standards and research contents for the assessment of resilience with regard to various temporal and spatial scales. However, as it is now, no such unifying resilience paradigm exists in research and practice (e.g., Coaffee and Lee, 2016). Quite the contrary, we observe an increasingly diverse "landscape" of research approaches and frameworks for assessing resilience. Conceptualizing social resilience as "matrix" of levels and capacities adds to this diversity through combining an inclusive approach with the analysis of multiple social levels and through reworking the set of resilience capacities (coping, participative, adaptive, and transformative capacity). We encourage students of resilience to consider this complexity. At the same time we are aware that some simplifications may be in order – especially with regard to the situational contingencies of resilience research and efforts of assessing this notion in practice. Therefore, the following should not be read as a recipe to apply our understanding of social resilience, but as a set of only a few selected suggestions that may serve as starting points in qualitative social science research on resilience.

- Designing a project on social resilience: Social science research on resilience can take on many different forms. For instance, it is helpful to distinguish between projects of basic research, collaborative research jointly by researchers and practitioners, doing an evaluation study, and action research (Van de Ven, 2007)[7]. It seems plausible to assume that researchers only have limited strategic choice over the form of a research project, especially in their early career. We therefore suggest that students of resilience should seek to clarify the "nature" of their project based on pragmatic as well as methodological considerations about the overall project design (purpose, project duration, form of project, resources of the project, and so forth). These considerations apply to projects with or without focus on social resilience. From a sociological viewpoint, human agency is not only a concept to analyze social resilience as the "object" of investigation, but also to understand past accomplishments, present circumstances, and future possibilities of research.

- Considering the context of multiple voices: Qualitative social science research often intends to provide rich accounts of context conditions and social processes related to the phenomenon of interest. Coaffee and Lee (2016) show that context and process mechanisms are crucial to analyze and assess (urban) resilience in practice. If researchers adopt the sociological concept of resilience introduced in this contribution to the textbook about vulnerability and resilience to natural hazards, it is plausible to expect that they will need rich accounts of context conditions, for instance, to provide a sufficiently detailed and valid analysis of intra-organizational, organizational, and inter-organizational processes with regard to coping and participation as well as adaptation and transformative efforts. Context and process mechanisms will involve and affect a diversity of actors (e.g., state actors, communities, citizens). We suggest that researchers are "strict" with regard to adopting a multi-level analysis to organizing for resilience and – at this stage of work – "open" for multiple voices that are potentially relevant to consider resilience under specific context conditions.
- Conceptualizing social resilience: Based on a rich context analysis and a tentative understanding how diverse actors act in social processes that are potentially relevant for resilience, researchers are encouraged to provide a conceptually convincing approach to how they understand social resilience and how they implement resilience analysis with regard to an inclusive and multi-level approach. We suggest that the whole matrix of social levels and capacities of resilience is only applicable under specific favorable conditions (e.g., large project of true collaborative research, Van de Ven, 2007). If a project is to be conducted under less favorable conditions, research usually has to focus on specific aspects of social resilience. We strongly encourage students of resilience to acknowledge and articulate these necessary pragmatic limitations and the respective claims of validity of their research. Furthermore, we would like to highlight another aspect with regard to the expanding landscape of social resilience research: In recent years a critical discourse on resilience, especially its alleged hidden neoliberal transcript emerged (e.g., Coaffe and Lee, 2016; Joseph, 2013; Zebrowski, 2013). While this discourse is an important corrective, it also tends to do away with the benefits and positive notions of resilience in natural hazard and disaster management. We see

the need for critical, but still balanced, concepts of resilience (see the next suggestion).

- Analyzing barriers to resilience: As just discussed, there is a pessimistic "scenario" to every optimism about organizing for resilience. Hence, analyzing barriers of resilience is crucial to reach a balance between optimism and pessimism. In the following we will emphasize some salient barriers with respect to the different social levels we used throughout this contribution. On the intra-organizational level, the lack of participative capacity may impede the development of new ideas for improving hazard management. Intra-organizational power structures may obstruct any common event experience and hinder the development of new solutions. People in "high places" or rigid organizational cultures can rule out any innovation and rather render risks and potential disasters unspeakable (e.g., Cerulo, 2008). The concept of resilience may become a politically charged concept that prohibits resilient responses in the face of natural hazards and disasters (Hempel and Lorenz, 2014). Thus, it is essential not only to analyze which resources are needed to achieve resilience, but also to consider how and why actors address "resilience" and how the idea of resilience is deployed to gain resources (Tierney, 2015). On the organizational level, certain cultural practices as well as organizational assumptions and routines may lead to organizational blindness in the face of vital risks, worst cases, and disasters (Cerulo, 2008). Recently, scholars (e.g., Matthewman, 2015; Nixon, 2014) have argued that the predominant view on natural hazards is too narrow and neglects slow and creeping attritional disasters, such as chronic environmental degradation. Furthermore, broader changes in the natural environment (e.g., global environmental change, climate change) and technological developments (e.g., digitalization, genetic engineering) transform potentials for risk, crises, and disaster resulting in much more complex risks and large-scale disasters with cascades of disastrous events that require the development of adequate coping capacities (e.g., Coaffee and Lee, 2016). On the inter-organizational level, adaptive capacity might not be achieved or, worse, diminish. The interaction of organizations from diverse societal realms is especially difficult, challenging, and prone to failure. Furthermore, shifts in the (global) social matrix such as globalization, urbanization, and technological developments (e.g., digitalization) transform not only the potentials for risk, crises, and disaster, but also human agency, interpretative patterns, power structures, and subjective notions of what resilience might be and how it can be achieved. To find new adequate forms of adaptive and transformative capacities under these dynamics will be an ongoing major task for the years to come.

- Assessing social resilience: Social resilience, at least in our understanding, arises out of the synthesis of diverse capacities at specific levels of social order. Social processes of synthesizing issues of interpretation, power, and participation, as well as issues of learning, adaptation, and transformation cannot be addressed through checklists, recipes that resemble recipes in cookbooks, and simple causal models. In our view and as laid out in our third learning aim, assessing social resilience requires balancing all aspects mentioned in this chapter in a highly context-sensitive manner. Taking the notion of human agency and social realms seriously, we cannot advocate a single approach to assess social resilience, but would like to encourage researchers and students to develop new and combine existing research approaches in fruitful ways in order to explore and understand social resilience with respect to their specific research question, their research conditions and their given limitations. In order to provide some conceptual guidance for this endeavor and for navigating the expanding landscape of resilience we proposed our "matrix" of social levels to distinguish between multiple social levels of organizing for resilience and as a set of capacities of actors in modern society that is characterized by social processes of perception, interpretation, and meaning-making. But assessing and even more implementing social resilience requires continuous efforts of agents that desire to enhance social resilience and that are able to consider such diverse issues as constructing and communicating meaning as well as playing power games in risky settings.

## 9.5 Conclusion

The idea of social resilience encourages us to pay ample attention to the capacities and resources of individual persons, teams, organizations, and networks to proactively adapt to and recover from unexpected disturbances and unusual experiences. We are confident that actors that desire to increase social resilience will provide important contributions to collective action that addresses complexity, uncertainty, and ambiguity related to natural hazards and risk (Klinke and Renn, 2002; Renn et al., 2011). Framing thoughts and beliefs in line with positive resilience thinking and expecting desired change may pave the way for actual change in team development, organizational strategies and routines, and network dynamics relevant for natural hazard management ("self-fulfilling prophecy," Weick, 1995, 2001). Of course, it is also possible to hear voices that "see" positive resilience thinking as somehow naïve, especially politically naïve. State actors may be more inclined to practice "resilience talk" (Hutter and Kuhlicke, 2013) for political reasons (e.g., "responsibilization," Coaffee and Lee, 2016: 39) than to improve natural hazard management through their own continuous efforts of capacity development (e.g., Weick and Sutcliffe, 2015).

This contribution to the textbook on vulnerability and resilience to natural hazards does acknowledge that "resilience" is and will be used for political and ideological reasons, in political processes, and in a tactical manner by self-interested actors in the public realm. Especially in a social science contribution to resilience research, it would be unwise to omit power and politics in modern society (e.g., Clegg et al., 2006; Weick et al., 2005). Therefore, our contribution does not wish to claim that achieving social resilience in the face of natural hazards is easy and highly likely and in fact widely spread in the social world. Quite the contrary, our contribution showed that actors desiring to achieve resilience need to synthesize a broad range of diverse capacities in the context of efforts of proactively and reactively dealing with disturbances and the unexpected. Actors truly interested in developing the capacities of social resilience need to consider issues of interpretation and meaning making, power and participation, learning from experience, and for imagined, even "strange" futures as well as issues of reflecting on deeply situated institutions and processes of institutional change and transformation.

Given that social resilience may exist and may develop on different social levels, we are hopeful that small change (for instance, in teams and single organizations) can significantly contribute to change in networks, social capital, and local communities, and even societies for managing effectively disturbances and the unexpected. However, research on social change in general, institutions and the change of "regimes" in particular, also shows that desired changes in line with positive resilience thinking and principles of sustainability may take decades to grow and spread (e.g., Egermann and Hutter, 2014; Markard et al., 2012; Poole and Van de Ven, 2004). Desired change may also require changes in fundamental structures of societies (e.g., changing the processes of the concentration of energy and people as well as of political and economic power, Tierney, 2014 based on Perrow, 2007). This underlines that social resilience from the perspective of human agency needs to be placed in a broader framework that considers the relations between agents of resilience and institutional change (see Scott, 2014; Tierney, 2014).

With reference to the learning aims of this chapter, we offer three take-home messages to the reader. We assume that the reader is more or less a novice in resilience research and practice. Hopefully, novices will advance their resilience understanding through extensive as well as intensive thinking on this concept and through action in accordance with resilience-inspired beliefs. If readers read this chapter in a library and wonder what messages they may take home to continue thinking and acting on resilience, we believe that the following can play a role – among further possibilities to use our arguments in research and practice:

- Clarify the social level of interest! It is easy to get lost in the literatures that are relevant for understanding resilience. One way to avoid unproductive confusion is to be clear about the social level (or levels) of interest. Are you concerned about how teams perform in the context of danger and natural hazards or are you involved in inter-organizational relations? Do you expect further advancements in social resilience only from interaction between social levels? All kinds of theorizing and assessing resilience should be aware of their respective level (or levels) and should not overstretch their explanatory potential. Therefore, clarifying the social level of interest is a good starting point to observe how people talk about possible ways to deal with an uncertain and, in a strict sense, unknowable future.
- Take perception and interpretation seriously! As researchers, whether natural scientists, engineers or social researchers, we tend to downplay the importance of context-specific perception and interpretation in practice. This is so for manifold reasons, for instance, reasons that are related to ongoing pressures to "publish or perish" which emphasizes how research is perceived and interpreted by reviewers and editors and less so by human actors in practice. However, our conceptualization emphasizes that social resilience as complex set of capacities can only be understood through an in-depth understanding of how "real people" perceive and interpret their context-specific situation. This may include an in-depth understanding of situations that emerge and develop in history and the acknowledgment that perceptions and interpretations may be conflicting and, thus, contributing to ambiguous resilience processes including secondary or tertiary effects on different social levels. Hence, we encourage the reader to consider the possibilities of "true" process research (Van de Ven, 2007).
- Beware of overconfidence in the context of resilience assessments! There is wide agreement in the diverse resilience literatures that a well-developed capacity of resilience does not mean "invulnerability". In contrast, individual and social resilience are actor- and context-specific capacities and context conditions are highly influenced by factors outside the control of human action (for instance, physical processes that lead to natural hazards). There is also wide agreement that confidence in being resilient in the future may also develop into overconfidence, if resilience assessments are used mainly to demonstrate "high resilience" to others and less so to initiate productive critical discussion and improvements. Considering that

social resilience involves different social levels and a complex set of capacities, you should prick up your ears if you hear people arguing for simple and static notions of "high resilience" and "easy ways" to achieve resilience.

To sum up, we encourage readers that have just left the library to think about how they can combine (1) confidence that social resilience makes a difference in the "real world" with (2) doubts that social resilience may be achieved through simple quantitative assessments to demonstrate "high resilience" to important stakeholders of natural hazard management.

# 10

# Economic Resilience

CARLOS DIONISIO PÉREZ-BLANCO, DAVID ADAMSON, AND ADAM LOCH

## 10.1 Introduction

Typical definitions of economic resilience relate to both the ability of a system to maintain function when shocked by an event (static economic resilience, Cavallo and Noy, 2011; Rose, 2009; Skoufias, 2003) and later efficient utilization of available resources on recovery, repair, and reconstruction to speed that system's recovery following a shock (dynamic economic resilience, Carter et al., 2007; Hallegatte et al., 2016; Le De et al., 2013; Rose and Krausmann, 2013). From a business perspective, static resilience makes the best possible use of available resources to reduce business interruption losses at a given moment in time; while dynamic resilience works after the event, relying on repair, and reconstruction to speed up recovery and on the ability of businesses to reschedule, or recapture, foregone output (Park et al., 2011; Wein and Rose, 2011). Recent studies also include a third category of resilience that refers to the capacity of the system to reduce vulnerability and therefore limit economic impacts during and after the event (OECD, 2016; US Economic Development Administration, 2016), although this is sometimes considered to fall in the mitigation sphere (Wein and Rose, 2011). An alternative classification distinguishes between the direct impacts of the event (lives and asset losses) and the "ability of the economy to cope, recover, and reconstruct and therefore to minimize aggregate consumption losses" (Hallegatte, 2014: 1), referred to as macroeconomic resilience – although it involves also actions at a micro- and meso-economic level. Mancini et al. (2012: 1) draw a threshold and define resilience as the "capacity of an ecosystem to tolerate disturbance without collapsing into a state controlled by a different set of environmental conditions and processes."

Economic resilience is thus distinct from other forms (e.g., engineering) through its focus on property damage, which takes place at a specific point in time and continues until the economy has recovered, or reaches a new normal. Any new normal state will be naturally complicated by individual behavioral considerations and public policy decisions, such that Kajitani and Tatano (2009) suggest economic resilience stems from any decisions that reduce business interruption from physical damage to structures from shock events. UNU Wider (2008) define resilience as a "policy-induced ability." Critically, the ability to be resilient can be "nurtured," which explains why a number of inherently vulnerable regions have attained relatively high levels of development (Briguglio et al., 2006).

Many economies are most often shocked by extreme weather events. Extreme weather events, coupled with exposed and vulnerable human and natural systems, can lead to disasters (IPCC, 2014a) that are economically, socially, and psychologically costly to avoid (ex-ante costs) and/or recover from (ex-post costs). Amplified by human-induced climate change, long term economic losses from natural disasters averaged US$314 billion per year in 2015 (UNISDR, 2015b) based on a consideration of impacts on man-made capital alone; it does not consider direct or indirect income losses nor environmental impacts. Despite this partial accounting, an economy capable of raising an annual income equivalent to the losses reported in United Nations Office for Disaster Risk Reduction (UNISDR) estimations would rank 36th worldwide in terms of GDP. Higher incidence of disaster event impacts on an economy will result in disruptions to economic growth opportunity costs through insufficient investment in, for example, health, education, and social protection measures. Where disaster events are large enough to threaten the replacement of the capital stock, then an economy will also transition toward a lower economic development growth due to the opportunity costs associated with replacing that capital stock. For example, the percentage of capital stock lost to natural disasters in Europe and Central Asia is expected to be 20 times lower than in small-island developing states, for which natural disasters and climate change are an existential threat (UNISDR, 2015b). Natural disasters can therefore be particularly challenging to developing nations[1] where the lack of institutions can impede responses to disaster events, and adverse capital stock losses can be harder to replace (OECD, 2015).

Increasing evidence on the economic repercussions of disaster shocks has propelled a more holistic and proactive approach to their management around the world, which aims to enhance the economic resilience to natural disasters in developed and developing economy contexts. The Yokohama Strategy (UNISDR, 1994) and especially the 2005–15 Hyogo Framework for Action (HFA; UNISDR, 2005) motivated many nations to adopt a systematic approach to reducing their vulnerabilities and risks to natural disaster events. The HFA highlighted broad international acknowledgment that efforts to reduce natural disaster risks must be systematically integrated into national policies, plans and programs, and called for participants to prepare their own Disaster Risk Reduction (DRR) strategy to expand current management objectives from disaster response to proactive disaster risk management approaches – thus "nurturing" resilience (UNISDR, 2005). In response, many national governments have adopted these suggestions, identifying more than 190 national focal points and crafting over 100 institutional arrangements to enhance disaster risk resiliency (UNISDR, 2015b).

Despite gains in the identification of priority risk areas and institutional change, the 2010 HFA mid-term review and regional and national progress reports noted little proactivity in changes to the policies implemented by participating governments (UNISDR, 2011, 2015a). In the main, DRR strategic approaches appeared to focus on disaster response activity through ex-ante hard engineering[2] solutions and ex-post contingency funds to be activated and disbursed after disaster events (WEF, 2016). They also noted that major commitments reflected in national laws and policies were yet to be deployed into practical DRR politics and/or principles (Surminski et al., 2015), such that present paradigms of risk management have spurred investments that boost economic growth at the expense of higher exposure, e.g., through floodplain development (Kiedrzyńska et al., 2014; Schindler et al., 2014).

As a result, disaster exposure has not been reduced toward levels anticipated under the HFA, and absolute economic losses were expected to continue their increase[3] (UNISDR, 2012a). In response to these issues, the recent zero-draft of the Sendai Framework for Disaster Risk Reduction (2015–30) highlighted the need to define targets that are operationally feasible, measurable, and achievable to properly incentivize effective DRR (United Nations, 2014). However, such resolute principles disappeared from the final outcome of the conference. None of the objectives finally set in the framework offered any quantitative targets against which progress could be measured; instead, the text relied on vague wording and called for "substantial" advancement. Further, the objectives defined had a largely global focus, and therefore did not provide any defined regional or national targets (UNISDR, 2015a).

Short-sighted strategies focused only on ex-post disaster response will clearly fail to address this challenge, and may lead to a tipping point beyond which any required economic recovery will exceed the capacity of future generations (UNISDR, 2015b). Urgent action is needed to translate DRR discourse into effective policies, arrangements and investments that enhance economic resiliency against natural hazards now and into the future (Zimmermann and Keiler, 2015). It is increasingly accepted that DRR is a matter of political and institutional capital, rather than a financial one (Goodwin and Vado, 2007; Mysiak et al., 2013). Yet, disasters are costly to manage prior to their occurrence, or after; and both forms of management naturally impact on economic resilience. Thus, economics has a key role to play in enhancing resilience more broadly.

This chapter provides an overview of the instruments and methods economics science relies upon to deal with disaster risk. Section 10.2 presents the instruments that policy makers have at their disposal to design plans that "nurture" resilience and programs for DRR and discusses its advantages and disadvantages. Section 10.3 assesses how economics can contribute to discerning what instruments have the potential to contribute to resilience and DRR in a cost-effective and/or efficient way. Section 10.4 illustrates our analysis with a case study of the 2011 Brisbane flood disaster in Australia. Section 10.5 concludes the chapter.

## 10.2 Instruments for Economic Resilience

Disaster risk management strategies are typically adapted to the specific challenges faced by the economy at risk, within the formal constraints imposed by the legal and institutional framework, and the informal constraints stemming from local customs, traditions and norms (UNISDR, 2012). Complexity across any of these constraints creates some degree of path dependency that will slow down transitions to more suitable or adaptive arrangements (Williamson, 2000). Public inability to appreciate the need for the current investment to guard against uncertain future outcomes leads to significant discounting effects – particularly where "visible" protections may not be possible from the proposed strategy. This explains why, despite the wide array of available options to manage disaster risk, governments may rely on highly visible and expensive hard engineering solutions, complemented by immediately visible and politically attractive contingency funds if those engineering solutions prove incapable of preventing damages (UNISDR, 2015b).

Yet by their very nature, hard engineering solutions also result in path-dependent outcomes that may reduce future options for managing disaster risk and event impacts. Reliance on such a strategy further reveals a narrow view of vulnerability focused on the physical resistance of engineering infrastructures (potentially) ignorant of other sources of vulnerability explained by environmental and social processes (IPCC, 2014a). It also favors uncontrolled economic development and increases disaster exposure at a time when climate change is increasing natural hazards in many regions (IPCC, 2014a). Finally, in the face of significant recovery costs following any natural disaster, budgetary constraints will often limit or politically challenge the ability of governments to address economic damages through the use of contingency funds (Standard and Poor's, 2015a). In this context, current strategic disaster risk management may be contributing to inflated disaster risk and limited economic resilience.

In the quest towards resilient economies, society must identify the set of instruments that best balances exposure and vulnerability, together with the costs for changing hazard patterns, so that economic and noneconomic targets are attained. Under such an approach, it may be necessary to balance the benefits and costs of reallocating existing resources away from vulnerable regions to prevent unnecessary intergenerational costs; yet ultimately total social welfare will increase if the best strategies are selected and vulnerable resources are not allowed to develop in areas of known high disaster risk. Disaster risk management therefore broadly entails a set of instruments aimed at preventing or minimizing economic damage in the first stage, and economic losses in the second. Despite these instruments, economic damages may not always be fully preventable, making it necessary (from a first-world perspective) to drive recovery or resilience with damage compensation instruments. Damage compensation instruments are subject to regulation on liability (tort law), and comprise the interplay between private insurance, state aid, nongovernment organization efforts, and public acts of charity. Damage compensation instruments assist recovery after a catastrophe, and are ultimately designed with the purpose of reducing the time to return to either the preexisting economic, social, and environmental conditions or towards a new system that is less exposed to these future adverse events. For example, recent research has shown that sufficiently insured natural hazards are inconsequential in terms of foregone output (Von Peter et al., 2012). We summarize these ideas in Figure 10.1, and discuss these issues further down.

### *10.2.1 Damage Prevention*

As stated above, damage prevention is typically focused on hard engineering and funded by the public sector. This process then transfers part of the hazard of risk burden from risk-exposed asset holders to taxpayers. This distortion of risk can alter the public's perception of the true nature, frequency, and consequences of future risk. Consequently the transfer of risk to the public purse can increase private welfare by encouraging net migration into an

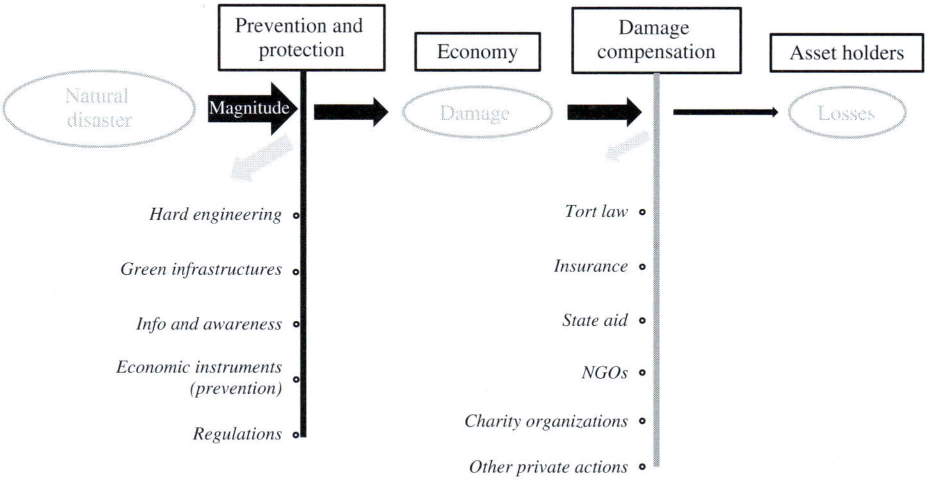


Fig. 10.1. Disaster risk management policies

area, leading to inflated property values in highly-exposed areas which have some comparative advantages (e.g., aesthetic values, better soils, accessibility; Schindler et al., 2014). Thus, rather than helping to reduce the realized consequences of disaster events, the true future cost may well be far higher under this trend toward more public distributions of risk.

This trend was particularly intense during the two decades of sustained economic growth prior to the 2008 financial crisis (Crichton, 2008). Instead of addressing potential risk impacts by deploying complementary instruments to reduce risk exposure and address environmental and social vulnerability processes, natural disaster management has become reactive and incremental. This has reinforced observed trends and led to a rapid increase in the marginal costs of damage prevention, as more exposed areas are increasingly expensive to protect (Botzen and van den Bergh, 2008). The 2008 financial crisis, which amplified the opportunity costs of these investments, and the growing frequency and intensity of the extreme events reported in several areas worldwide (UNISDR, 2012), added pressure on this unsustainable dynamic. Eventually, the marginal costs of hard engineering may reach a point where either the budgetary implications are prohibitive and/or the economic cost to mitigate these risks outweigh the benefits.

Complementary preventative measures can be applied in many contexts, addressing social and environmental processes at the same time. Green infrastructure[4] may offer a complementary and cost-effective replacement to hard engineering. Properly designed green infrastructure is flexible, effective, efficient, and incorporates no-regret measures that integrate human activities with natural processes and ecological systems (Dawson et al., 2011; Kates et al., 2012). The rationale behind the promotion of green infrastructure is that it has the potential to contribute to both risk reduction as well as providing positive externalities including recreational amenities, the provisioning of a wider range of other ecosystem services, and economic resilience. There is a wide array of green infrastructure-based solutions with potential attendant economic or social benefits that can be used to address natural disasters, including afforestation against landslides and runoff, wetlands and sustainable drainage systems to prevent floods, aquifer recharge against droughts, green roofs against heat stress and heavy precipitation, or dunes and salt marshes to mitigate storm surges, to name but a few (Demuzere et al., 2014; Forino et al., 2014; Wolch et al., 2014). Yet, their uptake is still limited. This can be explained by the low visibility of some of their benefits, which are generated outside of the market (externalities). There is also a lack of political and private experience or empirical evidence demonstrating the practical value of green infrastructures for DRR, and little understanding of the instruments that can support the process (Biesbroek et al., 2013).

Regulations can also contribute to effective DRR and economic resilience. Regulations are a relatively inexpensive option, when compared against hard and green infrastructure, and enable some focus on the social component of disaster risk management, setting specific targets that agents have to meet. In this respect, transaction costs analysis and comparative studies are useful to determine the true costs and values of various options. Norms and standards are often used in combination with other instruments, e.g., enacting building code requirements in floodplains, limiting aquifer abstractions and protecting wetlands so that green infrastructure can be maintained/restored/developed. Regulations can also be used to pass the true cost of risk management back to private individuals (i.e., cost of compliance) and, if designed well, reduce the cost of future risks to the public purse. The underlying shortcoming of regulatory approaches is their inability to change the driving forces behind agents' actions, which may result in strategic behavior to avoid compliance (Heyes, 2000; Winter and May, 2001). Furthermore, rebound effects in which performance is actually worsened as compared to the baseline scenario may also arise (Alcott, 2005, 2008). For example, if sufficient and adequate enforcement (which may be expensive) is not in place, restrictions on surface water-use may result in illegal groundwater abstractions of equal or greater amount, worsening drought preparedness and economic resilience to droughts (Pérez-Blanco and Gómez, 2014). For many of these issues, data availability and analysis is often a key constraint to effective planning and management.

Information and awareness campaigns remove some of the information constraints agents may face when making decisions. This can result in lower exposure, e.g., through the decision not to build on a floodplain. Although information and awareness campaigns have no direct impact on vulnerability, they contribute to raise responsiveness on disaster risk management at different decision levels (including individual responses) and may turn the tide in favor of actions that may effectively contribute to DRR (Spiekermann et al., 2015).

A straightforward way to address many of these problems is the use of economic incentives. In our previous surface water-use example, authorities could set sanctions proportional to the likelihood of risk exposure and impacts to penalize realized noncompliance by affected individuals/businesses. Alternatively, they could charge users for any withdrawals to reduce water use to targeted levels with adjustments to these approaches over time, as further information is revealed. Economic instruments are thus complementary to, rather than substitutes for, other DRR instruments. For example, they both complement and offer an edge over legislative/regulatory water policies through their ability to design incentives capable of adapting individual decisions to collectively agreed goals at least cost. Economic instruments such as prices (including the prices of "social bads," i.e., sanctions), subsidies, and market-based instruments (e.g., water trade) internalize the adverse impacts that stem from disaster risk into the behavior of economic agents, so that their response is compatible with policy goals. Payment for Ecosystem Services (PES) is another popular economic instrument (subsidy) to enhance green infrastructure supply, as happens with water purchase tenders for aquifer recharge in the semi-arid states of western USA (Hanak and Stryjewski, 2012). In the Po River basin in northern Italy, a combination of regulations (minimum environmental flows) and economic incentives (the possibility to decide on the restrictions to make) are used to encourage agreements among users during droughts (Mysiak et al., 2013). Economic instruments have thus been increasingly adopted to address DRR and climate change adaptation, and this has served to improve its design and performance (EEA and OECD, 2013).

Yet, doubts about the effectiveness of these instruments remain. They are often designed without a clear purpose; although their explicit objective might be to manage DRR, they may often pursue other goals such as revenue raising, and this dichotomy could hinder their public acceptability, uptake, and overall performance. They can also result in rebound effects that contribute to natural capital depletion, climate change, and higher disaster risk; these may occur with some subsidies to improve energy

(Greene et al., 1999) and irrigation efficiency (Adamson and Loch, 2014). In the end, the economic instruments discussed above are the result of designed tools whose performance is conditional upon the political capacity and institutional frameworks in which they develop.

### *10.2.2 Disaster Compensation*

Although incentives may motivate individuals and businesses to self-manage their risk, it may not be possible to fully remove future damages from flood, drought, and other risks in an era of climate change and intensification of natural hazards (IPCC, 2014a). DRR strategies therefore demand a comprehensive approach to hazard management not limited to damage prevention, but also inclusive of compensation. Damage compensation policies include arrangements such as tort law (liability) relief, ex-post state aid, insurance, nongovernment organization or charity assistance and public actions in response to disaster events that may all provide relief to those affected.

Tort laws address the economic compensation of disaster-related damages, caused or exacerbated intentionally or through negligence or omission that harm rights or protected interests of others (Sugarman, 2006). The applicability of tort law differs across countries: for example, English tort law is based on the duty of care, while German tort law applies in cases when the offender infringed a legal right of the victim (European Group on Tort Law, 2005). Tort law offers people affected by natural disasters the possibility to use the legal system to obtain financial recovery for the damages that are responsibility of a third party. However, victims of natural hazards cannot use this remedy if damages stem from wholly natural disasters – the so-called Act of God events.

Many countries also have provisions to supply state aid after a natural disaster event. State aid is typically used to rebuild foregone capital and compensate foregone income, and can adopt different forms including ad-hoc payments, tax exemptions, soft loans and reduced social security contributions. State aid is generally more affordable (compared to some insurance premiums), equitable (in principle, all victims are included), and solvent (conditioned to sovereign creditworthiness). However, the use of state aid has constraints. The World Trade Organization and many governments worldwide follow strict intervention rules to ensure an equitable allocation of resources, and state aid to make good the damage caused by natural disasters is only admissible if some prerequisites are met. For example, the EU defines de minimis aid and General Block Exemption Regulations that define ceilings and exemptions (European Commission). Another limitation to state aid refers to budgetary constraints, which is of increasing relevance to many countries after the 2008 financial crisis. Moreover, creditworthiness can be largely affected by natural disasters and limit the ability of states to cope with damages and act as the insurer of last-resort. For example, a one-in-250-year earthquake could downgrade national credit ratings by more than two levels (notches) in countries such as Japan and Chile. Alternatively, in developing countries like Bangladesh and the Dominican Republic a one-in-250-year tropical storm may lead to downgrades in credit ratings of 2 to 2.5 levels respectively (Standard and Poor's, 2015a), with far more significant national impacts. These figures are likely to be aggravated by climate change (Standard and Poor's, 2015b).

In this context, insurance arrangements that build partnerships between public and private sectors and include the regular payment of a premium by policyholders have received renewed attention (European Commission, 2013b; OECD, 2015; UNISDR, 2015b; WEF, 2016). Insurance is an economic instrument that redistributes the cost of risk from taxpayers back to asset holders, offering individual protection against the risk of losses caused by various perils through pooling of risks (Baltensperger et al., 2008). Insurance is a heterogeneous instrument, and can adopt different forms. For example, it can be mandatory (France) or voluntary (UK), and involve wholly public (Greek crop insurance), private (Australian hail insurance), or mixed public-private (as happens in Spain) arrangements. Insurance is thus again complementary to, rather than a substitute for, other disaster risk management policies.

Similar to tort law and state aid, insurance eases recovery after a natural disaster and thus limits economic losses. But unlike tort law it is widely applicable; and unlike ex-post state aid, it is (at least partially) funded by policyholders. Recent research also highlights its effectiveness. For example, using a sample of forty-eight countries and a 5 percent damage-to-value ratio, Standard and Poor's (2015a) shows that a 50 percent insurance coverage can reduce impacts on growth by 40 percent, as compared to a baseline scenario without insurance. Moreover, if the private sector participates in the insurance market at an average level, state aid can be limited to targeting residual risk impacts[5] (Melecky and Raddatz, 2011), and national credit ratings will be more resilient to natural disaster impacts (Standard and Poor's, 2015b). When complemented with other incentives such as risk-based pricing (e.g., flood insurance in the UK), insurance also provides disincentives for individual risky behavior (Surminski, 2009; Warner et al., 2009). This can contribute to reverting current trends towards higher risk exposure, and facilitate the transition towards resilient and adaptive economies. However, all these advantages are subject to the ability of insurers to settle claims. Solvency in natural hazard insurance is typically enhanced through regulations, reinsurance, and public support – where the latter two may come together, as is the case of the Spanish Insurance Compensation Consortium (ICC, 2014). Another insurance shortcoming relates to equity and affordability. For example, risky assets in disadvantaged areas may be relatively more expensive to insure, or even be uninsurable under private arrangements. The inclusion of solvency, affordability and equity issues in the design of insurance against natural disasters expands the role of state aid from residual risk management to a combination of ex-ante and ex-post subsidization. This in turn demands new and more sophisticated regulations, both at a national and supranational level. Public intervention has other negative by-products, especially those concerning the weakening of the linkage between risk and pricing and its negative impacts on private incentives for undertaking self-risk mitigation and/or adaptation measures (Holub and Fuchs, 2009; Surminski and Oramas-Dorta, 2014).

Finally, important assistance following any natural disaster can come from domestic and international nongovernmental organizations (NGOs), charities and private actions that support those affected. NGOs in particular may provide important resources toward environmental recovery and ecosystem service resilience where these aspects are critically linked to an affected economy. Examples include replanting riparian areas following damage to mitigate future damage or ecosystem losses, or the provision of native wildlife refuge sites that will enable recovery of impacted fauna or flora species. Additional recovery resources may arise from appeals and donations organized by private companies (e.g., offers by large company employees to house disaster affected individuals, arranged by their employer who may have many affected staff). As such, these organizations can often span gaps in public/private systems aimed at economic resilience, and provide valuable support or specialized expertise to address resilience requirements. They can be particularly effective at providing social connectedness and psychological benefits above those available from more distant arrangements associated with state aid or insurance measures. Such organizations thus reduce the total ex-post costs and time involved in recovery, driving improved resilience outcomes.

The wide array of options available for mitigating and managing the risks associated with natural disasters increases the complexity and difficulty with analyzing subsequent possible outcomes, and providing practical recommendations for public officials. We address these issues in the next section.

## 10.3 Economic Analysis

Multiple instruments, hazards, and climate change scenarios make disaster risk management a complicated matter demanding advanced data and information management. Managing these trade-offs poses relevant technical, operational, and coordination challenges (Pérez-Blanco et al., 2016). Economic models and research both at a microeconomic (Graveline, 2016; Heckelei et al., 2012; Loch et al., 2012; Pérez-Blanco et al., 2015) and macroeconomic level (Bosello et al., 2012; Eboli et al., 2010; Hallegatte, 2008; Hallegatte et al., 2016; Rose and Liao, 2005) are often coupled with physical models and representations (Carrera et al., 2015; Harou et al., 2009; Jongman et al., 2012; Pauw et al., 2011) to assess the outcomes of alternative risk and policy scenarios.

These models offer an increasingly detailed and comprehensive representation of abatement costs and benefits. Yet, their outcomes often conflict with actual decision-making. Many studies have been devoted to the difficulty of applying research-based knowledge in the policy advice sphere (Loch et al., 2014; Spiekermann et al., 2015). A growing literature explores how the existence of institutions condition the cost-benefit balance in conventional risk abatement analysis through the assessment of transaction costs; namely the costs of the resources used to (1) define, establish, maintain, use, and change institutions and organizations, and (2) define the problems that these institutions and organizations are intended to solve (Marshall, 2013). Due to the limited empirical evidence available, streaming transaction costs into policy assessments may be challenging, especially for ex-ante analysis. Nonetheless, recent research suggests transaction costs can be relevant and should be incorporated in policy assessment (Crase et al., 2013; Garrick et al., 2013; Ofei-Mensah and Bennett, 2013). We touch upon this issue again later.

### *10.3.1 Economic Analysis: Abatement*

When economists deal with environmental problems, they typically look at abatement costs and benefits. Abatement costs quantify the trade-offs involved in reducing an environmental negative, such as disaster risk (Vogt-Schilb and Hallegatte, 2014). Abatement costs of alternative policies can be compared with some expected benefits (cost-benefit analysis) or nonmonetary targets (cost-effectiveness). With cost-benefit analysis, all the significant benefits related to the policy are assumed to be known and capable of being expressed in monetary units. A value of time (discount rate) is used to express the flow of costs and benefits involved on a common basis; typically the net present value (see e.g., European Commission, 2014). In a similar fashion, cost-effectiveness analysis compares costs and outcomes, but does not assign a monetary value to the latter (WHO, 2014). The use of one or another method depends on a series of factors including the nature of the costs and benefits involved, their uncertainty, and the analytical ability to monetize them accurately.

The cost-benefit principle lies at the core of economic analysis, and implies we should take only those actions whose benefits exceed their costs (Frank, 2000). The logic inherent behind this assertion makes it difficult to disagree: if a 1-in-100-year flood is expected to hit your farm and the costs of available policies are higher than the implicit benefits, you should face the ensuing losses. However, if the same happens in an urban area and implicit benefits now include human lives, using cost-benefit analysis to justify inaction becomes controversial (Kuttner, 1999). The accuracy of estimates, even if technically feasible, is also a frequently-cited shortcoming of cost-benefit analysis (de Moel et al., 2012; Rabin and Köszegi, 2007). In particular, benefit estimations collected through nonmarket valuation methods often rely on soft data collected through surveys, and sources of bias are many. Alternatively, revealed preference methods that rely on hard data can be used, but this is contingent on information availability (Varian, 2012). Academia has devoted much research to design consistent contingent valuation methods (Arrow et al., 1993). Yet, even if these sources of bias can be successfully identified and resolved, the issue of substitutability still remains: monetizing two goods puts them on equal footing, but it may be wrong to protect only the good with the higher value if both are complementary; or even worse, if this leaves irreplaceable assets and related services in danger (e.g., a unique ecosystem and its functions, Daly, 1974, 1992). Finally, allotting a value to time in order to put all costs and benefits on a common temporal basis is challenging and may distort outcomes (Freeman, 2003).

In the face of such uncertainties, one may argue that we should protect all areas in which incommensurable losses such as human lives may emerge, and rely on cost-benefit analysis otherwise. Unfortunately, this is not always feasible in a world of scarce resources and economic trade-offs. Disasters will happen with a high degree of probability at some point in the future, but resources devoted to protect a specific area against floods could be used elsewhere in the interim, perhaps to save a larger number of lives. This is where cost-effectiveness enters. Cost-effectiveness defines measurable and collectively agreed targets that policies need to achieve at the least cost. By defining a specific target instead of aiming at maximizing the difference between monetized benefits and costs, cost-effectiveness analysis avoids the uncertainties surrounding benefit estimations. Cost-effectiveness analysis also overcomes other potential research gaps (in economics or other scientific areas), adopting a precautionary approach aiming at preventing potentially harmful outcomes. For instance, cost-effectiveness is typically used in health research (see e.g., Tsertsvadze et al., 2014) and environmental management (e.g., the EU Water Framework Directive, European Commission, 2000). Helpfully, the post-2015 agenda for DRR defines many cost effectiveness indicators, several of them nonmonetary: included are disaster mortality, number of affected people, economic loss, damage to critical infrastructure and disruption of basic services, number of national and local disaster risk reduction strategies, international cooperation to developing countries, and access to early warning systems and disaster risk information. However, it fails to set measurable targets (UNISDR, 2015a).

### *10.3.2 Economic Analysis: Changing Trajectory*

The way that economies face any crisis will be inherently conditioned by any past decision-making (Anderies et al., 2006). Reactive and incremental policy-making can work if selected instruments are adequate to the challenge at hand, while broader and more lasting transformation may be necessary when incremental policies prove insufficient (Kates et al., 2012). In either case, such arrangements involve transaction costs that can be thought of as the institutional (rules and regulation capacity) and organizational (people and knowledge capacity) investment needed by any economy to create/implement/support an operational policy or program aimed at addressing public problems. Once implemented, such programs usually adopt a production cost phase that operates as part of that economy's everyday business. To ensure success, however, we monitor performance and assess that against targets that provide background transaction costs. If as a result of that monitoring the decision is made to adjust or redirect the policy/program then a new transaction costs investment phase will be experienced (alongside the continuing production/operation costs) to achieve those changes. As we have already seen, transaction costs matter since incorrectly targeted public investments have serious consequences; evidenced by regular costly hard engineering solutions to insulate economic development from natural hazards, or the consistent use of emergency compensation funds to cope with disasters when the first line of defense collapses (UNISDR, 2011). Economic growth, increasing exposure to risk, and the looming threats of climate change demand innovative solutions that current disaster-risk management has failed to provide (OECD, 2015). A further risk to the effective implementation of innovative policy/programs are lock-in transaction costs (Marshall, 2013) that may limit the range of available options for managing risk and increase the total level of further investment required for adjustment/redirection. Limited options may in turn decrease public resilience and increase total recovery costs associated with natural disaster events. We now provide a case study from Australia that highlights many of the issues discussed in the preceding sections.

## 10.4 Case Study: The Millennium Drought and the 2011 Brisbane Flood

In January 2011 the city of Brisbane in the southeast of Queensland, Australia, experienced one of the worst flood disasters in its history. The city sits toward the mouth of the Brisbane River, which has several large water storage dams located along it and some of the larger rivers and creeks that flow into the river (Queensland Government, 2017; Figure 10.2).

These structures were built according to a reactionary policy of mitigation against the extremes of water supply: drought-related water scarcity and cyclone-induced flood events. Both of these extremes can coincide during the summer months (December–February), with cyclone risk typically extending until the end of April. The Lake Manchester Dam (completed 1916) provided much of the city's water supply until its capacity constraints were reached. Between 1927 and 1931 three flood events suggested that Lake Manchester was struggling to meet the growing demand for water in Brisbane, and that it was incapable of preventing floods. To overcome these failings, an 1893 proposal to build a dam at Somerset in the catchment's north was revisited. Somerset Dam was completed in 1953 and effectively saved Brisbane from being flooded in 1955 (Barnes, 2010). Somerset Dam was then further extended in 1959 to provide hydropower.

In 1963 two new sites were proposed to meet Brisbane's growing water demand in the face of uncertain supply: the first at Wivenhoe downstream from Somerset Dam and the second at Wolfendene along the Albert River south of Brisbane. By 1971 a decision was made to construct the dams in sequential order: Wivenhoe first and then Wolfendene (Queensland Government, 1988). However, in January 1974 before the Wivenhoe works had commenced, Brisbane was severely flooded by a severe degrading cyclonic system – still widely regarded as Brisbane's worst flood due to the number of fatalities and damage incurred.

The 1974 flooding prioritized Wivenhoe's construction between 1975 and 1985 (Queensland Government, 1988). The final storage was expected to hold five times Brisbane's annual potable water requirements, and mitigate future

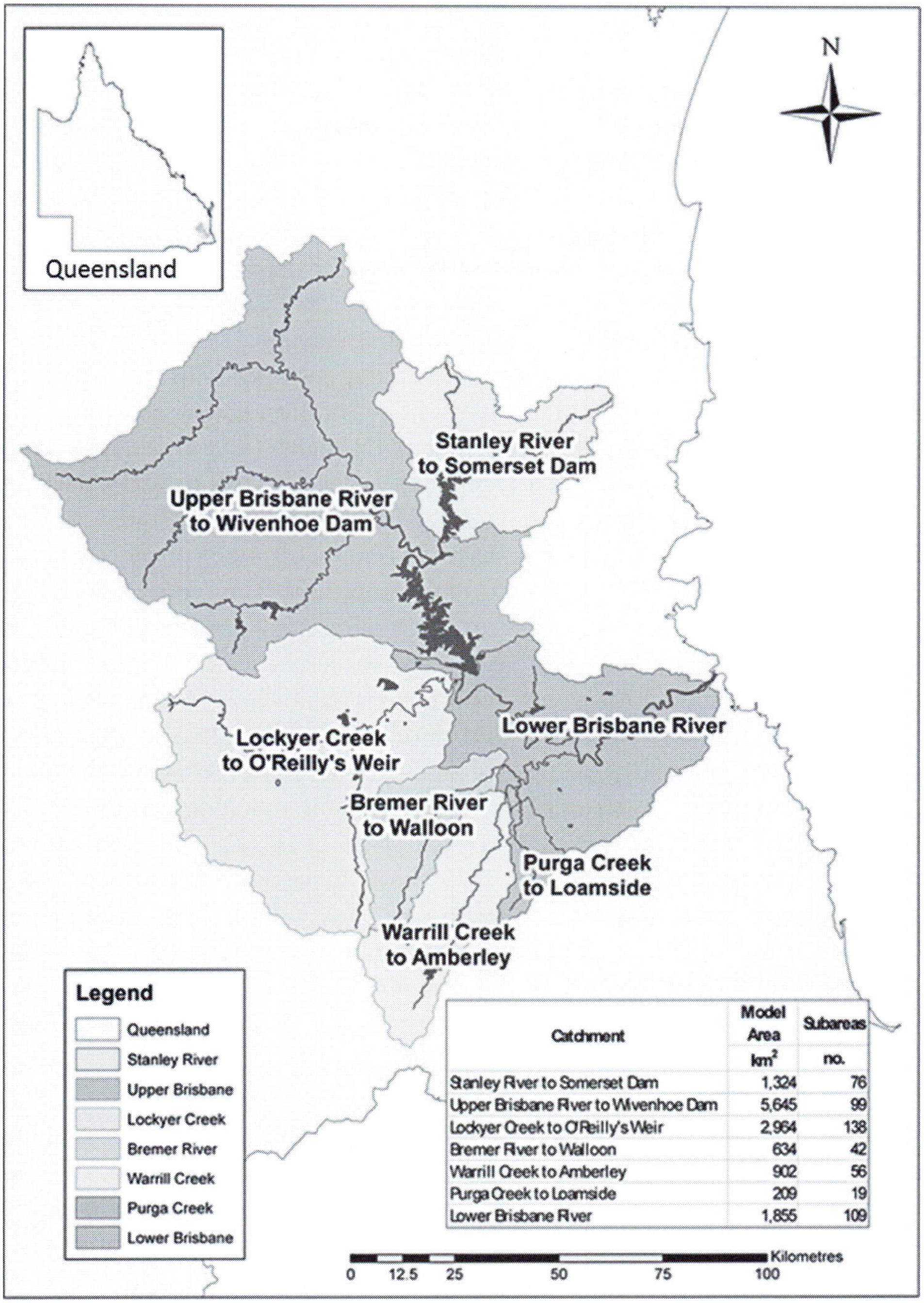


| Catchment | Model Area | Subareas |
|---|---|---|
| | $km^2$ | no. |
| Stanley River to Somerset Dam | 1,324 | 76 |
| Upper Brisbane River to Wivenhoe Dam | 5,645 | 99 |
| Lockyer Creek to O'Reilly's Weir | 2,964 | 138 |
| Bremer River to Walloon | 634 | 42 |
| Warrill Creek to Amberley | 902 | 56 |
| Purga Creek to Loamside | 209 | 19 |
| Lower Brisbane River | 1,855 | 109 |

Fig. 10.2. Southeast Queensland water supply grid. (A black and white version of this figure will appear in some formats. For the color version, please refer to the plate section.)

flood impacts. However, the city also undertook a major review and revision of local floodplain development policies, regulatory changes, flood management levies, and improved drainage systems followed in an effort to ensure such damage and impacts were never again experienced. These efforts, coupled with local campaigns against its construction, saw a delay on the commencement of Wolfendene Dam works – which at 1.1 million megaliters would have provided similar storage capacity to Wivenhoe.[6] Further, studies arguing no new water sources were required by Brisbane until 2050 saw such "unnecessary costs" become an election issue in 1989. The change of government that year led to the works finally being canceled. Through the 1990s Brisbane continued to grow in size, and its water demands with it.

Between 2000 and 2010, many parts of Australia experienced one of its worst droughts on record (the Millennium Drought), and Brisbane's water supply was severely affected. To combat the effects of this drought the state government tackled both the supply and demand for water. Around AU$6.9 billion was invested in new infrastructure to drought-proof southeast Queensland (Queensland Water Commission, 2010) by developing new sources of supply (e.g., upgrading existing dams and desalination options) and linking the supply network together. Brisbane residents were encouraged to reduce their demand for water from this new water grid by subsidizing the cost of domestic rainwater tanks and purchasing water efficient appliances, increasing the fixed and variable cost of accessing domestic water, the introduction of water-use restrictions, and a public campaign aimed at motivating the public to actively reduce their consumption. As a result of these efforts, water-use fundamentally changed in Brisbane households, with the average quantity of water consumed per person per day falling from 282 liters in 2004/05 to 143 liters in 2008/09 (Queensland Water Commission, 2010). Interestingly, despite the end of the drought and a return to more reliable supply of water, Brisbane's average water-use remains low to current periods. Over a 14-day period concluding February 10, 2016 (summer), it had only increased to an average of 168 liters per person (Seqwater, 2016; Figure 10.3). Cost impacts from the addition of expensive fixed-cost capital such as desalination plants into the grid also meant that these costs had to be passed back to consumers, which helped keep total water demand/use very low once the drought eventually broke. These demand/use changes thus became indoctrinated over a long period.

Past experience and limited flood incidence after 1974, significant public investment in hard engineering infrastructure, an integrated water supply network, and fundamentally changed water-use patterns led many to assume Brisbane was relatively safe from future flooding. Dam storage levels had

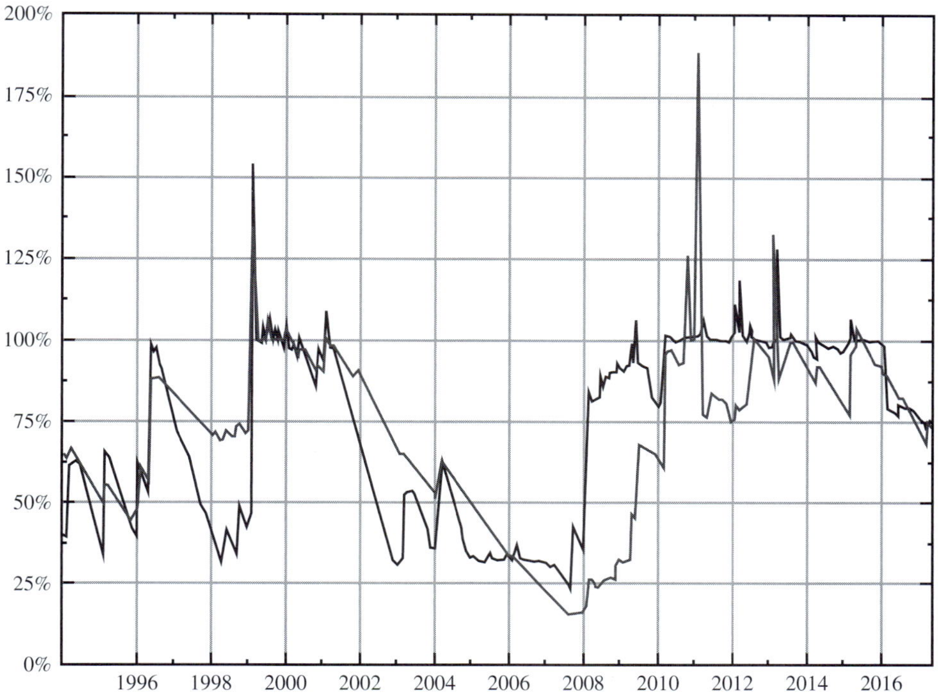


Fig. 10.3. Southeast Queensland dam storage levels 1995–2017. (A black and white version of this figure will appear in some formats. For the color version, please refer to the plate section.)

dropped continuously between 2000 and 2008, when three successive wet periods (2008–10) resulted in storage inflows driving storage levels to full capacity once again. The drought was ended and many water managers, politicians and citizens breathed a huge sigh of relief that the strategy for dealing with the drought had worked. Some water-use restrictions were relaxed, and dam management was focused on ensuring supply held for as long as possible until the next drought occurred – particularly in the summer months of December to April (Seqwater, 2017; see Figure 10.3).

In 2011, the combination of several factors led to one of the worst natural disasters Brisbane has ever experienced. Excellent accounts of the impacts of this event are available elsewhere (Box et al., 2016; van den Honert and McAneney, 2011), but a summary of the event is as follows:

- Soil moisture profiles were full after several consecutive years of rainfall, preventing their capacity to capture some of the runoff.
- A massive rainfall event over Toowoomba sent a wall of water over the range and down into the Bremner River and Lockyer Creek catchments

Table 10.1. *Brisbane flood event damage summary. After van den Honert and McAneney (2011)*

| | February 1893 | January 1974 | January 2011 |
|---|---|---|---|
| Deaths | 35 | 16 | 23 Locker Valley, 1, in Brisbane |
| Injury | 300 | 30 | — |
| Persons left homeless | 5,000 | 9,000 | 18,000 (Brisbane, Ipswich and Brisbane Valley) |
| Persons evacuated | | 9,000 | — |
| Buildings damaged | 5,000 | 7,000 | — |
| Cars damaged or destroyed | | 1,000 | — |

which feed into the Brisbane River. The water wiped out the small rural township of Grantham killing many residents, and caused significant damage to other rural areas in the catchment area.

- This water continued to flow downstream toward Brisbane.
- Further extensive rainfall associated with strong La Niña events across the Wivenhoe and Somerset Dam catchment areas north of Brisbane resulted in inflows to the dams. As the storages were operating at full-service levels (FSL) no further water could be contained. At risk of overtopping and destroying the dam wall structure, operators had little choice but to release the excess water into the Brisbane River.
- The two flows of water converged, causing damage to areas such as Ipswich and outlying suburbs west of Brisbane.
- As the Brisbane River rose and the flood waters reached the city's central areas, low-lying suburbs or those in proximity to the river's edge suffered flooding damage, as storm water systems could not discharge into the already swollen river. Silt and debris from the upstream rural areas were distributed over a large area, compounding damage impacts for affected residents and businesses.

Table 10.1 contrasts some of the economic and social damage in 2011 to earlier Brisbane flooding events. The relative cost to rural communities from the 2011 flooding event is especially poignant.

There are several lessons from the resilience perspective that arise from this case study. First, ex-ante much of the investment in water infrastructure during the drought was relatively high-cost to mitigate drought effects, rather than flood effects, when historically the two would have been managed together. The natural cycle between droughts and floods has driven the development of hard engineering programs and the management of those systems in Australia. These solutions are generally implemented during an extreme drought, or ex-post in response to a major flooding event when existing infrastructure and approaches have proven insufficient for managing water supply volatility. This "panic investment" increases the real costs of development. Once the immediate crisis has ended, the political need for reform and investments is forgotten until the next crisis (Connell, 2007). This in turn leads to a boom-bust infrastructure investment cycle.

In this case, such drivers likely altered the underlying management focus of the structures toward one of mitigating future drought events while their recent effects were fresh in people's minds – arguably increasing the probability of the 2011 flood outcome. Compounding this, Wivenhoe Dam was not designed for overtopping (i.e., where the water is permitted to spill over the dam wall to facilitate release), which meant that it had to be released to avoid such an outcome or risk complete dam failure. Fundamental shifts in water demand and incremental relaxation of water-use restrictions over a long period also meant that previous demand patterns, which may have drawn down storage levels, did not eventuate, further compounding the impacts. Infrastructure therefore provided some highly visible basis of mitigation against the flood event, which might have been increased had the Wolfendene Dam been built. Wolfendene would not have helped prevent the flooding of Brisbane; however, the presence of Wolfendene may have changed the way water was impounded and managed (Härtel and Latemore, 2011). Ultimately, the magnitude of this event was such that engineering solutions alone provided insufficient protection. As an aside, the self-insurance nature of the infrastructure repair/reconstruction arrangements (75 percent federal and 25 percent state in this case) may have provided further incentives for the state to free-ride on the rest of the nation should an event such as this occur. Thus, the true costs of mitigation and risk management were probably not considered prior to the event as an accurate driver of risk-minimization strategies.

Second, many Brisbane residents appreciated the risk of flooding (especially after the 1974 event) and had taken coverage against flood damage. However, following the 2011 flood, many insurance companies changed their description of the event and categorized it as a riverine flooding event, rather than a localized flooding event, for which many people were not covered – thus negatively impacting about 15 percent of policyholders when their claims were denied (van den Honert and McAneney, 2011). Public perceptions of the value of insurance cover and providers fell dramatically as a consequence, and many people reviewed their coverage and changed insurance companies. However, the diminished coverage from private providers necessitated public compensatory schemes in line with our earlier discussion. In response, the state and federal governments once again became the insurer of last resort and a national Temporary Flood and Cyclone Reconstruction Levy was imposed on high-income earners (i.e., >AU$50,000 p.a.) to fund the Brisbane flood recovery effort. This levy shifted the burden from private risk-management, which was (incorrectly) assumed by many to be in place, and backed it with wider public assistance to effect faster and more effective economic recovery.

Third, the local, state, and federal government all worked well together to coordinate the damage mitigation, cleanup, recovery, and eventual return to normal economic conditions. However, a significant part of the recovery relief was provided by charity, nongovernment organizations (NGOs) and the public "mud-army" who mobilized in support of flood victims to assist with the cleanup, removal of debris, and repairs[7]. This meant that the public purse avoided a significant proportion of the costs by, and especially increased the resilience of, those who were not ultimately covered by their insurance agencies. Private householders and some employers (e.g., The University of Queensland) also offered accommodation, interest-free loans and other resources to increase the ability of affected residents to recover from the event. At one point, the offers of assistance were so large that they needed to be turned down under an inability to fully coordinate those interested in helping. As a consequence, mental-health effects are likely far lower than would have been the case if these factors had not eventuated, resulting in reduced total recovery costs and higher total resilience for affected citizens/businesses. This highlights the requirements for complementary economic resilience tools, and the effective outcomes that can arise when they are incorporated successfully.

Finally, the flood event was followed by a formal inquiry into its causes, management, and outcomes. Improved dam-safety audits, communication, and infrastructure planning coordination were recommended as required changes. Sewage and storm-water systems were also redesigned to improve future floodwater handling and removal. Finally, public-access local flood mapping now provides citizens and businesses with greater access to risk-management tools to increase awareness about the future risk of property flooding (i.e., frequency of flooding and the expected height of water over that land). This information, if accessed, can then remove information asymmetry about risks and expectations about insurance costs for prospective property purchasers. This demonstrates the need for analysis and assessment both in the possible lead-up to, and aftermath of, any natural disaster event.

## 10.5 Conclusion

The economic repercussions of natural disasters are significant, and asymmetric among regional or national economies. Different factors explain the extent of income and capital damages, the speed of recovery, and losses after a catastrophe. Previous studies highlight exposure, vulnerability, and hazard risks, but most especially the way they are managed through policies and other instruments. In this chapter we offer an overview of the DRR policy and institutional landscape to assess the portfolio of instruments policy makers have at their disposal to cope with natural disasters, and present the economic rationale behind certain disaster risk management strategies. General concepts are illustrated with examples and a case study of the 2011 flood event in Brisbane, Australia.

The state of the play of disaster risk management and limited progress reported in assessments of DRR to date demand a more proactive approach that extends the scope of government intervention from response to planning and programming. This is far from being a straightforward task. The uncertainties surrounding complex socioecological systems are exacerbated in the case of DRR by the vagaries of climate change and intricate institutional settings. In this context, identifying an effective and/or efficient combination of instruments among the range of available policies and programs can be challenging.

In response, economic analysis has developed a series of tools that make possible an increasingly detailed and comprehensive assessment of DRR repercussions. Traditional focus on abatement has been expanded to account also for the role of institutions and transaction costs. This chapter offers a better understanding of the divergence between scientific discourse and actual political decisions, and outlines some economic tools that may be used to conduct assessments accounting for abatement costs and the benefits and costs stemming from institutional features and performance.

Further research is still necessary to enhance economic resilience:

- Empirical evidence of the impact of transaction costs and a better understanding of the propagation of uncertainty that comes with integrated frameworks remain limited.
- Despite relevant advances, research on abatement is often carried out within specific paradigms (e.g., macroeconomics versus microeconomics), where interdisciplinary approaches would provide significant value.
- Ex-ante and ex-post economic research reveals a myriad of policy combinations that can be used to address disaster risk depending on the context. Economic assessments of actual DRR policies implemented by

governments suggest that strategies need to be tailored to specific country, regional, and even local characteristics and challenges. Even strategies relying on similar instruments may present diverging features that make them unique, as happens with mandatory versus voluntary insurance. What is necessary is a consistent strategy that relies on appropriate instrument(s) that deliver desired DRR targets in a cost-effective and efficient manner, accounting for institutional dynamics and transaction costs, and a transparent register of sources of uncertainty – e.g., through error terms for performance evaluation. Ideally they should also increase economic resilience and recovery following any natural disaster event.

# 11

# Institutional Resilience

SAMUEL D. BRODY AND KAYODE ATOBA

## 11.1 Introduction

Despite efforts to mitigate the adverse impacts of flooding in the USA, losses from both acute and chronic events continue to rise, particularly in low-lying coastal areas that are afflicted by inundation from both precipitation and storm surge events. Coastal flood-prone areas have experienced rapid increases in population, pavement, and associated structures over the last several decades. As a result, the economic and human costs of floods are spiraling upward across the nation. From 1999 to 2009 alone, the USA suffered approximately US$33.5 billion in insured flood losses. Counties/parishes along the Gulf of Mexico coastline reported almost US$21.5 billion of this total (Brody et al., 2011). These damage estimates help solidify what has been generally understood for years: that floods pose a major risk to property and with increasing development in low-lying coastal areas the problem appears to be getting worse.

Given increasing adverse impacts from both acute and chronic storm events, it has become clear that the rising cost of floods is not solely a function of changing weather patterns or inflationary monetary systems. Rather, flood risk and associated losses can only be addressed by understanding built-environment patterns across flood-prone landscapes. Financial incentives, population growth, and associated sprawling development within and adjacent to the 100-year floodplain is placing more structures at risk to flooding (Brody et al., 2011). All mitigation strategies, both structural and nonstructural, must consider the cumulative impacts of rapid development and land use change in coastal margins well into the future. More than ever, there is a critical need to understand how to reduce flood risk at the local jurisdictional and organizational level and facilitate the establishment of more resilient communities in the USA and around the world.

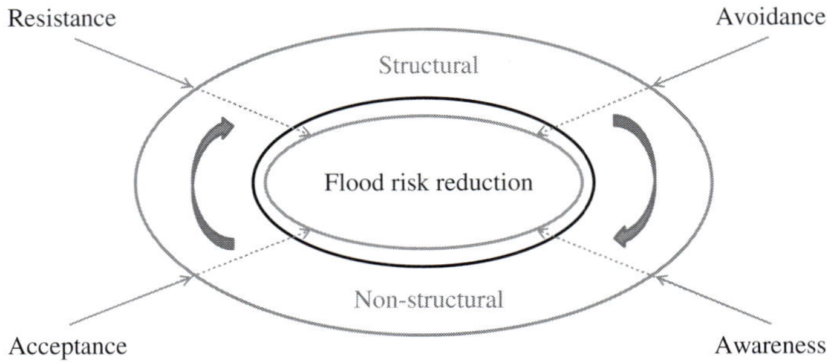


Fig. 11.1. Flood risk reduction framework

Current US flood risk reduction policy entails a range of mitigation techniques, often conceived and implemented in a haphazard manner across various scales and governmental jurisdictions. Experts and decision makers are usually divided between supporting a structural engineering or nonstructural land use planning approach to flood protection. Furthermore, ad-hoc strategies implemented as a reaction to specific storm events has resulted in a patchwork of flood defenses and development policies that ignore the broader systemic nature of flood risk nationwide.

Flood-resilient coastal communities will need to consider and adopt a range of mitigation strategies working synergistically over time. These range from drainage infrastructure to education and awareness programs. The specific portfolio of flood risk reduction techniques will, of course, depend on the composition and characteristics of each local jurisdiction, but will draw from a common set of approaches. In the following pages, we present a conceptual framework for understanding flood risk reduction strategies in the USA and helping localities pursue a more flood-resilient future. This framework rests on four categories of mitigation techniques: resistance, avoidance, acceptance, and awareness (Figure 11.1). Within each category, we review the major structural and nonstructural mitigation techniques available to decision makers at the local jurisdictional level. Specifically, we discuss the advantages and disadvantages of pursuing a specific strategy, under which conditions it might be most appropriate, and how various strategies may be combined to produce positive synergistic effects (Table 11.1). Finally, we outline major challenges facing local jurisdictions in the USA in implementing an integrated flood program that will reduce risk and associated losses in the future.

Table 11.1. *Flood risk reduction strategies*

| Mitigation strategy | Description | Saltwater | Freshwater |
|---|---|---|---|
| | ***Resistance*** | | |
| Dikes/Levees | Solid constructed walls that prevent elevated water levels from flooding interior lowlands. | X | X |
| Dams | Artificial barrier usually constructed across a stream channel to impound or store water. | | X |
| Flood gates/ Barriers | Adjustable gates that prevent storm surge from flooding coastal areas. | X | X |
| Breakwaters | Detached structures built parallel to the coast. | X | |
| Groins/Jetties | Typically short structures attached perpendicular to the shoreline, extending across at least part of the beach out into the surf zone. | X | |
| Bulkheads | Vertical retaining walls to hold or prevent soil from sliding seaward. | X | |
| Revetments | Armoring materials placed on an existing slope, embankment or seawall to protect the backside area from storm-driven waves. | X | |
| Artificial reefs | Construction of reefs in nearshore areas to reduce the impacts of storm surge and waves. | X | |
| Constructed dunes | Building or replacing dunes to protect communities from storm surge and wave action. | X | |
| | ***Avoidance*** | | |
| Freeboard/ Building elevation | Elevating structures above base flood to protect from inundation. | X | X |
| Fill | Elevating landscapes with compacted soil or dirt before construction of buildings to prevent inundation. | X | X |
| Buffers/Setbacks | A specific distance for which structures must be set back. | X | X |
| Clustering | Increasing the permissible development density in the least vulnerable areas within a specific property. | X | X |
| Density bonuses | Increasing development density and height requirements for specified parcels. | X | X |
| TDRs | Transfer of development rights from a vulnerable area to a less vulnerable or sensitive area. | X | X |
| Targeted public infrastructure | Invest in public utilities and other infrastructure in the least vulnerable areas. | X | X |

Table 11.1. (*cont.*)

| Mitigation strategy | Description | Saltwater | Freshwater |
|---|---|---|---|
| Acquisition | Purchase of some or all of property rights for open space protection for flood mitigation. | X | X |
| Relocation | Remove structures from a vulnerable location to a less vulnerable location | X | X |
| Drainage maintenance | Maintaining drainage devices (canals, ditches, storm drains, etc.) to ensure they operate effectively during a flood event. | | X |
| Protected areas/ Open space | Designating one or multiple parcels as protected open space for flood mitigation. | X | X |
| Local plans | Adoption of local planning instruments (e.g., floodplain plans, comprehensive plans, local mitigation strategies, etc.) that set forth a series of coordination policies aimed at mitigating flood impacts. | X | X |
| Low impact development | Development standards and techniques designed to work with ecological functions to manage storm-water as close to its source as possible (e.g., bioswales, rain gardens, permeable pavement). | | X |
| | ***Acceptance*** | | |
| Retention/ Detention | Either dry or wet holding areas/ponds that collected storm water | | X |
| Underground cisterns | Large storm-water holding areas underground. | X | |
| Breakaway walls | First story walls on elevated homes designed to break away during storm surges. | X | |
| Garage vents | Openings at the base of a garage that allow water to pass through the structure. | X | X |
| Protected open space | Designating protected open spaces or passive recreation sites for flood detention. | X | X |
| Constructed wetlands | Creating wetlands around structures or on vacant parcels. | | X |
| | ***Awareness*** | | |
| Flood risk information | Providing information about flood risks through multiple media outlets. | X | X |
| Education/ Training | Training through classes, workshops, certifications, etc. | X | X |
| Hazard disclosure | Disclosing a property's potential flood hazard to prospective buyers before the lender notifies them of the need for flood insurance. | X | X |

## 11.2 The Jurisdiction's Role in Institutional and Organizational Resilience

Discussions on the role of jurisdictional planning agencies in promoting socio-ecological resilience should begin with whether they are viewed as organizations or as institutions. Previous research highlight the characteristics of planning agencies as well as the quality of their plans in fostering resilience to natural hazards (Berke and Godschalk, 2009; Berke et al., 2012; Brody et al., 2010). However, additional discussions on whether planning agencies within jurisdictions promote organizational or institutional resilience are important.

Organizations have been defined from a "rational" perspective as being highly formalized and aimed at specific goals, and from an "open" perspective as consisting of multiple participants from varying backgrounds that function properly through informal linkages (Scott, 2008). Organizations can also be defined as "goal-directed, boundary-maintaining and socially constructed systems of human activity" (Aldrich and Ruef 2006: 4; Gupta et al., 2010). Examples of organizations in the context of these definitions include public agencies, donor groups, research institutes, or lobby groups (Garschagen, 2013). Goal setting, maintaining boundaries, and creating linkages are all characteristics of local planning agencies within jurisdictions responsible for flood reduction.

Institutions are defined as "systems of rules, decision-making procedures, and programs that give rise to social practices, assign roles to the participants in these practices, and guide interactions among the occupants of the relevant roles" (IDGEC, 1999: 14). This definition highlights the importance of rules and each participant within the institution understanding their roles through interactions. Institutions can be in the form of formal policies by government organizations and informal partners, which can be hidden or visible (Arts, 2006). Institutions also represent a system that is highly normative (Scott, 2008). Planning agencies have a system of rules guiding their policies and are primarily responsible for decision making in flood hazard mitigation within their local jurisdictions.

The question remains whether we view planning agencies as institutions or organizations. Although organizations represent a formal decision-making structure, they cannot be used interchangeably with institutions, which also have their own values and norms (Gupta et al., 2010; IDGEC, 1999; Young, 1989; Zijderveld, 2000). Hence, it is important to consider the linkages between institutions and organizations. The term "organizational institutionalism" describes these linkages (Garschagen, 2013; Greenwood et al, 2008).

Resilience theory also argues that the socio-ecological system itself is an institution trying to maintain stability (Gunderson and Holling, 2002; Holling, 1973). An institution is said to be resilient within the socio-ecological system if it can also reduce uncertainty, react to instability, increase institutional diversity, and manage continuity effectively (Herrfahrdt-Pähle and Pahl-Wostl, 2012; Young, 2010). In this chapter, we consider organizational resiliency as not simply how resilient the structure of the organizations responsible for flood reduction policies are, but as a broader institutionalized environment with the goal of building interactions and formulating policies that will reduce flood hazards. An examination of the proposed framework for flood risk reduction will highlight the role of the jurisdiction in flood risk reduction.

## 11.3 Establishing a Portfolio of Flood Mitigation Strategies

Communities genuinely interested in flood resilience must consider adopting programs where multiple techniques working synergistically to reduce flood losses. Comprehensive flood risk reduction lies at the intersection of resistance, avoidance, acceptance, and awareness. It is up to each community to decide their optimal portfolio of flood mitigation strategies based on specific local contextual characteristics and the amount of savings they want to accrue in the future. For example, a locality afflicted by storm-surge events may opt to construct a "dike in dune" barrier behind which new home construction requires freeboard of several feet. Alternatively, a community that experiences heavy rainfall episodes may buffer development along critical riparian areas, transfer these development rights to a nearby urban core, and at the same time protect naturally occurring wetlands that provide recreational opportunities to surrounding residents. Both examples could involve education and awareness programs to ensure residents are informed about potential flood risks.

The most notable program in the USA for encouraging localities to adopt synergistic flood strategies is FEMA's Community Rating System (CRS). The CRS was established in 1990 as a way to encourage communities to exceed the NFIP's minimum standards for floodplain management. Communities participating in the CRS adopt various flood mitigation measures in exchange for federal flood insurance premium discounts, ranging from 5 to 45 percent. The nonstructural orientation of the CRS program categorizes planning and management activities into the following four "series" containing 18 mitigation "activities": (1) public information (Series 300) activities informing residents about flood hazards, the availability of insurance, and household protection measures; (2) mapping and regulation activities (Series 400) containing both critical data needs and regulations that exceed NFIP minimum standards; (3) damage reduction (Series 500) activities focusing on reducing flood damage to existing buildings, and may entail acquiring, relocating, or retrofitting existing structures; and (4) flood preparedness (Series 600) activities implementing strategies associated with warning and response to minimize the adverse effects of floods (for more information on the CRS, see: http://training.fema.gov/EMIWeb/CRS/).

Credit points are assigned to participating CRS communities based on the degree to which different flood mitigation activities are implemented, but not all activities carry the same amount of points (FEMA, 2013). The total number of credit points obtained by a participating locality determines the size of insurance premium discounts. Credit points are aggregated into "classes," from 9 (lowest) to 1 (highest), where communities awarded a higher CRS class will have implemented a larger number and scope of flood mitigation measures. Insurance premium discounts range from 5 (class 9) to 45 percent (class 1), depending on the degree to which a community plans for the adverse impacts of floods. While the local jurisdiction takes responsibility for adopting and implementing each mitigation activity, the individual homeowner receives the discount on their national flood insurance premium.

Overall, the CRS program incentivizes local communities to focus on avoiding flood risk at the outset, increasing public awareness of this risk to encourage protective household behaviors, and preparing for the impact before a flood event occurs. By taking this approach, communities can become more proactive in addressing flood problems over time.

Observational research demonstrates the effectiveness of the CRS program in reducing property damage from floods. For example, two related cross-sectional studies examined all 67 counties in Florida and 37 coastal counties in Texas over a five-year period (1997–2001) to test the effect of the CRS program on observed flood loss. These studies found that communities in both states implementing mitigation activities under the CRS experienced significant reductions in flood damage, even when controlling for multiple natural environment, built environment, and socioeconomic contextual characteristics. Specifically, in Florida, a real-unit change in CRS class (moving in increments of 5 percent) equaled, on average, a US$303,525 decrease in the amount of damage per flood (Brody et al., 2007). Results from the coastal Texas study showed that from 1997 to 2001, a real-unit increase in CRS class translated into a US$38,989 reduction in the average property damage per flood event (Brody et al., 2008b).

## 11.4 Case Study: A Framework for Flood Risk Reduction Strategies

### *11.4.1 Resistance*

A major component of flood protection in the USA is to resist the intrusion of wave or rainfall-based floodwaters into human settlements. Resistance strategies most often involve structural measures, such as large-scale building and construction projects that actively protect communities situated in vulnerable areas. This "stand and fight" approach to flood risk reduction recognizes the importance of locating development in flood-prone areas for commerce, industrial production, recreation, and aesthetics. Initial flood mitigation efforts in the USA focused on large-scale structural projects, beginning with the Mississippi River flood in 1927 (Birkland et al., 2003). The Flood Control Act of 1930 dedicated federal funds to build structural flood control works, many of which are still in use today. As shown in Table 11.1, there exist multiple structural techniques aimed at resisting floodwaters.

A major aspect of structurally based resistance activities involves "armoring" the coastline. Typically, coastal armoring is considered a last resort and used where substantial human investments are at risk making it necessary to protect the upper portion of a beach profile from storm-induced erosion and flooding (Pope, 1997). One of the most popular options is the construction of dikes or levees that restrict rising waters in the "channel phase" of flooding (Alexander, 1993). These structures often consist of solid concrete walls, either above or below ground that prevent elevated water levels from flooding interior lowlands. Dikes are usually associated with eliminating wave-based flooding caused by storm surges along the coast. Levees are located along stream and river channels to prevent flooding from precipitation-based runoff or storm-surge that travels upstream along the floodplain. These structures are best used where there is existing heavy development or location of critical facilities, such as oil and gas production or power plants. Dikes and levees enable development to occur in flood-prone areas, particularly in the short term. However, these structures are often expensive, politically contentious, and can have adverse environmental impacts. Moreover, if they are not monitored and maintained properly, they can fail, causing catastrophic damage.

The construction of movable floodgates or storm barriers is increasingly being used as a resistance strategy to prevent widespread inundation of coastal communities. These structures consist of adjustable gates designed to prevent a storm surge or high tide from flooding the protected area behind the barrier. A surge barrier is usually integrated into a larger flood-protection system consisting of floodwalls, levees/dikes, and other mitigation strategies. Modern-day barriers situated at the mouth of estuaries or river outflows allow water to pass under natural conditions until the structure is closed due to an impending storm. Storm surge barriers and floodgates are most notable in the Netherlands (Jonkman et al., 2008), but have also been implemented in St. Petersburg (Russia), England, and New Orleans (USA), and in the near future Venice (Italy). Many of these structures are praised for their ability to eliminate the threat of storm surge-based flooding without significant environmental impacts, but can be difficult to implement due to time, expense, and engineering capabilities.

Revetments consist of erosion-resistant materials placed on an existing slope, embankment or seawall to protect the backside area from storm-driven waves. This structure is comprised of three components: (1) a stable armor layer, (2) a filter cloth or under-layer, and (3) toe protection. The filter and under-layer support the armor while allowing for the passage of water through the structure. Toe protection prevents undercutting and provides support for all the layer materials (USACE, 2008). Different types of materials are used to absorb wave action, including geotextiles, sandbags, concrete tetrapods, rock, or wood.

Revetments are generally low-cost coastal flood mitigation techniques that complement other structural approaches. For example, earthen dikes with stone revetments have been constructed to protect major petrochemical facilities in Texas City, Texas from the adverse effects of coastal storms and high tides. While revetments are ubiquitous for coastal mitigation across the USA, they are also prone to failure. For example, if the toe fails, the entire revetment can unravel. Also, armor layer damage can result in interior exposure and erosion. Finally, overtopping and loss of foundation material can negate the effectiveness of this armoring approach.

A dam is another ubiquitous structural flood mitigation measure consisting of an artificial barrier usually constructed across a stream channel to impound or store water. These structures are built from timber, rock, concrete, earth, steel or a combination of these materials. Spillway systems are constructed to convey normal stream and flood flows over, around, or through the dam. Over 80,000 dams mark the US-American landscape, many serving multiple functions, such as recreation, water supply, and power generation (Graf, 2001). These mitigation structures are very effective in modulating stream flows and shielding downstream coastal communities from flooding, but can result in reduced nutrient and sediment transport, channel obstruction, loss of floodplain area, and overall hydrologic fragmentation (Nilsson and Berggren, 2000). Also, failure of the dam structure can generate a sudden catastrophic pulse of water and debris that inundates downstream areas (as has happened periodically since the 1800s).

Another notable armoring technique is the construction of bulkheads, which are vertical retaining walls that hold or prevent soil from sliding seaward. However, their primary function is to reduce erosion and prevent shore-side loss of land, not to mitigate coastal flooding and wave damage. Bulkheads are either cantilevered or anchored sheet piles or gravity structures such as rock-filled timber cribs and gabions. The ability of a bulkhead to protect against wave attack stems almost solely by the adequacy of fill, and if this material is lost, the bulkhead is rendered ineffective (USACE, 2008). The two critical elements of a well-designed bulkhead are return walls at the alongshore ends of the structure to prevent high water from washing away material behind the structure, and geotextiles to allow some water to flow through the structure. Nevertheless, loss of the backfill at a bulkhead from rain and/or wave overtopping is a common cause of mechanism failure.

A second set of structural flood resistance activities involves “moderating” the impacts of coastal storms. Beach erosion control techniques are designed to reduce the rate of sediment loss and can moderate erosion damages (Pope, 1997). For example, near-shore breakwaters consist of detached structures built parallel to the coast that act to disperse or redirect wave energy and reduce impacts along the shoreline. The reduction in wave energy slows the littoral drift, produces sediment deposition and a shoreline bulge in the sheltered area behind the breakwater (USACE, 2008). In addition to protecting inland communities from storm-surge damage, breakwaters can help maintain beach width for recreational purposes, reduce erosion, and stabilize wetland systems.

Breakwaters have been used extensively in the USA, primarily along sediment-starved shores with fetch-limited wave climates (Chasten et al., 1993; Pope and Dean, 1986). These structures are typically constructed in high wave-energy environments using large armor stone or precast concrete units or blocks. In lower wave-energy environments, grout-filled fabric bags, gabions, and other proprietary units have been used. While breakwaters can reduce the adverse impacts of storm-driven waves on coastal communities, there are several issues that must be considered when constructing these structures for flood protection. First, natural sediment transport can be altered. Breakwaters can accelerate the littoral current up-drift, causing initial erosion of the beach on the up-drift side. The same occurs in the area immediately down-drift. Second, wave deflection can exacerbate impacts on unprotected areas. Third, a significant forerunner can raise water levels above the breakwater structure before the actual storm surge arrives, negating their effectiveness (Oumeraci, 1994).

Groins are typically short stabilization structures attached perpendicular to the shoreline. They extend across at least part of the beach and out into the surf zone. Once installed, accretion over time causes a positive increase in beach width up-drift of the groin. Groins are constructed to maintain a minimum beach width to buffer the impacts of wave-based storms. They are particularly useful in areas experiencing chronic erosion problems from diminished sediment supply and where it may be too expensive to maintain a functional beach by continuing to bring in large quantities of sand from outside sources (Pope, 1997).

According to the USACE (2008: V-3-59), groins are "the most misused and improperly designed of all coastal structures." Trapping sediments often causes adverse impacts to adjacent beaches if sand held in the project area would normally move through or accrete on the neighboring beaches. Groins can therefore produce erosion and a decrease in beach width on the down-drift side of the structure. Coastal zone management policy in the USA and other countries discourages the use of groins for shore protection due to a history of poorly designed and improperly sited structures stemming from a lack of understanding of their functional design, failure to implement the correct construction sequence, or improper cross-sectional shape. However, if properly installed, groins can help establish and maintain dry beaches that serve to protect coastal communities from storm events.

A third category of structural flood mitigation techniques aimed at resisting the impacts of storms is the construction of natural features. This approach receives much less attention among engineers and flood managers, but is increasing in popularity based on their ability to replicate ecosystem services found naturally. For example, artificial reefs can be established using a variety of materials to mimic the functions of natural formations (coral, oyster, or worms) in high wave energy environments. There are no general design rules for constructing artificial reefs, particularly for storm protection, but case studies are beginning to emerge in the scientific literature (Pattiaratchi, 1997). Construction or restoration of dune systems is also gaining favor and recently has been proposed as a way to protect communities in New Jersey impacted by Hurricane Sandy in 2012. Dunes can be constructed by themselves entirely of sand, or with a solid core to leverage the benefits of a dike. These techniques provide ecological and recreational benefits, are aesthetically preferable, and maintain the coastline in a more natural state. However, the feasibility of dunes is dependent on a convenient source of sand for renourishment, the success of establishing grasses/vegetation to hold the sand in place, and a commitment to maintain these features over the long term.

While structural flood mitigation projects first gained traction in the USA during the 1930s, this approach has returned in favor with recent large-scale projects involving seawalls, gates, and pumping stations to protect communities from storm-surge flooding. For example, New Orleans, Louisiana, recently completed a complex flood control system totaling over US$14.5 billion as a response to the damage caused by Hurricane Katrina in 2005. Discussions are also under way to erect surge suppression systems for coastal Texas and New York to prevent the reoccurrence of recent storm-surge based events in major metropolitan areas.

At the same time, however, it is important to understand the limitations inherent in structural approaches to flood management. These structures are generally very costly, involve long implementation timeframes, can impose negative impacts on the natural environment, and can actually exacerbate the adverse impacts of flood events. For example, when flood events exceed the capacity of a flood control structure, the resulting flood damages can be significantly higher than if the area had been unprotected and thus less populated (Larson and Pasencia, 2001; Stein et al., 2000). Also, structures such as levees can raise the normal level of a river and increase the velocity of water pulsing downstream. By constricting a waterway and hardening its banks, these structures can actually increase the probability of downstream flooding (Birkland et al., 2003).

### *11.4.2 Avoidance*

A second category of mitigation techniques that local jurisdictions can adopt to reduce the adverse impacts of floods over time focuses on avoidance. An avoidance approach to flood risk reduction entails removing development or steering it away from the most vulnerable areas, such as the 1-in-100-year floodplain or category 1 surge zone. While wholesale retreat from the coast is not politically or economically viable, the idea of avoiding specific areas that experience chronic inundation from either precipitation or tidal events is gaining acceptance across the USA. Avoidance can be a vertical proposition, where structures and people are lifted above areas of inundation. Or, the approach can involve horizontal movement, where critical assets are either pulled back or prevented from being placed in flood-prone areas from the start.

The most prominent structural technique to vertically avoid floodwaters is the elevation of buildings on pilings or some other support structure.

The height of structural elevation is usually expressed as freeboard, which is the number of feet the first floor of a building is raised above the 1-in-100-year level of inundation (Base Flood Elevation, or BFE). Freeboard is considered a factor of safety that buffers the many unknown factors that could contribute to flood heights greater than the height calculated for a selected size flood and floodway condition (www.fema.gov). While there is no federal standard, local jurisdictions can adopt specific requirements in their building codes. For example, Brazoria County located in the Gulf Coast region of Texas adopted a two-foot freeboard requirement for all new construction in the floodplain in 2007 (see FEMA, 2015). FEMA recommends at least one foot of freeboard for structures in the 1-in-100-year floodplain. Freeboard requirements can be costly to developers and homeowners and difficult if the structure is a "slab on grade" design. However, the up-front costs can be offset though avoided flood losses over time. A comprehensive analysis of Galveston County, Texas, showed that in a scenario of incremental elevation of all feasible structures over a ten-year period, the financial benefits of avoiding flood damage would start accruing from the seventh year onward (Shakra, 2016).

Another form of vertical avoidance is using fill to raise a structure above BFE. Fill, as an avoidance strategy, allows naturally occurring landscapes to be altered by increasing its elevation with dirt or sand. This technique is done for either individual or a series of parcels in a new subdivision before a building is put in place. A major advantage of fill is that it can be combined with additional avoidance measures, such as crawlspace foundation on compacted fill material, thereby providing a higher level of flood protection. Fill has been widely used in the USA; dating back to the 1980s, about 50 percent of development in floodplains are either filled or elevated (Bollens et al., 1988). Parcel fill is becoming increasingly used by developers, especially in sprawling urban areas where the pressure to build in floodplains is high. For example, between 1999 and 2014 in the Houston metropolitan area, almost 7,000 parcels have been filled or approved for fill. Despite its perceived advantages however, fill is not permitted along the floodway or in areas prone to high velocity storm surge events due to the risk of foundation failure (FEMA, 2001). There are also concerns that filling floodplains will increase the BFE, thereby increasing flood risk for adjacent areas that have not been filled (FEMA, 2013). These impacts can be reduced by making provisions for compensatory storage through construction of detention ponds in filled subdivisions.

Most flood avoidance strategies involve nonstructural mitigation techniques that fall squarely into the domain of land use planning at the local level.

These policies focus on horizontal avoidance in that they are meant to guide or pull development away from vulnerable areas, such as coastlines, floodplains, or river bottoms (Beatley, 2009). Strategies under this category of mitigation include both regulatory and incentive-based policies that can help facilitate more flood resilient development patterns over the long term (Brody et al., 2011).

Regulatory techniques involve planning-based policies, such as human-use restrictions or buffers around critical areas where development should not take place. Buffers or setbacks designated in local development codes are one of the most effective methods for achieving horizontal avoidance because they can literally peel back development along riparian areas, sensitive tidal surge zones, and other vulnerable locations. These tools usually already exist in local plans for protecting critical natural habitats and can easily be expanded or modified to cover flood mitigation as well. Buffer distances range widely across the USA, from 50 to over 1,000 feet, depending on the specific locational characteristics being considered.

Another common regulatory approach to horizontal avoidance is protecting one or multiple parcels through land acquisition techniques, such as fee-simple or development rights purchases. With this strategy, all or part of a parcel is purchased to remove or prevent development occurring in extremely vulnerable areas. Conservation easements are another type of land acquisition technique in which the property owner maintains possession, but parts of the parcel are restricted to development in exchange for tax benefits (Beatley et al., 1994). Acquiring a portion or partial development rights can be an effective way to more affordably prevent structures from being built in flood-prone areas than the wholesale purchase of the land. This approach also enables the protection of specific pockets of land versus removing the entire parcel from the tax rolls, such as is usually the case with protected areas and park designations. Buildings already in place can be relocated to higher ground as a part of the acquisition process. Finally, land acquisition for flood risk reduction that overlaps with critical natural resources and recreational/tourism opportunities can add value over the long run.

Incentive-based mitigation policies create a pull towards the least vulnerable areas for development rather than the regulatory push as just discussed. These techniques often allow for increased development intensity in desirable areas in exchange for relieving development on chronically flooded parcels. For example, clustering provisions that allow homes to be concentrated in a specific portion of a land parcel can help avoid flood loss. Density or building height bonuses in less vulnerable or more structurally protected areas within a community can help focus new development in a flood resilient manner. Also, transferring development rights from the floodplain to receiving areas in higher elevations provides developers incentives to locate structures where it is less likely to flood. Finally, targeting public infrastructure, such as sewer and water lines in the least flood-prone areas within a community will also encourage safer development over the long term. Proactive planning measures that focus development either outside of the 1-in-100-year floodplain or away from flood-prone watercourses are most effective at minimizing flood damage in low-lying coastal communities experiencing rapid growth.

### *11.4.3 Acceptance*

A third category of mitigation techniques within the flood risk reduction framework shown in Figure 11.1 involves the acceptance of floodwaters. In this instance, community-based strategies allow or even encourage flooding in specific areas or under certain conditions. The idea is that communities can coexist with periodic inundation and even provide a relief valve when there is an excessive buildup of storm water runoff.

The most commonly used acceptance mitigation strategy is the placement of retention or detention ponds that collect, hold, and slowly release storm water. Retention areas are dry except during times of flooding, while detention ponds are filled with water year-round (but both serve the same purpose). Retention areas are usually vegetated depressions strategically hidden behind houses or used as playing fields during dry periods. In contrast, detention ponds are often placed in full view as community amenities where residents can walk, fish, or relax by a water feature. While detention ponds can add more value to a community, greater attention must be paid to maintaining proper pond/lake levels during heavy rains. Both strategies can consume large areas that could otherwise be developed and rely on a well maintained neighborhood drainage system to function properly. Retention and detention ponds are most effective in well-planned communities where they can be strategically placed for maximum effect. A lesser used technique is the creation of underground cisterns for holding flood waters from surface runoff. This option serves the same purpose as retention/detention, but without consuming so much developable land. However, cisterns can be expensive to build and require viable soil substrates to properly maintain their function.

Protecting or constructing wetlands for flood risk reduction is another promising mitigation technique. Wetlands function as natural retention/detention devices by collecting, holding, and slowly releasing floodwaters.

In fact, naturally occurring wetlands have been shown to significantly reduce freshwater flooding and flood losses. For example, Brody et al. (2007) found that for all counties in Florida, the alteration of wetlands increased the average property damage per flood at the county level by over US$400,000. Based on this rate of change, wetland development cost the state over US$30 million per year in flood losses. Both natural and constructed wetlands can be coupled with ecological, aesthetic, and recreational values. Even if no functional wetlands are present on site, parcels designated as parkland or for passive recreation can also be designed to hold floodwaters.

A final type of acceptance-based flood mitigation strategy involves the design of buildings. Building codes can require the construction of home or businesses to allow inundation by floodwaters without compromising the integrity of the structure. For example, coastal structures elevated above the BFE can also be required to have breakaway walls. With this technique, storm surge can flow under the first floor without destroying the entire building. Similarly, vents along the lower walls of garages can accept floodwater from tidal events without resulting in a total loss. However, if valuables such as automobiles are stored on the bottom floor or in garages, significant damage can still occur during a surge event.

### *11.4.4 Awareness*

A fourth category within the flood risk reduction framework is awareness. When residents understand the risks of flooding and how best to mitigate the adverse effects of storm events on their property, overall losses at the community level can be significantly reduced. Outreach projects that educate residents about the probability of inundation in and around the 1-in-100-year floodplain will help them make more informed decisions when purchasing new homes. Information about the various options available to mitigate flood impacts to households, from purchasing insurance to dry-proofing basements, will help residents protect their investments.

Community outreach projects can be implemented through multiple media outlets, including written materials, web-content, and in-person workshops or training sessions. Another important strategy to increase awareness of flood risk is the promotion of hazard disclosure during the real estate transaction process. Communities can work with bankers, realtors, and other parties to make potential flood hazards more transparent to buyers at the beginning of the purchase process. Disclosure of whether a home is within or close to a 1-in-100-year floodplain or if the structure has been inundated in the past will help home buyers make more informed decisions. It is essential that prospective buyers understand the risk they will incur before purchasing a particular property.

## 11.5 Conclusion

In this chapter, we propose a framework for flood risk reduction based on the four dimensions of resistance, avoidance, acceptance, and awareness. There is no single approach or technique that can eliminate flood impacts within local jurisdictions. Instead, mitigation strategies within each dimension must work synergistically under an overall program to effectively reduce flood risk and foster the development of resilient communities over the long run. The exact mix or portfolio of mitigation activities will depend on the specific characteristics and issues associated with each locality. Whatever strategies are implemented, it is critical that decision makers think beyond normal political cycles and consider changing physical, socioeconomic and built environment conditions over time. This approach involves addressing small impacts that accumulate over years as well as the big events that provoke sudden shifts in the system. In the same way, the effectiveness of multiple mitigation strategies should be evaluated over large spatial and temporal scales. The value of these interventions must be measured cumulatively and for long-range flood probabilities.

While the need for effective flood risk reduction programs has never been more apparent, communities face many challenges when adopting and implementing mitigation strategies. Because flood resistance strategies are expensive, jurisdictions often depend on federal funding and grants for its implementation. The vast majority of funding for coastal flood-related issues is provided by the federal government, only after a disaster occurs through emergency supplemental appropriations (NRC, 2014). A recovery-based approach predicated upon federal payments is, at its essence, an acceptance of failure when it comes to avoiding adverse impacts from floods. A combination of payouts from large hazard events and the lack of actuarially sound insurance rates have forced FEMA to regularly borrow money from the federal treasury to cover its deficit. As of August, 2013, the NFIP has borrowed approximately US$24 billion (Kousky and Kunreuther, 2013). In contrast, a protected-based strategy focuses on mitigating flood risk before an event takes place, or eliminating it altogether. This more proactive approach to risk reduction assumes that residents should never bear the burden of inundation and associated loss, regardless of where they are located within coastal landscapes. Such an approach favors both systems-based structural interventions and land use planning techniques that seek to remove or avoid structures from areas most at risk.

Local jurisdictions also face the costs of maintaining and updating flood infrastructure to meet current flood risk standards. In many cases, local flood control districts play significant roles in recommending resistance strategies, and are sometimes charged with the responsibility of building flood infrastructure. Hence, it is important for such agencies to collaborate with jurisdictions and implement structural measures consistent with local land use plans and future development initiatives, and prevent residual impacts due to new flood infrastructure. For the most part, decision making for floods in the USA occurs within towns and cities leaving fragmented governance structures across watersheds – the ideal unit of analysis when dealing with floods. As a consequence, there is often a lack of policy consistency among jurisdictions sharing the same river body. Fragmented and ad hoc interventions can impose adverse impacts on neighboring communities, weaken the overall functionality of these systems over the long term, and lead to more flood damage overall.

Avoidance-based policies such as clustering development, transfer of development rights, density bonuses, etc., are mostly implemented through comprehensive plans adopted by local planning agencies within a jurisdiction. Therefore, the capacity of a jurisdiction's planning organization is critical in determining how effectively land use policies will be formulated and implemented over time. Capacity-related characteristics include budgetary resources, the amount of staff and expertise available, knowledge accumulated through past experiences, and the level of commitment and leadership associated with flood mitigation (Brody et al., 2010). Planning agency capacity is an important factor that determines the effective management of ecosystems and developing policies that are effective for future development within vulnerable areas (Brody et al., 2004; Burby and May, 1998). Plan quality is also linked to ecosystem protection, including the ability to work across multiple jurisdictions in a hydrologic region such as a watershed (Berke et al., 2013).

Local jurisdictions may have the capacity to implement flood reduction strategies, but can be limited by lack of incentives to pursue a flood mitigation program. The CRS provides an incentive for implementing nonstructural flood mitigation strategies by reducing federal flood insurance premiums. However, jurisdictions are still faced with the challenge of examining the cost-benefits of implementing avoidance and acceptance strategies, such as avoiding floodplain development, providing floodable open spaces in comparison to revenue that could be accrued from taxing developments that could otherwise occur in these places.

Reducing flood risk can involve expensive and technically complex solutions that can overwhelm the capability of small communities. Moreover, localities often must deal with other seemingly more pressing issues (e.g., economic, educational, transportation, etc.) that take precedence over what is perceived to be a rare, naturally occurring event. On the other hand, when jurisdictions are proactive and implement avoidance strategies, there are net benefits such as a reduction of cost of repair to infrastructure from flood events. This approach is far preferable to chronic and repetitive costs typical of a recovery-based system for coping with floods. Generally, the return on investment is higher over the long term when investment in the most effective mitigation measures is made at the outset.

Building awareness of flood risks and potential mitigation techniques is often a burden to local jurisdictions strapped for financial resources. In response, many localities have formed partnerships with nongovernmental organizations and other institutions as part of their social responsibility to inform the public. These partnerships often focus on public outreach, disclosure of flood conditions during the home-buying process, and technical assistance to residents. Educational institutions such as universities can also serve as powerful partners for building awareness of flood risk reduction by creating technical tools and web interphases, such as the Galveston Bay Coastal Atlas focusing on the region around Houston, Texas (see http://newcoastalatlas.tamug.edu/atlaswebpage/tricounty/tricountyatlas.html). Tools like this allow residents with Internet access to better understand the risk of flooding to their homes and pinpoint strategies that will reduce this risk in the future.

A summary of the strategies in our flood resiliency framework and challenges faced by local jurisdictions in implementing these strategies are as follows:

- Flood resistance: Flood resistance strategies are implemented through construction of large-scaled flood infrastructure that protects vulnerable communities. An example is coastline armoring involving the use of dikes, levees, moveable floodgates, revetments, dams, or bulkheads along coastlines as a last resort in reducing coastal storm surge impact. They also occur through constructing "moderate" coastal storm infrastructure such as breakwaters and groins. Construction of dunes and artificial reefs are also examples of "natural" features for flood resistance.
- Flood avoidance: This strategy requires staying away from flood prone areas either vertically or horizontally. Vertical avoidance occur through freeboard and fill, while horizontal avoidance involve parcel/land acquisition

techniques, clustering development outside floodplains, transfer of development rights, and restricting development in floodplains.

- Flood acceptance: This strategy acknowledges the reality that flooding will occur, thereby making provision for floodwaters. This is achieved through construction of retention ponds to provide compensatory storage, constructing artificial wetlands, protecting naturally existing wetlands, or constructing buildings that allow inundation to occur without damaging the physical structure.
- Flood awareness: Local jurisdictions can use multiple outlets of public outreach programs to educate residents of their flood risk. This strategy prompts the response and interest of these residents to available local mitigation strategies.
- Flood loss reduction portfolio: The intersection of resistance, avoidance, acceptance, and awareness strategies forms a portfolio of comprehensive flood risk reduction that can be applied within specific local contextual conditions.
- Challenges to flood resiliency: Challenges faced by local jurisdictions include the recovery-based approach to flood mitigation in the USA, high cost of maintaining and updating flood infrastructure, planning agency capacity, and inadequate incentives in applying flood mitigation strategies.

# 12

# Linkages between Vulnerability and Resilience

SUSAN L. CUTTER

Vulnerability and resilience are well known and often discussed concepts in disaster research. Depending on the disciplinary, methodological, or philosophical perspective, these two concepts have very distinct and contested meanings and research traditions as previous chapters have illustrated. Building on those contributions, this chapter takes a broader view of the evolution of hazard vulnerability and resilience concepts and their application to public policy and practice at local, national, and global scales.

## 12.1 Divergent and Convergent Conceptualizations

Early on, the disaster research community had divergent paths in the study of vulnerability and resilience (Cutter et al., 2008). Initially popularized by the social sciences, vulnerability examined the potential for harm, especially to people and things they valued as well as the differential susceptibility to harm by various groups of people, their assets, and environments. In contrast, resilience was promulgated by the ecological sciences in their approach to understanding disturbances in systems and the ability of such systems to absorb "shocks," recover from them, and return back to some steady-state condition.

As the understanding of human-environmental linkages became more sophisticated, interdisciplinary framings of socio-ecological systems, coupled human-natural systems, complex adaptive systems, and later climate change adaptation and sustainability science ensued. Each in their own way added conceptual confusion to vulnerability and resilience, despite efforts at clarification (Folke, 2006; Janssen et al., 2006; Klein et al., 2003; Manyena, 2006; Miller et al., 2010; Smit and Wandel, 2006; Vogel et al., 2007).

The lack of clarity is best seen in the early work of Folke et al. (2002) who define resilience as the flip side of vulnerability; yet also list it as one of the determinants (along with exposure and sensitivity) of vulnerability. As pointed out by Klein et al. (2003), such reasoning is circular – a system is vulnerable because it lacks resilience and it is not resilient because it is vulnerable. In other conceptualizations (most notably Cutter et al., 2008) vulnerability and resilience are seen as complementary to one another – some limited degree of overlap as an inherent property of communities, yet no overlap whatsoever in a post-event context. Kelman et al. (2016) also agree with this view – resilience and vulnerability contain some common characteristics, but each is a separate, distinguishable concept, not an exact opposite of one another.

The increasing prominence of the term adaptive capacity especially in socio-ecological systems (exemplified by climate change adaptation and sustainability science researchers) has led to the increased ambiguity and the need to clarify the terminology used in the varying research traditions. This is analogous to the usage of the term mitigation – it means one thing to the climate change science community (e.g., reduction in the sources or actions to enhance the sinks of greenhouse gas emissions, IPCC, 2014b) and something quite different within the disaster community (e.g., the lessening or minimizing of adverse impacts of a hazardous event (UNISDR, 2017). Table 12.1 provides a cursory overview of the definitional differences between the two intellectual traditions as viewed from the main policy documents for each community.

The lack of definitional clarity provides fodder for younger scholars who themselves are grappling with understanding the relationship between vulnerability and resilience (among other concepts) in managing disaster and climate risks (see Kelman et al., 2016; Lei et al., 2014; Matyas and Pelling, 2012). Yet, this ambiguity also has helped to open a wider policy window for resilience – a term that is increasingly used to connect science to policy and practice – in different venues from systems engineering to humanitarian aid to planning. The positive tone of resilience – taking actions to reduce impacts – helps explain its widespread acceptance in policy and practice, especially when contrasted to vulnerability and its presumed negative connotation. On the other hand, the difficulties in operationalizing the definitions of vulnerability and resilience have led to a wide array and often competing approaches in measurement – the basic foundation for monitoring baselines and progress towards achieving the disaster policy goals of reducing vulnerability and building resilience.

Table 12.1. *Different definitions of key concepts*

| Term | Hazard/disaster context[1] | Climate change adaptation/sustainability science context[2] |
|---|---|---|
| Resilience | "The ability of a system, community or society exposed to hazards to resist, absorb, accommodate, adapt to, transform and recover from the effects of a hazard in a timely and efficient manner, including through the preservation and restoration of its essential basic structures and functions through risk management." | "The capacity of social, economic and environmental systems to cope with a hazardous event or trend or disturbance, responding or reorganizing in ways that maintain their essential function, identify and structure, while also maintaining the capacity for adaptation, learning and transformation" (p. 127). |
| Vulnerability | "The conditions determined by physical, social, economic and environmental factors or processes which increase the susceptibility of an individual, a community, assets or systems to the impacts of hazards." | "The propensity or predisposition to be adversely affected. Vulnerability encompasses a variety of concepts and elements including sensitivity or susceptibility to harm and lack of capacity to cope and adapt" (p. 128). |
| Mitigation | "The lessening or minimizing of the adverse impacts of a hazardous event." | "A human intervention to reduce the sources or enhance the sinks of greenhouse gases (GHGs). This report also assesses human interventions to reduce the sources of other substances which may contribute directly or indirectly to limiting climate change, including, for example, the reduction of particulate matter emissions that can directly alter the radiation balance (e.g., black carbon) or measures that control emissions of carbon monoxide, nitrogen oxides, Volatile Organic Compounds and other pollutants that can alter the concentration of tropospheric ozone which has an indirect effect on the climate" (p. 125). |
| Adaptive Capacity | Only defines capacity: "The combination of all the strengths, attributes and resources available within an organization, community or society to manage and reduce disaster risks and strengthen resilience." | "The ability of systems, institutions, humans and other organisms to adjust to potential damage, to take advantage of opportunities, or to respond to consequences" (p. 118). |
| Coping Capacity | "Coping capacity is the ability of people, organizations and systems, using available skills and resources, to manage adverse conditions, risk or disasters. The capacity to cope requires continuing awareness, resources and good management, both in normal times as well as during disasters or adverse conditions. Coping capacities contribute to the reduction of disaster risks." | Not defined |

Table 12.1. (*cont.*)

| Term | Hazard/disaster context[1] | Climate change adaptation/sustainability science context[2] |
|---|---|---|
| Disaster Risk | "The potential loss of life, injury, or destroyed or damaged assets which could occur to a system, society or a community in a specific period of time, determined probabilistically as a function of hazard, exposure, vulnerability and capacity." | Only defines disaster: "Severe alterations in the normal functioning of a community or a society due to hazardous physical events interacting with vulnerable social conditions, leading to widespread adverse human, material, economic or environmental effects that require immediate emergency response to satisfy critical human needs and that may require external support for recovery" (p. 122).<br>Risk: "The potential for consequences where something of value is at stake and where the outcome is uncertain, recognizing the diversity of values. Risk is often represented as probability or likelihood of occurrence of hazardous events or trends multiplied by the impacts if these events or trends occur. In this report, the term risk is often used to refer to the potential, when the outcome is uncertain, for adverse consequences on lives, livelihoods, health, ecosystems and species, economic, social and cultural assets, services (including environmental services) and infrastructure" (p. 127). |
| Hazard | "A process, phenomenon or human activity that may cause loss of life, injury or other health impacts, property damage, social and economic disruption or environmental degradation … Several hazards are socionatural, in that they are associated with a combination of natural and anthropogenic factors, including environmental degradation and climate change." | "The potential occurrence of a natural or human-induced physical event or trend or physical impact that may cause loss of life, injury, or other health impacts, as well as damage and loss to property, infrastructure, livelihoods, service provision, ecosystems and environmental resources. In this report, the term hazard usually refers to climate-related physical events or trends or their physical impacts" (p. 124). |
| Risk Management | "Disaster risk management is the application of disaster risk reduction policies and strategies to prevent new disaster risk, reduce existing disaster risk and manage residual risk, contributing to the strengthening of resilience and reduction of disaster losses." | The plans, actions or policies to reduce the likelihood and/or consequences of risks or to respond to consequences" (p. 127). |

[1] UNISDR (2017)
[2] IPCC (2014)

## 12.2 The Emergence of Resilience as an Alluring Concept in Disaster Risk Reduction[1]

Resilience has been a familiar concept in ecology and emergency management for decades, but has rapidly emerged in the last twenty years as a concept that embodies not only the need to be prepared for emergencies, but also to be proactive in taking actions to reduce impacts and to find positive opportunities in response. The international framing for disaster resilience began with the International Decade for Natural Disaster Reduction (IDNDR; 1990–99). The UN General Assembly declaration promoted the broad commitment to disaster reduction through a focus on disaster relief, preparedness, and prevention. In the early part of the decade, the focal point was early disaster warning systems as the means for reducing the consequences of natural hazards. However, as the decade progressed, it became clear that a broader awareness of both the economic and social drivers of disaster risk and their adverse consequences was equally important. While no mention of resilience was made at this time, there was a clear emphasis on vulnerability. For example, the 1994 World Conference on Natural Disaster Reduction in Yokohama and its Strategy and Plan of Action recognized the role of vulnerability in the disaster risk and consequences calculus:

> Whilst the natural phenomena causing disasters are in most cases beyond human control, vulnerability is generally a result of human activity. Therefore, society must recognize and strengthen traditional methods and explore new ways to live with such risk, and take urgent actions to prevent as well as to reduce the effects of such disasters
>
> *(UNISDR, 2004: 9).*

At the end of the decade, the UN General Assembly formalized a secretariat for the International Strategy for Disaster Reduction (ISDR) to promote disaster reduction by emphasizing the physical and social dimensions of vulnerability in the assessment and management of disaster risks, as well as the integration of disaster risk reduction into sustainable development policies and practices. The latter was codified in the 2002 World Summit on Sustainable Development (WSSD) meeting in Johannesburg where reducing disaster risk and vulnerability were included as main targets for sustainable development planning and poverty reduction initiatives. In the policy world at this time, vulnerability was generally thought of as the flip side of resilience and vice versa.

Disaster resilience slowly became the global mechanism for integrating disaster risk reduction and sustainable development (Cutter, 2014). As clearly recognized on the heels of the conclusion of the IDNDR the purpose for implementing disaster reduction policies and measures is "to enable societies to be resilient to natural hazards and ensuring that development efforts do not increase vulnerability to those hazards" (UNISDR, 2004: 8). This was the first formal commitment to the principle of resilience and its linkage with vulnerability with resilience defined as:

> The capacity of a system, community, or society potentially exposed to hazards to adapt, by resisting or any changing in order to reach and maintain an acceptable level of functioning and structure. This is determined by the degree to which the social system is capable of organizing itself to increase its capacity for learning from past disasters for better future protection and to improve risk reduction measures
> *(UNISDR 2004: 16–7).*

Interestingly, the present definition of resilience (Table 12.1), has not dramatically changed within UNISDR since the original publication in 2004 (quote above). What has changed is the expansion of the definition to include the role of risk management and some indicator of a temporal dimension such as returning to normal in a timely fashion.

The subtle shift from disaster vulnerability reduction to enhancing resilience came with the international agreement on the Hyogo Framework for Action (HFA) in Kobe, Japan in 2005. It was at this time that two important reorientations took place. The first was the shift from using the term "natural disasters" to simply disasters, or in some cases natural hazards. As noted by Brisceño (2015), the term natural disaster conveys that disasters are natural with little that can be done other than preparing and responding to them, in other words, disasters are acts of God or some other deity. Disasters, on the other hand are human-centered constructs, involving the interaction between natural processes and human settlements and social processes. Disasters are acts of people and you cannot have a disaster without this interactive exchange. Disasters can be managed in multiple ways by focusing on risk reduction (including vulnerability reduction) and risk management (including improving resilience) policies and practices. The formal acceptance of the new terminology (and the de facto philosophy behind it), was a major policy accomplishment of the HFA.

The second shift was the increasing primacy of the concept of resilience. Viewed as pathway for risk reduction for a safer future, the HFA featured resilience in the subtitle of its major report, "Building the Resilience of Nations and Communities to Disasters" (UNISDR, 2005). Measureable goals and outcomes for 2005–2015 were developed to achieve three strategic goals, with one of them directly addressing resilience (goal (b) below):

(a) The more effective integration of disaster risk considerations into sustainable development policies, planning and programming at all levels, with a special emphasis on disaster prevention, mitigation, preparedness and vulnerability;

(b) The development and strengthening of institutions, mechanisms, and capacities at all levels, in particular at the community level, that can systematically contribute to building resilience to hazards;
(c) The systematic incorporation of risk reduction approaches into the design and implementation of emergency preparedness, response and recovery programs in the reconstruction of affected communities.

*(UNISDR, 2005: 3–4).*

The increasing inclusion of resilience as a primary theme and strategic goal began with the HFA and continued through multiple Global Platform meetings throughout the 2000s. Nearing the end of the HFA process (2015) and not seeing as much progress towards the original goal: "the substantial reduction of disaster losses, in lives and in the social, economic, and environmental assets of communities and countries" (UNISDR, 2005: 3), led to a heightened sense of urgency and re-commitment to disaster risk reduction and building resilience from local to global scales. This sense of urgency was prompted by increasing disaster losses including the seemingly endless number of extreme events resulting in billions of dollars in economic losses and/or thousands of lives lost throughout the world (see for example MunichRe's (2017) Topics Geo series or UNISDR's (2015b) Global Assessment Reports on global loss trends).

The Global Platform meeting in Sendai, Japan, in 2015 produced the Sendai Framework for Disaster Risk Reduction (SFDRR) – another recommitment by signees to advance disaster risk reduction and building resilience, only this time through more focused priorities for action: (1) understanding disaster risk, (2) strengthening disaster risk governance to manage risk, (3) investing in disaster risk reduction for resilience, and (4) enhancing disaster preparedness for effective response and building back better in recovery. There is a continued movement away from preparedness with SFDRR with more focused attention on risk assessment and reduction and on specific and measureable targets for action (Zimmermann and Keiler, 2015). It is precisely the targets and metrics used to assess concordance with the SFDRR that is driving the contemporary discussions on the linkages between vulnerability and resilience at global scales.

The 2017 Global Platform for DRR meeting in Cancún resulted in two significant outcomes. The first was agreement on the framework and specific metrics for systematic accounting of disaster losses by individual nations in order to quantitatively assess progress towards meeting targets under the SFDRR monitoring (UNISDR, 2014). The second and perhaps the one of greater importance, was need for coherent and complementary actions among the Paris Climate Agreement (UNFCCC COP 21), the 2030 Agenda for Sustainable Development (Sustainable Development Goals, or SDGs), and the Sendai Framework for DRR (International Institute for Sustainable Development, 2017). All of these efforts include elements of building resilience to reduce and manage disaster risk and alleviate poverty as well as recognizing the overlapping and mutually reinforcing policies and monitoring that is required to achieve these goals. Such a paradigm shift is long overdue given the inextricably linked nature of disaster vulnerability and resilience.

## 12.3 Moving from Concept to Measurement

Vulnerability has always been an elusive concept and the same can be said of resilience. The sensitivity of populations to the consequences of natural hazards has been observed around the world and well documented by relief and aid organizations. Wisner (2016) traces the evolution of the vulnerability approach and methods employed for assessing within the context of natural hazards. A similar review is found in Chapter 2 in this volume.

Attempts at constructing vulnerability assessments employ a range of techniques and spatial scales from local to global. Dichotomized into qualitative versus quantitative approaches, the literature on natural hazard vulnerability assessments is quite expansive (Birkmann, 2013b). However, many of the models provide very good descriptions and conceptualizations of vulnerability, but are not operational enough to actually provide measures of it (Birkmann, 2013a; Wisner, 2016). Others are less conceptually rich but provide robust methods for assessment (Pelling, 2013). In their meta-analysis of social vulnerability drivers using floods as the hazard du jour, Rufat et al. (2015) found vulnerability to be a mature research field where the definitional issues have been settled, and the metrics established. They point out, however, that context remains an important consideration to reflect "vulnerability as a state, but also as a situation" (Rufat et al., 2015: 480). As they conclude, "greater consideration of context, measurability, and interaction between drivers can help highlight not only what is reflected in resultant vulnerability indices, but also what is absent" (Rufat et al., 2015: 480).

Resilience metrics are equally diffuse and range from qualitative to quantitative approaches from local to global scales. Given the wide-range of application arenas for resilience, the development of indicators has progressed along divergent paths with little cross fertilization of methods, models, metrics, or tools. Indicators range from sector-specific applications to more broadly based community measures, again at a range of spatial scales and geographies.

A review of disaster resilience indicators in the USA found no dominant approach to measuring resilience (Cutter, 2016b). The existing measurement approaches included tools, scorecards, and indices that were characterized by a focus on attributes/assets of resilience or capacities for resilience. The former focused on material things, the latter on social ties, connectivity, and community well-being. In their review of the literature on disaster resilience assessment models and tools, Ostadtaghizadeh et al. (2015) found limited agreement between the use of the concept of resilience and the operationalization of it for measurement purposes. They did, however, find some consistency in literature insofar as the domains that were measured – social, economic, institutional, physical (infrastructure), and natural. They also found a range of approaches from qualitative (the majority) to quantitative, and distinctions between studies of existing status or conditions of resilience and those focused on processes and actions of programs. Lastly, Sharifi and Yamagata (2016) focused their review of the extant tools based on their suitability for guiding community resilience planning. Using the definition of resilience provided by the USA National Academies (NRC 2012: 1): "the ability to prepare and plan for, absorb, recovery from and more successfully adapt to adverse events," the authors examined studies highlighting community resilience assessment (CRA) tools. The specific focus was on how closely the tools adhered to the four abilities in the NRC definition (e.g., prepare and plan, absorb, recover, and adapt) and the relative performance of the tool on each of these aspects. The authors found the least addressed and measured abilities were attributes related to absorption and adaptation.

## 12.4 Empirically Relating Resilience to Vulnerability

While disaster resilience and vulnerability are associated conceptually, there is little research on the comparisons between them in terms of metrics and tools. A number of interesting findings appear when studies are compared using similar measures of social vulnerability and resilience. First, in the three studies examined (Bergstrand et al., 2015; Cutter et al., 2014; Sherrieb et al., 2010) all empirically demonstrated that vulnerability and resilience were indeed separate constructs, but they were linked. For example, Sherrieb et al. (2010) focusing on Mississippi counties found only a 14 percent overlap between vulnerability (as measured by the Social Vulnerability Index or SoVI) and resilience (using the Community Resilience Index or CRI using only two parameters, economic development and social capital). On the other hand, Cutter et al., 2014 found a 25 percent overlap between social vulnerability using SoVI and the Baseline Resilience Indicators for

Communities or BRIC, again confirming statistical independence of the two concepts. Second, vulnerability and resilience are negatively associated – as vulnerability increases, resilience decreases, but the strength of the correlation varies (in Sherrieb et al., 2010 it was $r = -.68$, Cutter et al., 2014 it was $r = -.05$; Bergstrand et al., 2015 $r = -.386)^2$. Lastly, the geographic patterns of vulnerability and resilience relationships are inconsistent (especially across the USA where two of the studies were focused), suggesting that the social and historical context of places is important and situationally determinant (Bergstrand et al., 2015; Cutter et al., 2014). While the computation of the metrics varied (use of two different resilience indexes – CRI and BRIC) as did the study areas (Mississippi counties versus all USA counties) there is remarkable consistency in the findings in terms of the direction of the association (negative), degree of overlap (slight), and geographic patterning (not uniform across all regions).

## 12.5 Moving from Measurement to Action

If resilience and vulnerability reduction are going to be powerful forces for action in the policy world they need to be amenable to evaluation. Is community X more or less resilient than community Y and is it becoming more or less resilient through time? What payoffs might be garnered from proposed investments in resilience or vulnerability reduction? Such questions have led to the growing demand for metrics that can be used for purposes ranging from (1) defining baseline conditions within a system and monitoring changes through time to (2) making decisions about the allocation of management and financial resources to promote resilience.

But efforts to respond to this demand have found the process frustratingly difficult, which accounts for the paucity of policy-relevant research on resilience metrics, in particular. In many ways the policy community is way ahead of the science community in this regard. For example, resilience is relative (Cutter, 2016a; Gall, 2013) and requires answers. For what? For whom? Why? Who decides? Developing a metric is more tractable if the scope is narrow, yet by its very nature resilience and vulnerability are deeply imbedded in interdependencies between systems, scales, and historical processes. Narrowness misses the point. Because of the localized nature of building community resilience, top-down imposed metrics may not adequately capture the range of goals and perspectives by local stakeholders and thus thwart attempts to impose interventions deemed untenable. Bottom-up participation by community stakeholders may be more meaningful, but such bottom-up processes produce metrics that are difficult to compare between communities. Therein lies the conundrum in moving resilience from a concept to action.

## 12.6 What Is Success in the Policy Marketplace?

As noted earlier, resilience assessment (compared to vulnerability assessment) is a new field and clearly lags behind policy. In order to be used in policy, especially at USA federal levels there must be clarity not only in definitions of resilience among agencies and frameworks for assessment, but also coordinated efforts to minimize overlaps and contradictory strategies for implementation (NRC, 2012). As noted by Larkin et al. (2015) in their study of seven US federal agencies, there is no comprehensive framework for any agency, let alone a comprehensive framework across agencies or interagency partnerships for resilience. Unfortunately, two of the major players in the resilience space in the federal government were not included in the Larkin et al. assessment – Department of Housing and Urban Development (HUD) and the National Institute of Standards and Technology (NIST).

Two of the most robust resilience programs in the US federal government are HUD and NIST. The NIST program is focused on community resilience from the perspective of buildings and infrastructure systems. They have developed an extensive (two-volume) resilience planning guide to assist communities in setting resilience goals, priorities, and resource allocations to enhance resilience following a six-step process (NIST, 2016).

The HUD program (HUD, 2015a) is an outgrowth of the aftermath of Hurricane Sandy. Partnering with nonprofits and the philanthropy sector, HUD established a Rebuild by Design competition to reconstruct areas damaged by the storm in the greater New York metropolitan area in ways to enhance future resilience, not to simply build back. The innovation and perceived success of the Rebuild by Design competition led to an expansion of the program, opened up to all states and communities that experienced a major disaster between 2011–13 identified by the receipt of a Presidential Disaster Declaration. Nearly US$1billion in funding was allocated to support housing and infrastructure post-disaster redevelopment (HUD, 2017).

It is clear that the level of USA policy engagement would not have been as robust without three coalescing factors: the National Academies August 2012 report on Disaster Resilience, Hurricane Sandy in October 2012, and the establishment of the Rockefeller Foundation's 100 Resilient Cities initiative begun in 2013. The latter is particularly important in terms of media attention and the commitment of philanthropic resources to enhance urban resilience by (a) developing a consistent resilience strategy with monitoring provisions, (b) providing funds to hire a chief resilience officer for each city to assist in the implementation of the strategy, and (c) nurturing a collaborative network of cities that learn from one another.

In addition to the policy frameworks outlined in the SFDRR, nations across the globe are implementing resilience programs and strategies. For example, the European Union places resilience at the core of their foreign policy strategy for development and humanitarian aid as the mechanism for disaster risk reduction and enhancing resilience so as to reduce humanitarian needs (European Commission 2017). The foundation for the UK strategy for disaster risk reduction is provided by the UK Foresight Report (UK Government Office for Science, 2012). The present strategy is an integrated emergency management system involving infrastructure, communities, and businesses, each containing specific guidance on planning and action plans (UK Government, 2014). The OECD is equally engaged in promoting resilience as a mechanism for disaster risk reduction within member states and providing the necessary guidelines and tools for resilience systems analysis (OECD, 2014). The Rockefeller Foundation's 100 Resilient Cities initiative is also helping to promote and implement community-level resilience strategies across the globe from Santiago, Chile, to Melbourne, Australia, to Haiyan, China.

## 12.7 Critical Challenges Ahead

The nexus of vulnerability and resilience and its measurement pose some critical challenges for researchers and practitioners alike. The nature of relationship between vulnerability and resilience whether viewed in static or dynamic terms is becoming less important than establishing baselines, irrespective of their qualitative or quantitative measurement. It is difficult to proceed with activities that intervene to produce vulnerability reduction or resilience enhancement if we do not have a starting point for such assessments.

The re-creation of vulnerability and the reduction in resilience depends on who, what, where, and for whom – all of which are socially constructed and temporally variant. In other words, it depends on who is vulnerable to what and where, and in trying to mediate the vulnerability for whom and based on whose knowledge (Manyena, 2014). The same can be said of resilience (who is resilient, to what, where, and for whom) as noted previously in this chapter. The challenge ahead is to make sure that such assessments are transparent in responding to such concerns.

Another challenge is to move beyond the characteristics of existing levels of vulnerability and resilience to understand the underlying drivers of global processes and dynamic pressures that increase vulnerability and affect capacities for enhancing resilience (Weichselgartner and Kelman, 2015). These dynamic pressures are well known – technological innovation and change, globalization and economic interdependencies, environmental change, especially climate, and population growth – but we lack a comprehensive understanding of how such processes influence the spatial and temporal patterns of vulnerability and resilience. This creates an inability to effectively engage in reduction or enhancement actions at a policy level. What policy prescriptions can you undertake when you don't know what you don't know?

As alluded to earlier in this chapter there are a myriad of new methods, new data, and transdisciplinary approaches for linking vulnerability and resilience and their assessment. New approaches to uncertainty will surely benefit the quantification of measurement, but may not address the fundamental issue of mediating the underlying drivers. From a policy perspective, the former may be preferable, but the latter is a more durable approach to vulnerability reduction and enhanced resilience.

Lastly, we are at the precipice of a transformation in our science and policy of disaster risk. Global changes, be they social, economic, political, or environmental are creating tipping points in many places and within many socio-ecological systems. While still a laudable goal, disaster risk reduction requires reductions in existing and future vulnerabilities and enhancements in resilience. Whether we are able to achieve such a goal depends on demanding new and innovative thinking and practice by policymakers in civil society, government, and the private sector. It also requires the implementation of such policies by stakeholders from individuals, to households, to communities, to businesses.

Pessimism aside, there are some overt actions that can be taken to move us along the path towards reducing vulnerabilities and enhancing resilience:

- Raise awareness of risks and interdependencies at every opportunity, but especially after a large disaster when media attention is the greatest.
- Encourage a sense of shared responsibility and build a culture of resilience from the individual to households, neighborhoods, communities, and regions. Resilience is everybody's business, not just the government's, or the private sector's.
- Align resilience with existing ongoing processes and practices such as spatial planning, disaster recovery, infrastructure reconstruction, and revisions

in existing codes and standards. In other words, make resilience part of everyday life.
- Provide public recognition of resilience achievements and showcase local to global efforts. Show that enhancing resilience offsets vulnerability and enables communities to be proactive and chart their own future paths.

Disaster risk cannot be completely eliminated but the reduction in the residual risks that remain can be managed, especially by changing the vulnerability/resilience calculus, even in the face of large-scale social and environmental transformations. Therein lies the challenge for the next generation of disaster professionals.

# 13

# Synthesis and Conclusion

SVEN FUCHS AND THOMAS THALER

This book provides an impression of the challenges within the vulnerability and resilience landscape facing natural hazards. Defining natural hazard risk as a function of hazard, exposure and consequences, the understanding of vulnerability and resilience is essential in order to develop adaptation, protection and mitigation strategies to natural hazards. Vulnerability can be summarized as a product of exposure of people, economic activities, political systems, and elements at risk to natural hazards. Resilience can be seen as the coping ability and adaptive capacity of individuals and communities in response to natural hazards, but similarly also as the physical resistance of elements at risk to the impact of hazards. As the individual contributions of this book have clearly shown, both the concept of vulnerability and the concept of resilience are pillared by multiple disciplinary theories resulting in a range of paradigms for quantitative and qualitative assessment. Acknowledging the different roots of these multiple theories and underlying methods used for assessment, it becomes apparent that only by a multi-dimensional approach can the overall aim of reducing natural hazard risk be achieved. Efforts to reduce susceptibility to hazards and to create disaster-resilient communities require intersections among different theories since human action cannot be seen independently from the environmental settings.

This book advances our understanding on vulnerability and resilience with respect to natural hazards through its cross-disciplinary approach. Previous chapters have profiled the principal theoretical models of vulnerability and resilience and the prominent construction approaches of such models. Yet, there are still research gaps to fill both with respect to a practical implementation of methods and concepts and in order to better understand the topic and to explore possible future challenges emerging from dynamics in hazards and in exposure of societies. However, several factors contributing to the complexity of vulnerability and resilience can be identified: (1) the prevalent lack of suitable data for reliable measurement in many regions of the world; (2) the strongly site-specific nature of exposure resulting in variable vulnerabilities and in unequal resilience of different actors; (3) the challenges in quantifying spatial and temporal dynamics of vulnerability and resilience; (4) the respective institutional settings; and (5) the overall scale-dependency of vulnerability and resilience in natural hazard and risk management. The future challenge of understanding risk, as recently emphasized in the Sendai Framework for Disaster Risk Reduction (UNISDR, 2015a), calls for more information about vulnerabilities and resilience, and for fostering knowledge transfer between actors involved in disaster risk reduction, including scientists, practitioners and political decision makers. Scholars place high priority on knowledge sharing and transfer, but are often impeded by the blind spot of disciplinary thinking with the consequence to recognize the importance of the results in natural hazards risk management from other disciplines, but often excluding them in their (disciplinary) assessment and modeling work. No matter whether matrices, indicators, or mathematical functions are used to measure vulnerability and resilience, we have to cross disciplinary divides and vocational boundaries because no individual assessment method is able to capture the entire dimension of exposure to hazards and coping capacities suitable for reducing risk. A particular challenge is to respond to the need for integrated assessment methods using qualitative and quantitative approaches, and to take into account system dynamics on different temporal and spatial scales. Spanning from qualitative expert judgment to empirical quantification, each method has inherent challenges and the success is, furthermore, dependent on the overall institutional setting.

## 13.1 Where We Stand

At the beginning of the twenty-first century, the natural hazards research community advanced in various aspects of understanding vulnerability and resilience, as various chapters within this book have shown. Chapters on physical, social, and economic vulnerability or physical resilience, as an example, reported a large amount of research happening during the past decades. Some concepts of vulnerability and resilience have developed quite advanced methods to assess, to evaluate, and to understand the challenges in risk management. In contrast, even if advanced approaches have been proposed and are accepted as mainstream in the practice of disaster risk reduction, shortcomings in qualitative or quantitative data for assessment and computation repeatedly result in rather pragmatic approaches than precise outcomes. For example, as highlighted by Thaler and Jongman in their

Chapter 5 on economic vulnerability, the evolution of stage-damage curves since 1936 in academic and professional discourse has been a landmark for quantitative risk assessment but still suffers from the key challenge of uncertainties in data collection. Notwithstanding these challenges, stage-damage curves have evolved and are nowadays important tools to support natural hazard management in practice.

Some central challenges in vulnerability and resilience research remain and have been identified by the contributors to this book, such as the open question of ex-post versus ex-ante evaluation of losses. As shown in Chapter 10 by Pérez-Blanco et al., ex-post and ex-ante evaluation reveals a myriad of policy combinations that can be used to address disaster risk depending on the context. There is a need to tailor such evaluation to specific country, regional, and even local characteristics and challenges. Even strategies relying on similar instruments may present diverging features that make them unique, as happens with mandatory versus voluntary insurance. Other open points address the choice of data collection methods, the use of qualitative or quantitative data, the application of relative or absolute values during computation, and the selection of appropriate assessment methods (e.g., matrices, curves or indicators) for vulnerability and resilience evaluation. To choose a specific method in analyzing vulnerability or resilience is highly sensitive and implies specific uncertainties within individual results, which in turn proves the complexity of quantifying vulnerability and resilience.

As shown by various contributions, a central issue reflects on the role of institutions and organizations in natural hazard research and operational risk management. Studies on institutional and organizational vulnerability and resilience have gotten less attention in the past decades than others, such as academic discussion and scholarly works on physical vulnerability or societal resilience. Hence, there is still a need for in-depth comprehensive research on the understanding of how institutions and organizations act and response to natural hazard events, and how hazard resistance, avoidance, acceptance, and awareness can be merged and consolidated in disaster risk reduction (Chapter 11 by Brody and Atoba). For example, as Emrich and Tobin claim in their Chapter 7, research lacks studies on social justice and equity when responsibilities in natural hazard management are to be shared, in particular with respect to decreasing vulnerability and increasing resilience of affected societies. Effective governance and institutions may enhance community cohesion, as shown by Papathoma-Köhle and Thaler in Chapter 6. Consequently dilemmas of justice emerge, especially with the conflicts that some may gain and some lose from certain envisaged protection schemes. These issues have recently become even more complex due to the negative impact of financial crises and fiscal squeeze in some European countries, such as Greece, but also England and Wales.

Decisions about whom to protect and whom not to protect have become a central conflict line in natural hazard management, above all regarding the question of who is vulnerable and where the most vulnerable people and societies are located. These issues have so far only been barely discussed in operational natural hazard management, as shown in Chapter 12 of Cutter. Moreover, questions of social justice and equity are not simply questions of fair allocation and distribution of resources (e.g., level of protection), but also include processes of engagement, empowerment and inclusion of various societal groups in vulnerability and resilience research (e.g., sexual and gender minorities in emergency management planning).

## 13.2 Where We Go

### *13.2.1 Through Various Lenses*

Perhaps the key challenge and future need in vulnerability and resilience research is the integration of two diverse existing perspectives, (1) physical (natural) and (2) social sciences. Natural scientists often see both concepts through computable engineering metrics, such as in terms of the degree of loss of elements at risk resulting from the impact of a specified natural hazard event with a given magnitude and frequency. Scholars from social sciences, in contrast, tend to focus their research questions on socioeconomic and cultural characteristics that define individual vulnerability and resilience. As an antipole, representatives of each school seem to analyze vulnerability and resilience from their respective point of view with all their limitations, drawbacks, and exclusions. In Chapter 9, Hutter and Lorenz summarized this fact by the suggestion to avoid "overconfidence" in the context of assessments because a well-developed capacity of resilience does not necessarily mean a low vulnerability. In contrast, both vulnerability and resilience are described as actor- and context-specific, which needs individual capacities and context conditions in order to be assessed in a balanced way.

To overcome these issues in current risk management research and practical application, some scholars tried to integrate approaches from physical and social sciences into a holistic vulnerability and resilience assessment framework. The main challenge for such coupled frameworks is rooted in the connected system characteristics driven by both geophysical and social dynamics: assessing vulnerability and resilience is based on the prevailing system conditions, namely on number of elements at risk exposed and their valuation, coping capacities developed within societies and among societies and their institutions, etc. Similarly, the qualitative identification of hazard processes requires an understanding of triggering mechanisms in relationship to the process characteristics, i.e., the relationship between geomorphology, hydrogeology, geology, technical failures (such as overtopping of defense structures), changes in actual climate conditions and vegetation cover. It is broadly accepted, however, that natural processes are subject to dynamics due to the variations of the triggering factors resulting from climate change, which may alter existing frequency-magnitude relationships for hazard scenarios. Furthermore, there is connectivity between different hazardous processes and different elements at risk exposed, leading to short-term as well as long-term dynamics of risk. The evolution of risk due to socioeconomic transformation in particular since the 1950s, but also due to changes in the frequency and magnitude of processes varies remarkably on different temporal and spatial scales as various incidents in the past years have shown. Long-term changes are superimposed by short-term fluctuations, and both have to be considered when evaluating risk. Additionally, the social system and therefore institutions and policies are not constant over time and within spatial entities. Socioeconomic development in the man-made environment has led to an asset concentration over time and a shift in urban and suburban population in the many regions. Long-term changes are a result of the significant increase in population numbers and economic values exposed, and are manifest globally. Short-term fluctuations in values at risk supplement the underlying long-term trend, for example with respect to temporary variations of persons at risk and of vehicles on a road network. Consequently, there is a strong need to better include such dynamics into the concepts and methods for assessing vulnerability and resilience in order to better understand the underlying risk pattern, and to tailor risk management options.

Based on the individual scholarly contributions in this book, three main aspects have been identified that may help to overcome the monodisciplinary impasse. These include

(1) the integration of different components most studies on vulnerability and resilience consider, such as exposure, sensitivity, and adaptive capacities, and how to integrate such components into an overall understanding, model or framework;
(2) the integration of different methods for assessment, spanning from participatory, inductive methods aiming at a broader understanding of perceptions and capacities to rather taxonomic, deductive approaches by relying on different indicators or indices; and

(3) different target dimensions of vulnerability and resilience, such as the focus on economic consequences, consequences for the built environment or on environmental and social aspects.

### *13.2.2 Towards Scale-Dependent Assessment*

The selection of scale from local to global and appropriate data with adequate resolution is crucial in natural hazard risk management. With respect to the scale-dependency of vulnerability and resilience studies, there is a particular need to analyze both spatial and temporal scales and scale-dependencies on much deeper levels than research has done so far. As outlined in the contributions of Fekete and Montz (Chapter 2) as well as Fuchs et al. (Chapter 3), factors reflecting multiple scales and local characteristics of a place should be integrated into vulnerability and resilience modeling. It is important to consider how changes in scale might influence the results of the assessment, as reported by Burton et al. in Chapter 4. The lack of appropriate spatial scale vulnerability assessments translates into limited guidance in actionable mitigation that as such serves to create social and environmental justice issues within the context of risk reduction. Two aspects of scale are prominent with respect to these needs, (1) the interaction of varying and often also multiple spatial scales in vulnerability and resilience research, and (2) the integration of temporal scales into the assessment of vulnerability and resilience.

An assessment of vulnerability and resilience over multiple spatial and temporal scales is directly linked to hazard and risk management since sharing responsibilities – and thus, given power and policy structures – is differently implemented on global, national, regional, community, or individual household level. As stated in Chapter 7 by Emrich and Tobin, diverse challenges emerge from these issues, as questions of mitigation implementation and adaptive management are concerned. More specifically, Burton et al. highlight in their Chapter 4 that some theoretical schools in vulnerability assessment are able to provide a good understanding of vulnerability at a system level, but not at local level. However, a challenge of inter-scale and intra-scale transferability remains open, as for example local-level analyses in developed countries can often make use of high-resolution exposure datasets with detailed information on elements at risk, enabling the application of a range of vulnerability functions, while in less-developed countries such data is not available in the necessary spatial or temporal resolution and thus, more qualitative or even rudimentary approaches need to be applied. It is important to consider how changes in scale influence the assessment results since especially the evaluation of vulnerability and resilience request a cross-scalar perspective of natural hazards and their possible impact on the different sectors, such as economy, society, or policy. At minimum, research is needed to better understand the association between vulnerability and resilience metrics and impacts at different scales.

### *13.2.3 Towards Spatio-Temporal Dynamics*

So far, mainly static risk concepts were developed and applied with respect to natural hazards, neglecting any past risk levels and the history of evolution to the current situation under consideration as well as possible future levels of vulnerability and resilience. As a result, considerable gaps with respect to a possible adaptation of the risk concept applied remain open: Risk related to natural hazards is subject to spatio-temporal changes since the risk-influencing factors are variable over time and within space. Therefore, further studies focused on identifying, analyzing and modeling the spatiotemporal development of vulnerability and resilience are needed in order to provide information of short- as well as long-term changes of risk, and to better understand the underlying risk pattern. The implication of these dynamics on multiple spatial and temporal scales to integrated risk management has different roots, and should be emphasized in future research as exemplified by Bozza et al. in Chapter 8. Socioeconomic developments in the man-made environment have led to an asset concentration over time and a shift in urban and suburban population in many regions of the world. Thus, the temporal variability of elements at risk is an important key variable in the assessment of risk. For example, an increasing surface of settled area and the concentration of highly valuable assets (e.g., residential and nonresidential buildings) in exposed areas lead to higher probability of losses. These dynamics are generally attributed to the historical shift from traditional economies, based on cattle husbandry and extensive agriculture, to an industry- and service-based society in many areas, favored by a type of clustering which reflects a low awareness of planners towards natural hazard risk.

The evolution of risk due to socioeconomic transformation, but also due to climate change triggering changes in the frequency and magnitude of hazard processes varies remarkably on different temporal and spatial scales. Long-term trends are superimposed by short-term fluctuations, and both have to be considered when evaluating vulnerability and resilience. These long-term trends in risk could be regarded as the basic disposition of a society, contributing to the inherently complex and dynamic nature of vulnerability and resilience. Short-term fluctuations in risk supplement this continuing development within a specific spatial and temporal range, and are by definition highly variable. The challenge in policy development is to deal with these nonlinear relations between hazard triggers, exposure, and socio-economic development in order to better understand the causes and effects of increasing vulnerability or decreasing resilience. Only then will it be possible to tackle the tipping points responsible for vulnerability and resilience in many places and within many socio-ecological systems. As outlined in Chapter 4 by Burton et al. and simultaneously by Hutter and Lorenz in Chapter 9, however, there is limited available information of how to operationalize conceptual frameworks, on empirical evidence of theoretical concepts, and measurement methods. All kinds of theorizing and assessing vulnerability and resilience should be aware of their respective explanatory potential with respect to spatio-temporal dynamics.

## 13.3 Conclusion

Vulnerability and resilience, broadly defined as the potential for loss and the ability of a system to cope with change, respectively, are essential concepts in hazard and risk research and are central to the development of adaptation and mitigation strategies in natural hazard risk management. Moreover, as the individual contributions in this book have shown, both terms are a phenomenon that has been very influential in stimulating research and fostering understanding of the nature and distribution of impacts originating from hazards. This has repeatedly been acknowledged on the international level by UN agencies and programs, as recently in the Sendai Framework for Disaster Risk Reduction, which was adopted at the Third UN World Conference on Disaster Reduction in 2015. Globally, the Sendai Framework is the successor instrument to the Hyogo Framework for Action (HFA) 2005–15 "Building the Resilience of Nations and Communities to Disasters," and was conceived to give further impetus to the global work under the International Framework for Action for the "International Decade for Natural Disaster Reduction" of 1989, and the "Yokohama Strategy for a Safer World" and its Plan of Action, adopted in 1994 and the "International Strategy for Disaster Reduction" of 1999. Among others, the Sendai Framework clearly articulates the need for improved understanding of disaster risk in all its dimensions of exposure, vulnerability, resilience and hazard characteristics, and calls in particular for the strengthening of disaster risk governance in terms of delivering people-centered risk management, such as preparedness to "Build Back Better," recognition of stakeholders and their roles, and mobilization of risk-sensitive investment to avoid the creation of new risks. As such, on the international level of organizations for decades action has been undertaken and will further be undertaken to better prepare communities for natural hazards, and to decrease losses during the upcoming years.

On the level of practical implementation of these goals, however, several immanent challenges have to be considered when addressing vulnerability reduction and the increase of resilience. As shown in the individual chapters of this book, the heterogeneity of definitions used often impedes inter- and transdisciplinary approaches and sometimes even delays necessary action-taking, but it may also serve to broaden the scientific debate and the practical perspectives of disaster risk reduction. Bridging disciplinary gaps and demonstrating possible conceptual models and methodological approaches is important with respect to the implementation of improved management strategies. An issue of equal importance is the communication of risk to stakeholders involved, and the target-oriented delivering of information on natural hazard risk and possible mitigation alternatives. Both vulnerability and resilience are deeply linked with the conceptual differentiation between internal and external factors, that is, between a system and its "environment". Because risk management principles often only address one or the other, studies that consider the multidimensional nature of vulnerability and resilience and roots in science as well as in practical implementation are limited. One reason lies in the centralized intuitional organization of risk management actors in many countries, which hampers local authorities and communities to participate in necessary actions, to act self-organized and – most important – to mobilize the affected population. Moreover, a fair share of benefits and costs in natural hazard management could better contribute to raise awareness, to develop coping capacities, and to take responsibilities, all of which are connected to the accountability of authorities on different institutional levels. To conclude, integrating knowledge and developing methods rooted in both natural sciences and social sciences, pillared by the concept of human-environment interaction and directed towards applied research in terms of applicability in operational hazard risk management, are needed.

To achieve this goal, several issues are important and have to be considered in future research activities, but simultaneously in practical implementation of disaster risk reduction:

- Vulnerability and resilience cannot be seen as separate and independent from each other. Besides, both terms not being counterparts, but interacting and complementing opponents. Nevertheless, this interplay is far more complex and future research has to acknowledge this complexity.

- Vulnerability and resilience are scale-dependent. At minimum, research should be conducted to better elucidate the relation between evaluation indexes, metrics and mathematical relations, and the resulting impacts for disaster risk reduction on different scales.
- Vulnerability and resilience both have different conceptualizations and assessment methods in different scientific disciplines. Acknowledging these different roots, and calling for mutual respect among different scientific disciplines, a step forward in the development of methods and theories is to search for coupled scientific models that allow for a comprehensive assessment of vulnerability and resilience. Once either the internal or the external side is neglected, or only an approach from sciences or humanities is considered, results will be imperfect with respect to an overall understanding of natural hazard risk.
- Both vulnerability and resilience should be studied from a broad perspective, including physical, social, economic, ecological, and political dimensions. Available models so far mostly are centered on one or two of these issues and are therefore missing to provide an overall picture of vulnerability and risk.
- Vulnerability and resilience are temporally and spatially dynamic and at the same time context-specific. Context includes the hazard type, degree of exposure, scale of impacts and responses, geographic setting, and preexisting social, economic and political conditions, all of which are not static. The inevitable existence of change and therefore of uncertainties should be accepted in research, policy development and practical implementation. Dynamics in vulnerability and resilience act in an often nonlinear way, which shifts policy from an attempt to control change and create stability to managing the capacity of systems to cope with, adapt to, and shape change.

The individual contributions in this book showed a variety of approaches, theories, and methods for the assessment and evaluation of both vulnerability and resilience. Over time, important advances have been made in contextualizing and quantifying mutually both terms. Potential challenges for future research topics have been addressed, and we leave it to the ongoing activities of scholars from science and practice to further advance the topic in order to create more disaster-resilient communities, and to achieve the goals outlined in the Sendai Framework for Disaster Risk Reduction.

# Notes

## Chapter 5: Economic Vulnerability

1. Economic theory foresees an increasing of adaptation and mitigation efforts, when the marginal benefits of additional risk-reduction is equal to the marginal costs (Varian, 1992).
2. Economic vulnerability also plays an important role for the insurance sector to develop optimal insurance/re-insurance portfolio (Froot, 2001; Kunreuther and Pauly, 2006).
3. Economic vulnerability can be seen as the exposure of national, regional and local economies (macro level) and economic losses of individuals (micro level) to an exogenous shock, such as a natural hazard event (Briguglio et al., 2009). Consequently, the losses highly depend on the magnitude of economy gross national production or individual income (Noy, 2009).
4. Benefits were understood as the reduction in future losses (Green, 2003).
5. The damage assessment should focus on the losses occurred from an event instead of on the aspect of costs. "Losses" includes a broader definition, such as the direct losses generated from an environmental hazard event as well as the indirect consequences from the destruction. In contrast, costs mainly refer to the cost pay-outs from public authorities (through disaster funds or disaster aid payments) or insurance payments (NRC, 1999).
6. For the economic vulnerability assessment, the focus lies on the monetary values instead of the assessment of the physical vulnerability concentrated the main attention on loss ratio (see Chapter 3 and also Fuchs et al., 2007 or Papathoma-Köhle et al., 2011).
7. Within the economic literature, scholars distinguish between different terms, such as stocks and flows. Stocks include the quantity of goods at a certain time, whereas flows includes a dynamic perspective. Flows include output of stocks over time (usually the production of goods in its lifetime), for example to calculate the losses based on out of service of a machine over time and not only the reparation costs (ECLAC, 2003; Rose, 2004). Therefore, direct damages assess the losses of the stocks, in contrast to indirect losses mainly focusing on flow of the economy (ECLAC, 2003).
8. Monetary losses were calculated from observation of market values. Non-monetary losses cannot be measured from a market value and therefore need preference approaches (for example Willingness-to-pay (WTP) or Willingness-to-accept (WTA)), in which the aim is to assess the individual utility function as a means of estimating the welfare losses from future environmental hazard events (Green and Tunstall, 1991; Hanely and Spash, 1994). The methodological approach of data for non-monetary values were therefore conducted by surveys with all its limitation (Lo and Spash, 2013; Ryan and Spash, 2011).
9. Example for a What-if-question: "Which losses would you expect if the water depth was 2m?" (Penning-Rowsell et al., 2005).
10. CGE models allow short-, medium- and long-run time perspectives in contrast to input-output models which focus on a short-run perspective (Rose, 2004).

## Chapter 9: Social Resilience

1. While Section 9.3.5 will also present a rather optimistic scenario, Section 9.4 on resilience analysis and assessment seeks to provide a more balanced view that considers reasons to be optimistic as well as pessimistic about the prospects of organizing for future surprising disturbances.
2. Barriers to resilience organizing are embedded in social structures of modern society (e.g., centralized economic structures, Perrow, 2007) and in institutional constraints of human agency (Scott, 2014). An example would be the often mentioned "false" sense of security of citizens due to structural measures implemented to protect against natural hazards and due to deeply held and maintained beliefs about the responsibility of the state for natural hazard management. In this context, the term "institution" does not only refer to regulative aspects of institutional constraints, but also to normative and cultural-cognitive aspects (e.g., Scott, 2014).
3. In contrast to Pelling (2011) and Tierney (2014) who adopt a macro-perspective on social resilience in the context of social change and transformation, our analysis is more micro- and meso-oriented. However, the conceptual distinction between macro-, meso-, and micro-analysis should not be exaggerated (e.g., Hutter and Kuhlicke, 2013; Weick, 2001).
4. There is a widespread consensus in organizational studies that an organization is defined as a goal-directed, partly formalized, boundary maintaining social entity that integrates actions of individual persons (e.g., Baum, 2002).
5. For instance, low-probability/high-impact events like sudden sea level rise in the context of Flood Risk Management (FRM) for the city of London (Lonsdale et al., 2008, see also Hutter, 2007; Hutter and Schinke, 2016, on FRM in the city of Dresden).
6. In future research it would be worthwhile to explore how this stepwise evolvement could be conceptualized with regard to the different interlinked elements of human agency being iteration, projectivity, and practical evaluation as introduced in Section 9.2.
7. Van de Ven (2007: 271) compares these four forms of research based on two dimensions: Firstly, the perspective of researcher (outside/inside); secondly, the main purpose of research (describe + explain/control + evaluate). Often, research projects of graduate or PhD students will focus on description and explanation ("basic research"). However, even under these conditions, Van de Ven recommends to include continuous pragmatic communication between research and practice into the project. Presumably, true collaborative research will be rather rare (insider perspective and the purpose of description and explanation). In the real world, projects may be characterized by multiple forms of research ("hybrid research project"), especially in the case of a large research project with the aim of implementing results.

## Chapter 10: Economic Resilience

Collaborative funding for this paper was provided by the University of Adelaide's EU Centre for Global Affairs Annual Grant Program Building lasting collaborative arrangements on water scarcity management between Australia and the EU, an ARC DECRA Grant DE160100213 Optimising National Benefits from Restoring Environmental Water Flows, and an ARC DECRA Grant DE150100328 Transaction costs in Murray-Darling Basin water reform.

1. Economic resources and natural disasters are unevenly distributed among countries, and often they do not match. For example, average annual losses in low-income countries represent 22 percent of social expenditure as compared to 1.5 percent in high-income countries; and some of these low-income countries will be increasingly and disproportionally impacted by natural hazards – e.g., Small Islands Developing States (UNISDR, 2015b).
2. Hard engineering projects involve the construction of artificial structures that prevent the damages from natural disasters (e.g., reservoirs, fire-breaks, dykes).

3. Although the increase in the relative economic losses (i.e., accounting for economic growth) is arguably irrelevant globally, the looming threat of climate change may change this impact; and indeed for some regions this is already the case.
4. There is no unambiguous definition of what constitutes green infrastructure and the concept overlaps partly with nature-based solutions and ecosystem based approaches. The total operational and maintenance costs associated with such projects must be considered over time to ensure their cost-benefit accuracy.
5. Residual risk falls in the tail end risk; it is the uninsurable risk with a very small though unpredictable likelihood and a potentially high though unpredictable damage. This uncertainty may be too high for private insurance markets to develop without public support (Sugarman, 2006).
6. The construction of Wolfendene Dam might have alleviated some of the eventual 2011 flood damage: it may have enabled Wivenhoe's storage level to be kept low for flood mitigation – in line with original policies and planning – while Wolfendene could have met Brisbane's water supply needs from an (arguably by some) far more reliable source.
7. Mud Army participants included local Brisbane and outlying rural townships residents in the main. But many people from other Australian towns and cities, and even some overseas residents and disaster recovery specialists, travelled into the area to lend their support.

## Chapter 12: Linkages between Vulnerability and Resilience

1. I am indebted to Tom Wilbanks (1939–2017) for his insightful thoughts on this aspect of resilience.
2. Bergstrand et al. (2015) computed a *negative* Community Resilience Index so their paper reports a Pearson correlation coefficient of 0.386, which was inverted here for comparison purposes.

# References


Adamson, D., Loch, A. (2014). Possible negative feedbacks from "gold-plating" irrigation infrastructure. *Agricultural Water Management* 145, 134–44.

Adger, W. N. (2000a). *Indicators of social and economic vulnerability to climate change in Vietnam*. Norwich: CSERGE Working Paper GEC.

Adger, W. N. (2000b). Social and ecological resilience: Are they related? *Progress in Human Geography* 24 (3), 347–64.

Adger, W. N. (2006). Vulnerability. *Global Environmental Change* 16 (3), 268–81.

Adger, W. N., Hughes, T. P., Folke, C., Carpenter, S. R., Rockström, J. (2005). Social-ecological resilience to coastal disasters. *Science* 309 (5737), 1036–9.

Aerts, J. C. J. H., Botzen, W. J. W., Emanuel, K., Lin, N., de Moel, H. (2014). Evaluating flood resilience strategies for coastal megacities. *Science* 344 (6183), 473–5.

Akbas, S., Blahut, J., Sterlacchini, S. (2009). Critical assessment of existing physical vulnerability estimation approaches for debris flows. In J. Malet, A. Remaître, T. Bogaard (eds.), *Landslide processes: From geomorphological mapping to dynamic modelling*. Strasbourg: CERG Editions, pp: 229–33.

Akçar, S. (2001). *Grassroots women's collectives – roles in post-disaster effort: Potential for sustainable partnership and good governance (lessons learned from the Marmara Earthquake in Turkey)*. New York: United Nations Division for the Advancement of Women.

Alcott, B. (2005). Jevons' paradox. *Ecological Economics* 54, 9–21.

Alcott, B. (2008). Historical overview of the Jevons paradox in the literature. In J. M. Polimeni, M. Kozo, M. Giampietro, B. Alcott (eds.), *The Jevons paradox and the myth of resource efficiency improvements*. London: Earthscan, pp: 7–78.

Aldrich, D. P. (2011). The power of people: Social capital's role in recovery from the 1995 Kobe earthquake. *Natural Hazards* 56 (3), 595–611.

Aldrich, D. P. (2012). *Building resilience. Social capital in post-disaster recovery*. Chicago: The University of Chicago Press.

Aldrich, H. E., Ruef, M. (2006). *Organizations evolving*: London: SAGE.

Aleksandrova, M., Gain, A. K., Giupponi, C. (2016). Assessing agricultural systems vulnerability to climate change to inform adaptation planning: An application in Khorezm, Uzbekistan. *Mitigation and Adaptation Strategies for Global Change* 21 (8), 1263–87.

Alexander, D. (1993). *Natural disasters*. New York: Chapman & Hall.

Alexander, D. (1997). The study of natural disasters, 1977–97: Some reflections on a changing field of knowledge. *Disasters* 21 (4), 284–304.

Alexander, D. (2002). *Principles of emergency planning and management*. Guildford: Terra Publishing.

Alexander, D. (2005). Vulnerability to landslides. In T. Glade, M. Anderson, M. Crozier (eds.), *Landslide hazard and risk*. Chichester: John Wiley & Sons, pp: 175–98.

Alexander, D. (2013). Resilience and disaster risk reduction: An etymological journey. *Natural Hazards and Earth System Sciences* 13 (11), 2707–16.

Amadio, M., Mysiak, J., Carrera, L., Koks, E. (2016). Improving flood damage assessment models in Italy. *Natural Hazards* 82 (3), 2075–88.

Amirebrahimi, S., Rajabifard, A., Mendis, P., Ngo, T. (2016). A framework for a microscale flood damage assessment and visualization for a building using BIM–GIS integration. *International Journal of Digital Earth* 9 (4), 363–86.

Anderies, J. M., Ryan, P., Walker, B. H. (2006). Loss of resilience, crisis, and institutional change: Lessons from an intensive agricultural system in southeastern Australia. *Ecosystems* 9 (6), 865–78.

Anderson, M. B., Woodrow, P. J. (1989). *Rising from the ashes: Development strategies in times of disaster*. Boulder: Westview Press.

Ansell, Ch., Gash, A. (2008). Collaborative governance in theory and practice. *Journal of Public Administration Research and Theory* 18 (4), 543–71.

Apel, H., Aronica, G., Kreibich, H., Thieken, A. (2009). Flood risk analyses – how detailed do we need to be? *Natural Hazards* 49 (1), 79–98.

Arnold, M. (2008). *The role of risk transfer and insurance in disaster risk reduction and climate change adaptation*. Stockholm: Commission on Climate Change and Development.

Arnould, M. (1976). Geological hazards – Insurance and legal and technical aspects. *Bulletin of the International Association of Engineering Geology* 14, 263–74.

Arnstein, S. (1969). A ladder of citizen participation. *Journal of the American Institute of Planners* 35 (4), 216–24.

Arrow, K., Solow, R., Portney, P. R., Leamer, E. E., Radner, R., Schuman, H. (1993). Report of the NOAA panel on contingent valuation. *Federal Register* 58 (10), 4601–14.

Arts, B. J. M. (2006). *Forests, institutions, discourses: A discursive-institutional analysis of global forest politics*. Wageningen: Wageningen University.

Asprone, D., Manfredi, G. (2015). Linking disaster resilience and urban sustainability: A global approach for future cities. *Disasters* 39 (s1), s96–s111.

Atkisson, A., Petak, W., Alesch, J. (1984). Natural hazards exposures, losses and mitigation costs in the United States 1970–2000. *Transactions – Wisconsin Academy of Sciences, Arts and Letters* 72, 106–12.

Avenarius, C. (2003). *The structure of constraints: Social networks of immigrants from Taiwan*. Unpublished dissertation, Colonge: University of Cologne.

Avenarius, C., Johnson, J. C. (2004). *Recovery from natural disasters and the "Lack of weak ties"*. Portoroz, Slovenia: International Network for Social Network Analysis.

Ayyub, B. M. (2014). Systems resilience for multihazard environments: Definition, metrics, and valuation for decision making. *Risk Analysis* 34 (2), 340–55.

Bahadur, A. V., Ibrahim, M., Tanner, T. (2010). *The resilience renaissance? Unpacking of resilience for tackling climate change and disasters*. Brighton: Institute of Development Studies.

Baker, C. R. (2014). Breakdowns of accountability in the face of natural disasters: The case of hurricane Katrina. *Critical Perspectives on Accounting* 25 (7), 620–32.

Baker, D., Refsgaard, K. (2007). Institutional development and scale matching in disaster response management. *Ecological Economics* 63, 331–43.

Balica, S. F., Wright, N. G., van der Meulen, F. (2012). A flood vulnerability index for coastal cities and its use in assessing climate change impacts. *Natural Hazards* 64 (1), 73–105.

Baltensperger, E., Buomberger, P., Luppa, A. A., Wicki, A., Keller, B. (2008). *Regulation and intervention in the insurance industry – fundamental issues.* Zurich: The Geneva Reports – Risk and Insurance Research.

Bankoff, G. (2001). Rendering the world unsafe: "Vulnerability" as western discourse. *Disasters* 25 (1), 19–35.

Bankoff, G. (2007). Living with risk, coping with disasters. Hazard as a frequent life experience in the Philippines. *Education About Asia* 12 (2), 26–9.

Bankoff, G., Frerks, G., Hilhorst, D. (eds.) (2004). *Mapping vulnerability: Disasters, development, and people.* London: Earthscan.

Barnes, A. (2010). *Heritage award nomination form: Nomination document for the Somerset Dam.* Brisbane: Engineering Heritage Australia.

Barnett, J., Lambert, S., Fry, I. (2008). The hazards of indicators: Insights from the environmental vulnerability index. *Annals of the Association of American Geographers* 98 (1), 102–19.

Barredo, J. (2007). Major flood disasters in Europe: 1950–2005. *Natural Hazards* 42 (1), 125–48.

Barredo, J. (2009). Normalised flood losses in Europe: 1970–2006. *Natural Hazards and Earth System Sciences* 9 (1), 91–104.

Barrios, R. E. (2014). "Here, I'm not at ease": Anthropological perspectives on community resilience. *Disasters* 38 (2), 329–50.

Bateman, I. J., Brouwer, R., Ferrini, S., Schaafsma, M., Barton, D. N., Dubgaard, A., Hasler, B., Hime, S., Liekens, I., Navrud, S., de Nocker, L., Šceponavičiute, R., Semeniene, D. (2011). Making benefit transfer work: Deriving and testing principles for value transfers for similar and dissimilar sites using a case study of the non-market benefits of water quality improvements across Europe. *Environmental and Resource Economics* 50 (3), 365–87.

Battista, F., Baas, S. (2004). *The role of local institutions in reducing vulnerability to recurrent natural disasters and in sustainable livelihoods development.* Rome: FAO.

Batty, M. (2008). The size, scale, and shape of cities. *Science* 319 (5864), 769–71.

Baum, J. A. C. (eds.) (2002). *The Blackwell companion to organizations.* Oxford: Blackwell.

Bazerman, M., Watkins, M. D. (2008). *Predictable surprises. The disasters you should have seen coming and how to prevent them.* Boston: Harvard Business School Press.

Beatley, T. (2009). *Planning for coastal resilience: Best practices for calamitous times.* Washington: Island Press.

Beatley, T., Brower, D., Schwab, A. (1994). *An introduction to coastal zone management.* Washington: Island Press.

Bec, A., McLennan, C. L., Moyle, B. D. (2016). Community resilience to long-term tourism decline and rejuvenation: A literature review and conceptual model. *Current Issues in Tourism* 19 (5), 431–57.

Beccari, B. (2016). A comparative analysis of disaster risk, vulnerability and resilience composite indicators. *PLoS Currents Disasters*. Edition 1. doi: 10.1371/currents.dis.453df025e34b682e9737f95070f9b970.

Becker, A., Grünewald, U. (2003). Flood risk in central Europe. *Science* 300, 1099.

Béné, C., Frankenberger, T., Nelson, S. (2015). *Design, monitoring and evaluation of resilience interventions: Conceptual and empirical considerations*. Brighton: Institute of Development Studies.

Béné, C., Headey, D., Haddad, L., von Grebmer, K. (2016). Is resilience a useful concept in the context of food security and nutrition programmes? Some conceptual and practicial considerations. *Food Security* 8 (1), 123–38.

Benson, M. H., Craig, R. K. (2014). The end of sustainability. *Society & Natural Resources* 27 (7), 777–82.

Bergstrand, K., Mayer, B., Brumback, B., Zhang, Y. (2015). Assessing the relationship between social vulnerability and community resilience to hazards. *Social Indicators Research*, 122, 391–409.

Berke, P., Godschalk, D. (2009). Searching for the good plan: A meta-analysis of plan quality studies. *Journal of Planning Literature* 23 (3), 227–40.

Berke, P., Smith, G., Lyles, W. (2012). Planning for resiliency: Evaluation of state hazard mitigation plans under the disaster mitigation act. *Natural Hazards Review* 13 (2), 139–49.

Berke, P., Spurlock, D., Hess, G., Band, L. (2013). Local comprehensive plan quality and regional ecosystem protection: The case of the Jordan Lake watershed, North Carolina, USA. *Land Use Policy* 31, 450–59.

Berkes, F. (2007). Understanding uncertainty and reducing vulnerability: Lessons from resilience thinking. *Natural Hazards* 41 (2), 283–95.

Berkes, F., Colding, J., Folke, C. (2003). *Navigating social-ecological systems: Building resilience for complexity and change*. Cambridge: Cambridge University Press.

Berrittella, M., Hoekstra, A. Y., Rehdanz, K., Roson, R., Tol, R. S. J. (2007). The economic impact of restricted water supply: A computable general equilibrium analysis. *Water Research* 41 (8), 1799–813.

Bertrand, D., Naaim, M., Brun, M. (2010). Physical vulnerability of reinforced concrete buildings impacted by snow avalanches. *Natural Hazards and Earth System Sciences* 10 (7), 1531–45.

Bettencourt, L. M. (2013). The origins of scaling in cities. *Science* 340 (6139), 1438–41.

Bettencourt, L. M., Lobo, J., Helbing, D., Kühnert, C., West, G. B. (2007). Growth, innovation, scaling, and the pace of life in cities. *Proceedings of the National Academy of Sciences of the United States of America* 104 (17), 7301–6.

Bettencourt, L. M., West, G. B. (2010). A unified theory of urban living. *Nature* 467 (7318), 912–13.

Bettencourt, L. M., West, G. B. (2011). Bigger cities do more with less. *Scientific American* 305 (3), 52–3.

Biesbroek, G. R., Klostermann, J. E. M., Termeer, C. J. A. M., Kabat, P. (2013). On the nature of barriers to climate change adaptation. *Regional Environmental Change* 13 (5), 1119–29.

Biljecki, F., Stoter, J., Ledoux, H., Zlatanova, S., Cöltekin, A. (2015). Applications of 3D city models: State of the art review. *ISPRS International Journal of Geo-Information* 4 (4), 2842–89.

Bin, O., Polasky, S. (2004). Effects of flood hazards on property values: Evidence before and after hurricane Floyd. *Land Economics* 80 (4), 490–500.

Birkland, T. A., Burby, R. J., Conrad, D., Cortner, H., Michener, W. K. (2003). River ecology and flood hazard mitigation. *Natural Hazards Review* 4 (1), 46–54.

Birkmann, J. (2006a). Measuring vulnerability to promote disaster-resilient societies: conceptual frameworks and definitions. In J. Birkmann (ed.), *Measuring vulnerability to natural hazards: Towards disaster resilient societies*. Tokyo: United Nations University Press, pp: 7–54.

Birkmann, J. (2006b). Indicators and criteria for measuring vulnerability: Theoretical bases and requirements. In J. Birkmann (ed.), *Measuring vulnerability to natural hazards: Towards disaster resilient societies*. Tokyo: United Nations University Press, pp: 55–77.

Birkmann, J. (2013a). Measuring vulnerability to promote disaster-resilient societies: Conceptual frameworks and definitions. In J. Birkmann (ed.), *Measuring vulnerability to natural hazards: Towards disaster resilient societies*. 2nd edition. Tokyo: United Nations University Press, pp: 9–54.

Birkmann, J. (ed.) (2013b). *Measuring vulnerability to natural hazards: Towards disaster resilient societies*. 2nd edition. Tokyo: United Nations University Press.

Birkmann, J., Cardona, O. D., Carreno, L., Barbat, A., Pelling, M., Schneiderbauer, S., Kienberger, S., Kelier, M., Alexander, D., Zeil, P., Welle, T. (2013). Framing vulnerability, risk and societal responses: The MOVE framework. *Natural Hazards* 67 (2), 193–211.

Birkmann, J., Cutter, S. L., Rothman, D. S., Welle, T., Garschagen, M., van Ruijven, B., O'Neill, B., Preston, B. L., Kienberger, S., Cordona, O. D., Siagian, T., Hidayati, D., Setiadi, N., Binder, C. R., Hughes, B., Pulwarty, R. (2013). Scenarios for vulnerability: Opportunities and constraints in the context of climate change and disaster risk. *Climatic Change* 133 (1), 53–68.

Birkmann, J., Kienberger, S., Alexander, D. (eds.) (2014). *Assessment of vulnerability to natural hazards: A European perspective*. San Diego: Elsevier.

Birol, E., Karousakis, K., Koundouri, P. (2006). Using economic valuation techniques to inform water resources management: A survey and critical appraisal of available techniques and an application. *Science of the Total Environment* 365 (1–3), 105–22.

Birol, E., Koundouri, P., Kountouris, Y. (2011). Assessing the economic viability of alternative water resources in water-scarce regions: Combining economic valuation, cost-benefit analysis and discounting. *Ecological Economics* 69 (4), 839–47.

Blaikie, P. M., Cannon, T., Davis, I., Wisner, B. (1994). *At risk: Natural hazards, people's vulnerability and disasters*. London: Routledge.

BMLFUW – The Austrian Federal Ministry of Agriculture, Forestry, Environment and Water Management (2017). *Austrian Service for Torrent and Avalanche Control*. www.die-wildbach.at/ [last accessed: January 15, 2017].

Bohle, H. G. (2001). Vulnerability and criticality: Perspectives from social geography. *IHDP Update* 2, 3–5.

Bohle, H. G. (2008). Förderung von Resilience als Herausforderung für die Risikowelten von morgen. In C. Felgentreff, T. Glade (eds.), *Naturrisiken und Sozialkatastrophen*, Berlin: Springer Spektrum, pp: 435–41.

Bohle, H. G., Downing, T. E., Watts, M. J. (1994). Climate change and social vulnerability: Toward a sociology and geography of food insecurity. *Global Environmental Change* 4 (1), 37–48.

Bohle, H. G., Etzold, B., Keck, M. (2009). Resilience as agency. *IHDP Update* 2, 8–13.

Boin, A., Comfort, L. K., Demchak, C. C. (2010). The rise of resilience. In L. K. Comfort, A. Boin, C. C. Demchak (eds.), *Designing resilience. Preparing for extreme events*. Pittsburgh: University of Pittsburgh Press, pp: 1–12.

Boin, A., McConnell, A. (2007). Preparing for critical infrastructure breakdowns: The limits of crisis management and the need for resilience. *Journal of Contingencies and Crisis Management* 15 (1), 50–9.

Bollens, S. A., Kaiser, E. J., Burby, R. J. (1988). Evaluating the effects of local floodplain management policies on property owner behavior. *Environmental management* 12 (3), 311–25.

Boon, H. J., Cottrell, A., King, D., Stevenson, R. B., Millar, J. (2012). Bronfenbrenner's bioecological theory for modelling community resilience to natural disasters. *Natural Hazards* 60 (2), 381–408.

Bosello, F., Nicholls, R. J., Richards, J., Roson, R., Tol, R. S. J. (2012). Economic impacts of climate change in Europe: Sea-level rise. *Climate Change* 112 (1), 63–81.

Bosher, L., Dainty, A., Carrillo, P., Glass, J. (2007). Built-in resilience to disasters: A pre-emptive approach. *Engineering, Construction and Architectural Management* 14 (5), 434–46.

Botzen, W. J. W., Aerts, J. C. J. H., van den Bergh, J. C. J. M. (2009). Willingness of the homeowners to mitigate climate risk through insurance. *Ecological Economics* 68, 2265–77.

Botzen, W. J. W., van den Bergh, J. C. J. M. (2008). Insurance against climate change and flooding in the Netherlands: Present, future, and comparison with other countries. *Risk Analysis* 28 (2), 413–26.

Bourdieu, P. (1984). *Distinction: A social critique of the judgement of taste*. Boston: Harvard University Press.

Box, P., Bird, D., Haynes, K., King, D. (2016). Shared responsibility and social vulnerability in the 2011 Brisbane flood. *Natural Hazards* 81 (3), 1549–68.

Boyd, E., Folke, C. (eds.) (2012). *Adapting institutions: Governance, complexity and social-ecological resilience*. Cambridge: Cambridge University Press.

Boyd, R., Ibarraran, M. E. (2009). Extreme climate events and adaptation: An exploratory analysis of drought in Mexico. *Environment and Development Economics* 14 (3), 371–95.

Bozza, A., Asprone, D., Manfredi, G. (2015). Developing an integrated framework to quantify resilience of urban systems against disasters. *Natural Hazards* 78 (3), 1729–48.

Briceño, S. (2015). Looking back and beyond Sendai: 25 years of international policy experience on disaster risk reduction. *International Journal of Disaster Risk Science* 6 (1), 1–7.

Briguglio, L., Cordina, G., Farrugia, N., Vella, S. (2006). Conceptualising and measuring economic resilience. In L. Briguglio, G. Cordina, E. J. Kisanga (eds.), *Building the economic resilience of small states*. London: Formatek, pp: 265–88.

Briguglio, L., Cordina, G., Farrugia, N., Vella, S. (2009). Economic vulnerability and resilience: Concepts and measurements. *Oxford Development Studies* 37 (3), 229–47.

Brody, S. D., Davis, S. E. III, Highfield, W. E., Bernhardt, S. (2008a). A spatial-temporal analysis of section 404 wetland alteration in Texas and Florida: Thirteen years of impact along the coast. *Wetlands* 28 (1), 107–16.

Brody, S. D., Zahran, S., Highfield, W. E., Grover, H., Vedlitz, A. (2008b). Identifying the Impact of the built environment on flood damage in Texas. *Disasters* 32 (1), 1–18.

Brody, S. D., Grover, H., Lindquist, E., Vedlitz, A. (2010). Examining climate change mitigation and adaptation behaviours among public sector organisations in the USA. *Local Environment* 15 (6), 591–603.

Brody, S. D., Highfield, W., Carrasco, V. (2004). Measuring the collective planning capabilities of local jurisdictions to manage ecological systems in southern Florida. *Landscape and Urban Planning*, 69 (1), 33–50.

Brody, S. D., Highfield, W. E., Kang, J. E. (2011). *Rising waters: Causes and consequences of flooding in the United States.* Cambridge: Cambridge University Press.

Brody, S. D., Kang, J. E., Bernhardt, S. P. (2010). Identifying factors influencing flood mitigation at the local level in Texas and Florida: The role of organizational capacity. *Natural Hazards* 52 (1), 167–84.

Brody, S. D., Zahran, S., Maghelal, P., Grover, H., Highfield, W. (2007). The rising costs of floods: Examining the impact of planning and development decisions on property damage in Florida. *Journal of the American Planning Association* 73 (3), 330–45.

Bronfenbrenner, U. (2005). *Making human beings human: Bioecological perspectives on human development*. London: Sage.

Brooks, N. (2003). Vulnerability, risk and adaptation: A conceptual framework. *Tyndall Centre for Climate Change Research Working Paper* 38, 1–16.

Brouwer, R., Schaafsma, M. (2013). Modelling risk adaptation and mitigation behaviour under different climate change scenarios. *Climate Change* 117 (1), 11–29.

Brouwer, R., van Ek, R. (2004). Integrated ecological, economic and social impact assessment of alternative flood control policies in the Netherlands. *Ecological Economics* 50 (1–2), 1–21.

Browne, M. J., Hoyt, R. E. (2000). The demand for flood insurance: Empirical evidence. *Journal of Risk and Uncertainty* 20 (3), 291–306.

Bründl, M., Fuchs, S., Sandersen, F., Kveldsvik, V., Eidsvig, U., Bischof, N., Korup, O., Rheinberger, C., Romang, H., Barbolini, M. (2008). *Technical report, Deliverable 5.2, Integral risk management of snow avalanches, rock avalanches and debris flow in Europe*. Davos: IRASMOS Consortium.

Bruneau, M., Chang, S., Eguchi, R., Lee, G., O'Rourke, T., Reinhorn, A., Shinozuka, M., Tierney, K., Wallace, W., von Winterfeldt, D. (2003). A framework to quantitatively assess and enhance seismic resilience of communities. *Earthquake Spectra* 19 (4), 733–52.

Buchanan, J. M. (1975). The Samaritan's dilemma. In E. A. Phelps (ed.), *Altruism, morality and economic theory*. New York: Russel Sage Foundation, pp: 71–85.

Büchele, B., Kreibich, H., Kron, A., Thieken, A. H., Ihringer, J., Oberle, P., Merz, B., Nestmann, F. (2006). Flood-risk mapping: Contributions towards an enhanced assessment of extreme events and associated risks. *Natural Hazards and Earth System Sciences* 6 (4), 485–503.

Burby, R. J. (2006). Hurricane Katrina and the paradoxes of government disaster policy: Bringing about wise governmental decisions for hazardous areas. *The Annals of the American Academy of Political and Social Science* 604 (1), 171–91.

Burby, R. J., May, P. J. (1998). Intergovernmental environmental planning: Addressing the commitment conundrum. *Journal of Environmental Planning and Management* 41 (1), 95–110.

Burrell, A. (2011). Evaluating policies for delivering agri-environmental public goods. *OECD Workshop on the Evaluation of Agri-environmental Policies*. Braunschweig: OECD.

Burton, C., Cutter, S. L. (2008). Levee failures and social vulnerability in the Sacramento-San Joaquin Delta area, California. *Natural Hazards Review* 9 (3), 136–49.

Burton, I., Kates, R. W. (1964). The floodplain and the seashore: A comparative analysis of hazard zone occupance. *Geographical Review* 54 (3), 366–85.

Burton, I., Kates, R. W., White, G. F. (1978). *The environment as hazard*. 1st edition. New York: Oxford University Press.

Burton, I., Kates, R. W., White, G. F. (1993). *The environment as hazard*. 2nd edition. New York: Guilford Press.

Buskens, V. (1998). The social structure of trust. *Social Networks* 20 (3), 265–89.

Butts, C. T., Acton, R. M., Marcum, C. (2012). Interorganizational collaboration in the Hurricane Katrina response. *Journal of Social Structure* 13 (1), 1–36.

Cammerer, H., Thieken, A. H., Lammel, J. (2013). Adaptability and transferability of flood loss functions in residential areas. *Natural Hazards and Earth System Sciences* 13 (11), 3063–81.

Cannon, T. (1994). Vulnerability analysis and the explanation of "natural" disasters. In A. Varley (ed.), *Disasters, Development, and Environment*. Chichester: John Wiley & Sons Ltd., pp: 13–30.

Cannon, T. (2006). Vulnerability analysis, livelihoods and disasters. In W. J. Ammann, S. Dannenmann, L. Vulliet (eds.), *Coping with risks due to natural hazards in the 21st Century*. London: Taylor & Francis, pp: 41–9.

Cannon, T., Twigg, J., Rowell, J. (2003). *Social vulnerability, sustainable livelihoods and disasters Investigating methodologies for vulnerability assessments and their links with the sustainable livelihoods approach*. www.eldis.org/go/home&id=21628&type=Document#.WRxe_MakKHs [last accessed: January 10, 2017].

Cardillo, A., Scellato, S., Latora, V., Porta, S. (2006). Structural properties of planar graphs of urban street patterns. *Physical Review E* 73 (6), 066107.

Cardona, O. D. (2004). The need for rethinking the concepts of vulnerability and risk from a holistic perspective: A necessary review and criticism for effective risk management. In G. Bankoff, G. Frerks, D. Hilhorst (eds.), *Mapping vulnerability: disasters, development and people*. London: Earthscan, pp: 37–51.

Cardona, O. D. (2005). *Indicators of disaster risk and risk management: Program for Latin America and the Caribbean*. Washington: Inter-American Development Bank.

Cardona, O. D., van Aalst, M. K., Birkmann, J., Fordham, M., McGregor, G., Perez, R., Pulwarty, R. S., Schipper, E. L. F., Sinh, B. T. (2012). Determinants of risk: Exposure and vulnerability. In C. B. Field, V. Barros, T. F. Stockeret, D. Qin, D. J. Dokken, K. L. Ebi, M. D. Mastrandrea, K. J. Mach, G. K. Plattner, S. K. Allen, M. Tignor, P. M. Midgley (eds.), *Managing the risks of extreme events and disasters to advance climate change adaptation. A special report of Working Groups I and II of the Intergovernmental Panel on Climate Change (IPCC)*. New York: Cambridge University Press, pp: 65–108.

Carpenter, S., Walker, B., Anderies, J. M., Abel, N. (2001). From metaphor to measurement: Resilience of what to what? *Ecosystems* 4 (8), 765–81.

Carrera, L., Standardi, G., Bosello, F., Mysiak, J. (2015). Assessing direct and indirect economic impacts of a flood event through the integration of spatial and computable general equilibrium modelling. *Environmental Modelling & Software* 63 (1), 109–22.

Carter, M. R., Little, P. D., Mogues, T., Negatu, W. (2007). Poverty traps and natural disasters in Ethiopia and Honduras. *World Development* 35 (5), 835–56.

Casagrande, D. G., McIlvaine-Newsad, H., Jones, E. C. (2015). Social networks of help-seeking in different types of disaster responses to the 2008 Mississippi River Floods. *Human Organization* 74 (4), 351–61.

Cavallaro, M., Asprone, D., Latora, V., Manfredi, G., Nicosia, V. (2014). Assessment of urban ecosystem resilience through hybrid social-physical complex networks. *Computer-Aided Civil and Infrastructure Engineering* 29 (8), 608–25.

Cavallo, E., Noy, E. (2011). Natural disasters and the economy — A survey. *International Review of Environmental and Resource Economics* 5 (1), 63–102.

Cerulo, K. (2008). *Never saw it coming: Cultural challenges to envisioning the worst*. Chicago: University of Chicago Press.

Chakraborty, J., Tobin, G. A., Montz, B. E. (2005). Population evacuation: Assessing spatial variability in geophysical risk and social vulnerability to natural hazards. *Natural Hazards Review* 6 (1), 23–33.

Chambers, R. (1989). Editorial introduction: Vulnerability, coping, and policy. *IDS Bulletin* 20, 1–7.

Chang, S. E. (2001). Structural change in urban economies: Recovery and long-term impacts in the 1995 Kobe Earthquake. *The Kokumin Keizai Zasshi* 183, 47–66.

Chasten, M. A., Rosati, J. D., McCormick, J. W., Randall, R. E. (1993). *Engineering design guidance for detached breakwaters as shoreline stabilization structure*. Vicksburg: U.S. Army Engineer Waterways Experiment Station.

Chatterton, J. B., Pirt, J., Wood, T. R. (1979). The benefits of flood forecasting. *Journal of the Institution of Water Engineers* 33 (3), 237–52.

Ciscar, J. C., Feyen, L., Soria, A., Lavalle, C., Raes, F., Perry, M., Nemry, F., Demirel, H., Rozsai, M., Dosio, A., Donatelli, M., Srivastava, A., Fumagalli, D., Niemeyer, S., Shrestha, S., Ciaian, P., Himics, M., van Doorslaer, B., Barrios, S., Ibáñez, N., Forzieri, G., Rojas, R., Bianchi, A., Dowling, P., Camia, A., Libertà, G., San Miguel, J., de Rigo, D., Caudullo, G., Barredo, J. I., Paci, D., Pycroft, J., Saveyn, B., van Regemorter, D., Revesz, T., Vandyck, T., Vrontisi, Z., Baranzelli, C., Vandecasteele, I., Silva, F., Ibarreta, D. (2014). *Climate impacts in Europe. Results from the JRC PESETA II project*. Como: JRC scientific and policy reports.

Ciurean, R. L., Schröter, D., Glade, T. (2013). Conceptual frameworks of vulnerability assessments for natural disasters reduction. In J. Tiefenbacher (ed.), *Approaches to disaster management – examining the implications of hazards, emergencies and disasters*. www.intechopen.com/books/approaches-to-disaster-management-examining-the-implications-of-hazards-emergencies-and-disasters/conceptual-frameworks-of-vulnerability-assessments-for-natural-disasters-reduction [last accessed: May 22, 2017].

Clark, G. E., Moser, S. C., Ratick, S. J., Dow, K., Meyer, W. B., Emain, S., Jin, W., Kasperson, J X., Kasperson, R. E., Schwarz, H. E. (1998). Assessing the vulnerability of coastal communities to extreme storms: The case of Revere, Massachusetts, US. *Mitigation and Adaptation Strategies for Global Change* 3 (1), 59–82.

Clegg St., R., Courpasson, D., Phillips, N. (2006). *Power and organizations*. London: Sage.

Coaffee, J., Lee, P. (2016). *Urban resilience. Planning for risk, crisis and uncertainty.* London: Palgrave.
Cochrane, H. C. (1974). *Predicting the economic impact of earthquakes.* Boulder: University of Colorado.
Cochrane, H. C. (2004). Indirect losses from natural disasters: Measurement and myth. In Y. Okuyama, S. E. Chang (eds.), *Modeling spatial and economic impacts of disasters.* Berlin, Heidelberg: Springer, pp: 37–52.
Cohen, A., Siegelman, P. (2010). Testing for adverse selection in insurance markets. *The Journal of Risk and Insurance* 77 (1), 39–84.
Cohen, C., Werker, E. D. (2008). The political economy of "natural" disasters. *Journal of Conflict Resolution* 52 (6), 795–819.
Cole, S. (1995). Lifelines and livelihood: A social accounting matrix approach to calamity preparedness. *Journal of Contingencies and Crisis Management* 3 (4), 1–11.
Collier, S. J., Lakoff, A. (2008). The vulnerability of vital systems: How "critical infrastructure" became a security problem. In M. A. Dunn, K. S. Kristensen (eds.), *Securing "the Homeland": Critical infrastructure, risk and (in)security.* London: Routledge, pp: 17–39.
Collins, T. W. (2008). The political ecology of hazard vulnerability: Marginalization, facilitation and the production of differential risk to urban wildfires in Arizona's White Mountains. *Journal of Political Ecology* 15 (1), 21–43.
Collins, T. W., Grineski, S. E., de Lourdes Romo Aguilar, M. (2009). Vulnerability to environmental hazards in the Ciudad Juárez (Mexico)–El Paso (USA) metropolis: A model for spatial risk assessment in transnational context. *Applied Geography* 29 (3), 448–61.
Comfort, L. K., Boin, A., Demchak, C. C. (eds.) (2010). *Designing resilience. Preparing for extreme events.* Pittsburgh: University of Pittsburgh Press.
Commons, J. R. (1931). Institutional economics. *The American Economic Review* 21, 648–57.
Congress, U. S. (1936). Flood control act of 1936. *Public Law 74-738*, 49.
Connell, D. (2007). *Water politics in the Murray-Darling basin.* Riverwood: The Federation Press.
Cote, M., Nightingale, A. J. (2012). Resilience thinking meets social theory: Situating social change in socio-ecological systems (SES) research. *Progress in Human Geography* 36 (4), 475–89.
Cova, T. J., Theobald, D. M., Norman III, J. B., Siebeneck, L. (2013). Mapping wildfire evacuation vulnerability in the western US: the limits of infrastructure. *Geo Journal* 78 (2), 273–85.
Crane, T. A. (2010). Of models and meanings: Cultural resilience in social-ecological systems. *Ecology and Society* 15 (4), 19.
Crase, L., O'Keefe, S., Dollery, B. (2013). Talk is cheap, or is it? The cost of consulting about uncertain reallocation of water in the Murray-Darling Basin, Australia. *Ecological Economics* 88, 206–13.
Crawford, S. E. S., Ostrom, E. (1995). A grammar of institutions. *The American Political Science Review* 89 (4), 582–600.
Crichton, D. (2008). Role of insurance in reducing flood risk. *The Geneva Papers on Risk and Insurance – Issues and Practice* 33 (1), 117–32.
Crossan, M. M., Lane, H. W., White, R. E. (1999). An organizational learning framework: From intuition to institution. *The Academy of Management Review* 24 (3), 522–37.

Crowley, JR, Elliott, J. (2012). Earthquake disasters and resilience in the global North: Lessons from New Zealand and Japan. *The Geographical Journal* 178 (3), 208–15.

Cutter, S. L. (1996). Vulnerability to environmental hazards. *Progress in Human Geography* 20 (4), 529–39.

Cutter, S. L. (ed.) (2001). *American hazardscapes: The regionalization of hazards and disasters*. Washington: Joseph Henry Press.

Cutter, S. L. (2003). The vulnerability of science and the science of vulnerability. *Annals of the Association of American Geographers* 93 (1), 1–12.

Cutter, S. L. (2014). Building disaster resilience: Steps toward sustainability. *Challenges in Sustainability* 1 (2), 72–9.

Cutter, S. L. (2016a). Resilience to what? Resilience for whom? *The Geographical Journal* 182 (2), 110–13.

Cutter, S. L. (2016b). The landscape of disaster resilience indicators in the USA. *Natural Hazards* 80 (2), 741–58.

Cutter, S. L., Ash, K. D. Emrich, C. T. (2014). The geographies of community disaster resilience. *Global Environmental Change* 29, 65–77.

Cutter, S. L., Barnes, L., Berry, M., Burton, C., Evans, E., Tate, E., Webb, J. (2008). A place-based model for understanding community resilience. *Global Environmental Change* 18, 598–606.

Cutter, S. L., Boruff, B. J., Shirley, W. L. (2003). Social vulnerability to environmental hazards. *Social Science Quarterly* 84 (2), 242–61.

Cutter, S. L., Burton, C. G., Emrich, C. T. (2010). Disaster resilience indicators for benchmarking baseline conditions. *Journal of Homeland Security and Emergency Management* 7 (1), 51.

Cutter, S.L. and C. Corendea (eds.) (2013). *From social vulnerability to resilience: Measuring progress toward disaster risk reduction*. SOURCE No. 17/2013. Bonn: United Nations University Institute for Environment and Human Security (UNU-EHS).

Cutter, S. L., Emrich, C. T. (2006). Moral hazard, social catastrophe: The changing face of vulnerability along the hurricane coasts. *The Annals of the American Academy of Political and Social Science* 604, 102–12.

Cutter, S. L., Finch, C. (2008). Temporal and spatial changes in social vulnerability to natural hazards. *Proceedings of the National Academy of Sciences of the United States of America* 105 (7), 2301–6.

Cutter, S. L., Mitchell, J. T., Scott, M. S. (2000). Revealing the vulnerability of people and places: A case study of Georgetown County, South Carolina. *Annals of the Association of American Geographers* 90 (4), 713–37.

Dabla-Norris, E., Gündüz, Y. B. (2014). *Exogenous shocks and growth crises in low-income countries: A vulnerability index*. Washington: International Monetary Fund.

Dai, F., Lee, C., Ngai, Y. (2002). Landslide risk assessment and management: An overview. *Engineering Geology* 64 (1), 65–87.

Daly, H. E. (1974). The economics of the steady state. *The American Economic Review* 64 (2), 15–21.

Daly, H. E. (1992). Allocation, distribution, and scale: Towards an economics that is efficient, just, and sustainable. *Ecological Economics* 6 (3), 185–93.

Dalziell, E. P., McManus, S. K. (2004). Resilience, vulnerability, and adaptive capacity: Implications for system performance. Presented at the international forum for engineering decision making (IFED), Stoos, Switzerland, December

6–8, 2004. https://ir.canterbury.ac.nz/handle/10092/2809 [last accessed: May 22, 2017].

Daniel, V. E., Florax, R. J. G. M., Rietveld, P. (2009). Flooding risk and housing values: An economic assessment of environmental hazard. *Ecological Economics* 69 (2), 355–65.

Daniell, J. E., Khazi, B., Wenzel, F., Vervaeck, A. (2011). The CATDAT damaging earthquake database. *Natural Hazards and Earth System Sciences* 11 (8), 2235–51.

Davidson, R. A., Lambert, K. B. (2001). Comparing the hurricane disaster risk of U.S. coastal counties. *Natural Hazards Review* 2 (3), 132–42.

Davidson, R., Shah, H. C. (1997). A multidisciplinary urban earthquake disaster risk index. *Earthquake Spectra* 13 (2), 211–23.

Davies, M., Béné, C., Arnall, A., Tanner, T., Newsham, A., Coirolo, C. (2013). Promoting resilient livelihoods through adaptive social protection: Lessons from 124 programmes in South Asia. *Development Policy Review* 31 (1), 27–58.

Davis, I., Bellers, R. (1995). The context of community vulnerability assessment (CVA) and summary of workshop issues. Paper presented at The Assessment of Community Vulnerability in Hazard Prone Areas, Workshop, London 31 March, Committee of UN IDNDR.

Davoudi, S., Shaw, K., Haider, L. J., Quinlan, A. E., Peterson, G. D., Wilkinson, C., Fünfgeld, H., McEvoy, D., Porter, L. (2012). Resilience: A bridging concept or a dead end? "Reframing" resilience: challenges for planning theory and practice interacting traps: resilience assessment of a pasture management system in Northern Afghanistan urban resilience: what does it mean in planning practice? Resilience as a useful concept for climate change adaptation? The politics of resilience for planning: a cautionary note. *Planning Theory & Practice* 13 (2), 299–333.

Dawson, R. J., Ball, T., Werritty, J., Werritty, A., Hall, J. W., Roche, N. (2011). Assessing the effectiveness of non-structural flood management measures in the Thames Estuary under conditions of socio-economic and environmental change. *Global Environmental Change* 21 (2), 628–46.

de Moel, H., Aerts, J. C. J. H. (2011). Effect of uncertainty in land use, damage models and inundation depth on flood damage estimates. *Natural Hazards* 58 (1), 407–25.

de Moel, H., Asselman, N. E. M., Aerts, J. C. J. H. (2012). Uncertainty and sensitivity analysis of coastal flood damage estimates in the west of the Netherlands. *Natural Hazards and Earth System Sciences* 12 (4), 1045–58.

de Moel, H., Jongman, B., Kreibich, H., Merz, B., Penning-Rowsell, E., Ward, P. J. (2015). Flood risk assessments at different spatial scales. *Mitigation and Adaptation Strategies for Global Change* 20 (6), 865–90.

de Sherbinin, A. (2014). *Mapping the unmeasurable? Spatial analysis of vulnerability to climate change and climate variability*. Unpublished dissertation, Enschede: University of Twente.

Dempsey, J., Robertson, M. M. (2012). Ecosystem services: Tensions, impurities, and points of engagement within neoliberalism. *Progress in Human Geography* 36 (6), 758–79.

Demuzere, M., Orru, K., Heidrich, O., Olazabal, E., Geneletti, D., Orru, H., Bhave, A. G., Mittal, N., Feliu, E., Faehnle, M. (2014). Mitigating and adapting to climate change: Multi-functional and multi-scale assessment of green urban infrastructure. *Journal of Environmental Management* 146, 107–15.

Dequech, D. (2006). Institutions and norms in institutional economics and sociology. *Journal of Economic Issues* 40 (2), 473–81.
Di Baldassarre, G., Viglione, A., Carr, G., Kuil, L., Yan, K., Brandimarte, L., Blöschl, G. (2015). Debates-perspectives on socio-hydrology: Capturing feedbacks between physical and social processes. *Water Resources Research* 51, 4770–81.
Dibben, C., Chester, D. K. (1999). Human vulnerability in volcanic environments: The case of Furnas, Sao Miguel, Azores. *Journal of Volcanology and Geothermal Research* 92 (1–2), 133–50.
Dittmer, C., Lorenz, D. F., Reiter, J., Wenzel, B. (2016). Drei Jahre nach dem Deichbruch – Über die Gegenwart einer nicht abgeschlossenen Katastrophe. *Notfallvorsorge* 4, 17–25.
Doerfel, M. L., Chewning, L. V., Lai, C. H. (2013). The evolution of networks and the resilience of interorganizational relationships after disaster. *Communication Monographs* 80 (4), 533–59.
Douglas, J. (2007). Physical vulnerability modelling in natural hazard risk assessment. *Natural Hazards and Earth System Sciences* 7 (2), 283–8.
Dovers, S. R., Handmer, J. W. (1992). Uncertainty, sustainability and change. *Global Environmental Change* 2 (4), 262–76.
Downton, M., Miller, J., Pielke, R. (2005). Reanalysis of U.S. National Weather Service flood loss database. *Natural Hazards Review* 6, 13–22.
Downton, M., Pielke, R. (2005). How accurate are disaster loss data? The case of U.S. flood damage. *Natural Hazards* 35 (2), 211–28.
Drabek, T. E. (2012). *Human system responses to disaster: An inventory of sociological findings*. New York: Springer.
Drabek, T. E., McEntire, D. A. (2002). Emergent phenomena and multiorganizational coordination in disasters: Lessons from the research literature. *International Journal of Mass Emergencies and Disasters* 20 (2), 197–224.
Dutt, A. K. (2011). Institutional change and economic development: Concepts, theory and political economy. *Journal of Institutional Economics* 7 (4), 529–34.
Dutta, D., Herath, S., Musiake, K. (2003). A mathematical model for flood loss estimation. *Journal of Hydrology* 277 (1–2), 24–49.
Eakin, H., Luers, A. L. (2006). Assessing the vulnerability of social-environmental systems. *Annual Review of Environment and Resources* 31, 365–94.
Ebert, A., Kerle, N., Stein, A. (2009). Urban social vulnerability assessment with physical proxies and spatial metrics derived from air- and spaceborne imagery and GIS data. *Natural Hazards* 48 (2), 275–94.
Eboli, F., Parrado, R., Roson, R. (2010). Climate-change feedback on economic growth: Explorations with a dynamic general equilibrium model. *Environment and Developments Economics* 15 (5), 515–33.
ECLAC – Economic Commission for Latin America and the Caribbean (2003). *Handbook for estimating the socio-economic and environmental effects of disasters*. Santiago de Chile: Economic Commission for Latin America and the Caribbean.
EEA – European Environment Agency (2016). *Corine Land Cover*. land.copernicus.eu/pan-european/corine-land-cover [last accessed: January 10, 2017].
EEA, OECD – European Environment Agency-Organization for Economic Co-operation and Development (2013). *European environment agency-organization for economic co-operation and development database on economic instruments*. www2.oecd.org/ecoinst/queries/ [last accessed: September 30, 2013].

Egermann, M., Hutter, G. (2014). Perspektiven für die Annäherung der Raum- und Transitionforschung – am Beispiel des EU-Projekts ARTS. *Planung Neu Denken* 3, 1–9.

Egli, T. (1996). *Hochwasserschutz und Raumplanung-Schutz vor Naturgefahren mit Instrumenten der Raumplanung dargestellt am Beispiel von Hochwaser und Murgängen*. Unpublished dissertation, Zürich: ETH Zürich.

Elliott, J. R., Pais, J. (2006). Race, class, and hurricane Katrina: Social differences in human responses to disaster. *Social Science Research* 35 (2), 295–321.

Elmer, F., Kreibich, H., Seifert, I., Thieken, A. H. (2010). A delphi method expert survey to derive standards for flood damage data collection. *Risk Analysis* 30 (1), 107–24.

Embleton-Hamann, C. (2007). *Geomorphological hazards in Austria. Geomorphology for the future–joint meeting of the Commission on Geomorphology of the Austrian geographical society and the IAG working group on geomorphology and global environmental change*. Obergurgl Austria: Innsbruck University Press.

Emirbayer, M., Mische, A. (1998). What is agency? *American Journal of Sociology* 103 (4), 962–1023.

Emrich, C. T. (2005). *Social vulnerability in US metropolitan areas: Improvements in hazard vulnerability assessment*. Unpublished dissertation, Columbia, South Carolina: University of South Carolina.

Emrich, C. T., Cutter, S. L. (2011). Social vulnerability to climate-sensitive hazards in the southern United States. *Weather, Climate, and Society* 3 (3), 193–208.

Enarson, E., Chakrabarti, P. D. (eds.) (2009). *Women, gender and disaster: Global issues and initiatives*. London: SAGE.

Enarson, E., Morrow, B. H. (eds.) (1998). *The gendered terrain of disaster: Through women's eyes*. Westport: Praeger.

Eriksen, S. H., Brown, K., Kelly, P. M. (2005). The dynamics of vulnerability: Locating coping strategies in Kenya and Tanzania. *The Geographical Journal* 171 (4), 287–305.

Erikson, K. (1976). *Everything in its path. Destruction of community in the Buffalo Creek Flood*. New York: Simon and Schuster.

Erikson, K. (1994). *A new species of trouble. The human experience of modern disaster*. New York: W.W. Norton and Company.

Escaleras, M., Anbarci, N., Register, C. (2006). *Public sector corruption and natural disasters: A potentially deadly interaction*. Boca Raton: Department of Economics, College of Business, Florida Atlantic University.

Etkin, D., Haque, C., Brooks, G. (eds.) (2003). *An assessment of natural hazards and disasters in Canada*. Kluwer: Dordrecht.

European Commission (2000). *Directive 2000/60/EC of the European Parliament and of the Council establishing a framework for the Community action in the field of water policy*. Brussels: European Commission.

European Commission (2007). *EU Directive 2007/60/EC of the European Parliament and of the European Council on the assessment and management of flood risks*. Brussels: European Commission.

European Commission (2010). *Commission staff working paper – risk assessment and mapping guidelines for disaster management*. Brussels: European Commission.

European Commission (2013a). *Council Regulation (EU) No 733/2013 of 22 July 2013 amending Regulation (EC) No 994/98 on the application of Articles 92 and 93 of the Treaty establishing the European Community to certain categories of horizontal State aid*. Brussels: European Commission.

European Commission (2013b). *Green paper on the insurance of natural and man-made disaster*. Brussels: European Commission.

European Commission (2014). *Guide to cost-benefit analysis of investment projects (Report). Economic appraisal tool for cohesion policy 2014–2020*. Brussels: European Commission.

European Commission. (2017). *Resilience. European civil protection and humanitarian aid operations*. http://ec.europa.eu/echo/what/humanitarian-aid/resilience_en [last accessed: July 03, 2017].

European Group on Tort Law (2005). *Principles of European Tort Law*. Vienna, New York: Springer.

Evans, T., Ostrom, E., Gibson, C. (2002). Scaling issues with social data in integrated assessment modelling. *Intergrated Assessment* 3 (2–3), 135–50.

Eves, C. (2002). The long-term impact of flooding on residential property values. *Property Management* 20 (4), 214–27.

Eyre, A. (2006). Remembering: Community commemoration after disaster. In H. Rodríguez, E. L. Quarantelli, R. Dynes (eds.), *Handbook of disaster research*. New York: Springer, pp: 441–55.

Faas, A. J., Jones, E. C., Whiteford, L. M., Tobin, G. A., Murphy, A. D. (2014). Gendered access to formal and informal resources in postdisaster development in the Ecuadorian Andes. *Mountain Research and Development* 34 (3), 223–34.

Fell, R. (1994). Landslide risk assessment and acceptable risk. *Canadian Geotechnical Journal* 31, 261–72.

FEMA – Federal Emergency Management Agency (2001). *Ensuring that structures built on fill in or near special flood hazard areas are reasonably safe from flooding in accordance with the National Flood Insurance Program*. www.ct.gov/demhs/lib/demhs/hmgp/guidance/tb10_01_fill_in_sfha.pdf [last accessed: May 31, 2016].

FEMA – Federal Emergency Management Agency (2013). *National Flood Insurance Program community rating system coordinator's manual*. www.fema.gov/media-library/assets/documents/8768?id=2434 [last accessed: September 23, 2013].

FEMA – Federal Emergency Management Agency (2015). *Brazoria County adopts freeboard regulation in combatting flood loss. Mitigation best practices portfolio*. www.fema.gov/media-library-data/1441822561672-254e32a71d6e179b8d32ecfbbb1358d4/08-Brazoria-County-Adopts-Freeboard-Regulation-in-Combatting-Flood-Loss_web.pdf [last accessed: May 22, 2017].

Fekete, A. (2009). Validation of a social vulnerability index in context to river-floods in Germany. *Natural Hazards and Earth System Sciences* 9 (2), 393–403.

Fekete, A. (2012). Spatial disaster vulnerability and risk assessments: Challenges in their quality and acceptance. *Natural Hazards* 61 (3), 1161–78.

Fekete, A., Damm, M., Birkmann, J. (2010). Scales as a challenge for vulnerability assessment. *Natural Hazards* 55 (3), 729–47.

Fekete, A., Hufschmidt, G. (2014). From application to evaluation: Addressing the usefulness of resilience and vulnerability. *International Journal of Disaster Risk Science* 5 (1), 1–2.

Fekete, A., Hufschmidt, G., Kruse, S. (2014). Benefits and challenges of resilience and vulnerability for disaster risk management. *International Journal of Disaster Risk Science* 5 (1), 3–20.

Fekete, A., Hufschmidt, G. (eds.) (2016). *Atlas der Verwundbarkeit und Resilienz – Pilotausgabe zu Deutschland, Österreich, Liechtenstein und Schweiz; Köln & Bonn | Atlas of Vulnerability and Resilience – Pilot version for Germany,*

*Austria, Liechtenstein and Switzerland.* www.kavoma.de/atlas-vr [last accessed: May 22, 2017].

Fekete, A., Tzavella, K., Armas, L., Binner, J., Garschagen, M., Giupponi, C., Mojtahed, V., Pettita, M., Schneiderbauer, S., Serre, D. (2015). Critical data source; tool or even infrastructure? Challenges of geographic information systems and remote sensing for disaster risk governance. *ISPRS International Journal of Geo-Information* 4 (4), 1848–69.

Field, C. B., Barros, V. R., Dokken, D. J., Mach, K. J., Mastrandrea, M. D., Bilir, T. E., Chatterjee, M., Ebi, K. L., Estrada, Y. O., Genova, R. C., Girma, B., Kissel, E. S., Levy, A. N., MacCracken, S., Mastrandrea, P. R., White, L. L. (eds.) (2014) *Climate change 2014: Impacts, adaptation, and vulnerability*. Cambridge: Cambridge University Press.

Finch, C., Emrich, C. T., Cutter, S. L. (2010). Disaster disparities and differential recovery in New Orleans. *Population & Environment* 31 (4), 179–202.

Fletcher, S., Cox, R. S., Scannell, L., Heykoop, C., Tobin-Gurley, J., Peek, L. (2016). Youth creating disaster recovery and resilience: A multi-site arts-based youth engagement research project. *Children, Youth and Environments* 26 (1), 148–63.

Folke, C. (2006). Resilience. The emergence of a perspective for social-ecological systems analyses. *Global Environmental Change* 16 (3), 253–67.

Folke, C., Carpenter, S. Elmqvist, T., Gunderson, L., Holling, C., Walker, B., Bengtsson, J., Berkes, F., Colding, J., Danell, K., Falkenmark, M., Gordon, L., Kasperson, R. E., Kautsky, N., Kinzig, A., Levin, S., Maler, K. G., Moberg, F., Ohlsson, L., Olsson, P., Ostrom, E., Reid, W., Rockstrom, J., Savenije, H., and Svedin, U. (2002). *Resilience and sustainable development: Building adaptive capacity in a world of transformations. Scientific background paper on resilience for the process of the world summit on sustainable development.* Stockholm, Sweden: Environmental Advisory Council to the Swedish Government.

Folke, C., Carpenter, S. R., Walker, B., Scheffer, M., Chapin, T., Rockström, J. (2010). Resilience thinking: Integrating resilience, adaptability and transformability. *Ecology and Society* 15 (4), 20.

Folke, C., Colding, J., Berkes, F. (2003). Synthesis. Building resilience and adaptive capacity in social-ecological systems. In F. Berkes, J. Colding, C. Folke (eds.), *Navigating social-ecological systems. Building resilience for complexity and change*. Cambridge: Cambridge University Press, pp: 352–87.

Folmer, H., Johansson-Stenman, O. (2011). Does environmental economics produce aeroplanes without engines? On the need for an environmental social science. *Environmental and Resource Economics* 48 (3), 337–61.

Forino, G., von Meding, J., Brewer, G., Gajendran, T. (2014). Disaster risk reduction and climate change adaptation policy in Australia. *Procedia Economics and Finance* 18, 473–82.

Fotheringham, A. S., Brundson, C., Charlton, M. (2002). *Geographically weighted regression: The analysis of spatially varying relationships*. Chichester: John Wiley & Sons Ltd.

Frank, R. H. (2000). Why is cost-benefit analysis so controversial? *The Journal of Legal Studies* 29 (S2), 913–30.

Frazier, T. G., Wood, N., Yarnal, B. (2010). Stakeholder perspectives on land-use strategies for adapting to climate-change-enhanced coastal hazards: Sarasota, Florida. *Applied Geography* 30 (4), 506–17.

Frazier, T. G., Wood, N., Yarnal, B., Bauer, D. H. (2010). Influence of potential sea level rise on societal vulnerability to hurricane storm-surge hazards, Sarasota County, Florida. *Applied Geography* 30 (4), 490–505.

Frazier, T. G., Thompson, C. M., Dezzani, R. J., Butsick, D. (2013). Spatial and temporal quantification of resilience at the community scale. *Applied Geography* 42, 95–107.

Frazier, T. G., Thompson, C. M., Dezzani, R. J. (2014). A framework for the development of the SERV model: A Spatially Explicit Resilience-Vulnerability model. *Applied Geography* 51, 158–72.

Freeman, A. M. (2003). *The measurement of environmental and resource values: Theory and methods*. Washington: Resources for the Future.

Freudenberg, M. (2003). *Composite indicators of country performance: A critical assessment*. Paris: Organisation for Economic Co-operation and Development.

Frey, B. S. (1990). Vergleichende Analyse von Institutionen: Die Sicht der politischen Ökonomie. *Staatswissenschaft und Staatpraxis* 2, 158–75.

Froot, K. A. (2001). The market for catastrophe risk: A clinical examination. *Journal of Financial Economics* 60 (2–3), 529–71.

Fuchs, S. (2009). Susceptibility versus resilience to mountain hazards in Austria – Paradigms of vulnerability revisited. *Natural Hazards and Earth System Sciences* 9 (2), 337–52.

Fuchs, S., Heiss, K., Hübl, J. (2007). Towards an empirical vulnerability function for use in debris flow risk assessment. *Natural Hazards and Earth System Sciences* 7 (5), 495–506.

Fuchs, S., Keiler, M., Sokratov, S., Shnyparkov, A. (2013). Spatiotemporal dynamics: The need for an innovative approach in mountain hazard risk management. *Natural Hazards* 68 (3), 1217–41.

Fuchs, S., Keiler, M., Zischg, A. (2015). A spatiotemporal multi-hazard exposure assessment based on property data. *Natural Hazards and Earth System Sciences* 15 (9), 2127–42.

Fuchs, S., Kuhlicke, C., Meyer, V. (2011). Editorial for the special issue: Vulnerability to natural hazards – the challenge of integration. *Natural Hazards* 58 (2), 609–19.

Fuchs, S., McAlpin, M. C. (2005). The net benefit of public expenditures on avalanche defence structures in the municipality of Davos, Switzerland. *Natural Hazards and Earth System Sciences* 5 (3), 319–30.

Fuchs, S., Röthlisberger, V., Thaler, T., Zischg, A., Keiler, M. (2017). Natural hazard management from a co-evolutionary perspective: Exposure and policy response in the European Alps. *Annals of the American Association of Geographers* 107 (2), 382–92.

Fuller, R. B., Snyder, J. (1969). *Operating manual for spaceship earth*. Carbondale: Southern Illinois University Press.

Furedi, F. (2007). The changing meaning of disaster. *Area* 39 (4), 482–9.

Füssel, H. M. (2007). Vulnerability: A generally applicable conceptual framework for climate change research. *Global Environmental Change* 17, 155–67.

Gaillard, J. C. (2007). Resilience of traditional societies in facing natural hazards. *Disaster Prevention and Management: An International Journal* 16 (4), 522–44.

Gaillard, J. C. (2015). *People's response to disasters in the Philippines: Vulnerability, capacities and resilience*. Palgrave Macmillan: London.

Gaillard, J. C., Clavé, E., Vibert, O., Dedi, A., Denain, J. C., Efendi, Y., Grancher, D., Liamon, C. C., Sari, D. R., Setiawan, R. (2008). Ethnic groups' response to

the 26th December 2004 earthquake and tsunami in Aceh, Indonesia. *Natural Hazards* 47 (1), 17–38.
Gaillard, J. C., Maceda, E. A., Stasiak, E., Le Berre, I., Espaldon, M. V. O. (2009). Sustainable livelihoods and people's vulnerability in the face of coastal hazards. *Journal of Coastal Conservation* 13 (2–3), 119–29.
Gaillard, J. C., Texier, P. (2010). Religions, natural hazards, and disasters: An introduction. *Religion* 40 (2), 81–4.
Gain, A. K., Giupponi, C. (2015). A dynamic assessment of water scarcity risk in the Lower Brahmaputra River Basin: An integrated approach. *Ecological Indicators* 48, 120–31.
Gain, A. K., Giupponi, C., Renaud, F. G. (2012). Climate change adaptation and vulnerability assessment of water resources systems in developing countries: A generalized framework and a feasibility study in Bangladesh. *Water* 4 (2), 345–66.
Gall, M. (2013). *From Social Vulnerability to Resilience: Measuring Progress toward Disaster Risk Reduction.* InterSecTions (Interdisciplinary Security Connections) 13/2013. Bonn: United Nations University, Institute for Environment and Human Security.
Garmezy, N. (1991). Resilience in children's adaptation to negative life events and stressed environments. *Pediatrics* 20 (9), 459–66.
Garrick, D., Whitten, S., Coggan, A. (2013). Understanding the evolution and performance of water markets and allocation policy: A transaction costs analysis framework. *Ecological Economics* 88, 195–205.
Garschagen, M. (2013). Resilience and organisational institutionalism from a cross-cultural perspective: An exploration based on urban climate change adaptation in Vietnam. *Natural Hazards* 67 (1), 25–46.
Gerl, T., Kreibich, H., Franco, G., Marechal, D., Schröter, K. (2016). A review of flood loss models as basis for harmonization and benchmarking. *PLoS One* 11 (7), e0159791.
Gibson, C. C., Ostrom, E., Ahn, T. K. (2000). The concept of scale and the human dimensions of global change: A survey. *Ecological Economics* 32, 217–39.
Gill, J. C., Malamud, B. D. (2014). Reviewing and visualizing the interactions of natural hazards. *Reviews of Geophysics* 52 (4), 680–722.
Giupponi, C., Biscaro, C. (2015). Vulnerabilities—bibliometric analysis and literature review of evolving concepts. *Environmental Research Letters* 10 (12), 123002.
Giupponi, C., Mojtahed, V., Gain, A., Balbi, S. (2015). Integrated risk assessment of water-related disasters. In P. Paron, G. Di Baldassarre, J. F. Shroder (eds.), *Hydro-meteorological hazards, risks, and disasters*. Amsterdam: Elsevier, pp: 163–200.
Godschalk, D. R. (2003). Urban hazard mitigation: Creating resilient cities. *Natural Hazards Review* 4 (3), 136–43.
Goklany, I. M. (2007). Integrated strategies to reduce vulnerability and advance adaptation, mitigation, and sustainable development. *Mitigation and Adaptation Strategies for Global Change* 12 (5), 755–86.
Goldstein, B. E. (ed.) (2012). *Collaborative resilience. Moving through crisis to opportunity*. Cambridge/London: The MIT Press.
Goodwin, B. K., Vado, L. A. (2007). Public responses to agricultural disasters: Rethinking the role of government. *Canadian Journal of Agricultural Economics* 55 (4), 399–417.

Google Maps (2017a). *Picture of Rome and the city map, showing an analogy with the circular shape.* www.google.at/maps/place/Rom,+Italien/@41.909986,12.3959155,11z/data=!3m1!4b1!4m5!3m4!1s0x132f6196f9928ebb:0xb90f770693656e38!8m2!3d41.9027835!4d12.4963655 [last accessed: May 22, 2017].

Google Maps (2017b). *Picture of Philadelphia and a map of the city showing its typically rectangular shape.* www.google.at/maps/place/Philadelphia,+Pennsylvania,+USA/@40.0046684,-75.2581155,11z/data=!3m1!4b1!4m5!3m4!1s0x89c6b7d8d4b54beb:0x89f514d88c3e58c1!8m2!3d39.9525839!4d-75.1652215 [last accessed: May 22, 2017].

Google Maps (2017c). *Picture of the city of Palma Nuova and its typically star shaped planimetry.* www.google.at/maps/place/33057+Palmanova,+Udine,+Italien/@45.9055387,13.3056509,16z/data=!3m1!4b1!4m5!3m4!1s0x477bb089f331a869:0x5d2129dbb7a04c79!8m2!3d45.909424!4d13.3057286 [last accessed: May 22, 2017].

Google Maps (2017d). *Map of Naples' historical center.* www.google.at/maps/search/Map+of+Naples%E2%80%99+historical+center/@26.136492,-81.8074225,15z/data=!3m1!4b1 [last accessed: May 22, 2017].

Gotham, K. F., Campanella, R. (2011). Coupled vulnerability and resilience: The dynamics of cross-scale interactions in post-Katrina New Orleans. *Ecology and Society* 16 (3), 12.

Graf, W. L. (2001). Damage control: Restoring the physical integrity of America's rivers. *Annals of the Association of American Geographers* 91 (1), 1–27.

Graveline, N. (2016). Economic calibrated models for water allocation in agricultural production: A review. *Environmental Modelling & Software* 81, 12–25.

Green, C. (2003). *Handbook of water economics. Principles and practice.* Chichester: John Wiley & Sons Ltd.

Green, C. (2004). The evaluation of vulnerability to flooding. *Disaster Prevention and Management: An International Journal* 13 (4), 323–29.

Green, C., Tunstall, S. M. (1991). Is the economic evaluation of environmental resources possible? *Journal of Environmental Management* 33 (2), 123–41.

Green, R. (2008). Informal settlements and natural hazard vulnerability in rapid growth cities. In L. Bosher (ed.), *Hazards and the built environment: Attaining built-in resilience.* Oxon: Taylor & Francis, pp: 218–37.

Greene, D. L., Kahn, J. R., Gibson, R. C. (1999). Fuel economy rebound effect for U.S. household vehicles. *The Energy Journal* 20 (3), 1–31.

Greenwood, R., Oliver, C., Suddaby, R., Sahlin-Andersson, K. (2008). *The Sage handbook of organizational institutionalism.* London: Sage.

Gruber, M. (2008). Alternative solutions for public and private catastrophe funding in Austria. *Natural Hazards and Earth System Science* 8 (4), 603–16.

Guillard-Gonçalves, C., Cutter, S. L., Emrich, C. T., Zezere, J. L. (2015). Application of Social Vulnerability Index (SoVI) and delineation of natural risk zones in Greater Lisbon, Portugal. *Journal of Risk Research* 18 (5), 651–74.

Gunderson, L. H. (2000). Ecological resilience—in theory and application. *Annual Review of Ecology and Systematics* 31 (1), 425–39.

Gunderson, L. H. (2003). Adaptive dancing. Interactions between social resilience and ecological crises. In F. Berkes, J. Colding, C. Folke (eds.), *Navigating social-ecological systems. Building resilience for complexity and change.* Cambridge: Cambridge University Press, pp: 33–52.

Gunderson, L. H. (2010). Ecological and human community resilience in response to natural disasters. *Ecology and Society* 15 (2), 18.

Gunderson, L. H., Holling, C. S. (eds.) (2002). *Panarchy: Understanding transformations in human and natural systems*. Washington, Covelo, London: Island press.

Gunderson, L., Light, S. S. (2006). Adaptive management and adaptive governance in the everglades ecosystem. *Policy Sciences* 39 (4), 323–34.

Gunderson, L. H., Holling, C. S., Light, S. S. (eds.) (1995). *Barriers and bridges to the renewal of ecosystems and institutions*. New York: Columbia University Press.

Gupta, J., Termeer, C., Klostermann, J., Meijerink, S., van den Brink, M., Jong, P., Nooteboom, S., Bergsma, E. (2010). The adaptive capacity wheel: A method to assess the inherent characteristics of institutions to enable the adaptive capacity of society. *Environmental Science & Policy* 13 (6), 459–71.

Hagan, P., Maguire, B. (2007). Disasters and communities. Understanding social resilience. *The Australian Journal of Emergency Management* 22 (2), 16–20.

Hahn, M. B., Riederer, A. M., Foster, S. O. (2009). The livelihood vulnerability index: A pragmatic approach to assessing risks from climate variability and change—A case study in Mozambique. *Global Environmental Change* 19 (1), 74–88.

Haimes, Y., Horowitz, B., Lambert, J., Santos, J., Lian, C., Crowther, K. (2005). Inoperability input-output model for interdependent infrastructure sectors. I: Theory and methodology. *Journal of Infrastructure Systems* 11 (2), 67–79.

Hall, B., Kerr, M. L. (1991). *1991-1992 green index*. Washington: Island Press.

Hall, J. W., Brown, S., Nicholls, R. J., Pidgeon, N. F., Watson, R. T. (2012). Proportionate adaptation. *Nature Climate Change* 2 (12), 833–34.

Hallegatte, S. (2008). An adaptive regional input-output model and its application to the assessment of the economic cost of Katrina. *Risk Analysis* 28 (3), 779–99.

Hallegatte, S. (2014). *Economic resilience: Definition and measurement*. Washington: The World Bank.

Hallegatte, S., Bangalore, M., Bonzanigo, L., Fay, M., Kane, T., Narloch, U., Rozenberg, J., Treguer, D., Vogt-Schilb, A. (2015). *Shock waves: Managing the impacts of climate change on poverty*. Washington: The World Bank.

Hallegatte, S., Bangalore, M., Vogt-Schilb, A. C. (2016). *Assessing socioeconomic resilience to floods in 90 countries*. Washington: The World Bank.

Hallegatte, S., Green, C., Nicholls, R. J., Corfee-Morlot, J. (2013). Future flood losses in major coastal cities. *Nature Climate Change* 3 (9), 802–6.

Hallegate, S., Henriet, F., Corfee-Morlot, J. (2011). The economics of climate change impacts and policy benefits at city scale: A conceptual framework. *Climatic Change* 104 (1), 51–87.

Hamilton, M. J. (2007). Coastal landscape and the hedonic price of accommodation. *Ecological Economics* 62 (3–4), 594–602.

Hammond, M. J., Chen, A. S., Djordjevic, S., Butler, D., Mark, O. (2015). Urban flood impact assessment: A state-of-the-art review. *Urban Water Journal* 12 (1), 14–29.

Hanak, E., Stryjewski, E. (2012). *California's water market. By the numbers – update 2012*. San Francisco: Public Policy Institute of California.

Handmer, J. W. (1986). *ANUFLOOD in New Zealand: Part 2, Background to flood loss measurement*. Canberra: Australian National University.

Hanley, N. D., Spash, C. L. (1994). *Cost-benefit analysis and the environment*. Edward Elgar, Hampshire: Aldershot.

Harou, J. J., Pulido-Velazquez, M., Rosenberg, D. E., Medellín-Azuara, J., Lund, J. R., Howitt, R. E. (2009). Hydro-economic models: Concepts, design, applications, and future prospects. *Journal of Hydrology* 375 (3–4), 627–43.

Härtel, C. E. J., Latemore, G. M. (2011). Mud and tears: The human face of disaster – A case study of the Queensland floods, January 2011. *Journal of Management and Organization* 17 (6), 864–72.

Hausmann, P., Kurz, C., Rebuffoni, G. (2012). *Floods in Switzerland – an underestimated risk*. Zurich: Swiss Re.

Hawkins, R. L., Maurer, K. (2010). Bonding, bridging and linking: How social capital operated in New Orleans following Hurricane Katrina. *British Journal of Social Work* 40 (6), 1777–93.

Healey, P. (2009). In search of the "strategic" in spatial strategy making. *Planning Theory & Practice* 10 (4), 439–57.

Heckelei, T., Britz, W., Zhang, Y. (2012). *Positive mathematical programming approaches – Recent developments in literature and applied modelling*. www.fupress.net/index.php/bae/article/view/10567 [last accessed: May 22, 2017].

Heesen, J., Lorenz, D. F., Nagenborg, M., Voss, M., Wenzel, B. (2014). Blind spots on Achilles' heel. On the limitations of Vulnerability and resilience mapping in research. *International Journal of Disaster Risk Science* 5 (1), 74–85.

Hempel, L., Lorenz, D. F. (2014). Resilience as an element of a sociology of expression. *Behemoth* 7 (2), 26–72.

Henkel, K. E., Dovidio, J. F., Gaertner, S. L. (2006). Institutional discrimination, individual racism and Hurricane Katrina. *Analyses of Social Issues and Public Policy* 6 (1), 99–124.

Herman, O. W. (1969). Econometrics as pioneering in nonexperimental model building. *Econometrica* 37 (3), 369–81.

Herrfahrdt-Pähle, E., Pahl-Wostl, C. (2012). Continuity and change in social-ecological systems: The role of institutional resilience. *Ecology and Society* 17 (2), 8.

Hewitt, K. (1983). The idea of calamity in a technocratic age. In K. Hewitt (ed.), *Interpretations of calamity from the viewpoint of human ecology*. Boston: Allen & Unwin Inc., pp: 3–32.

Hewitt, K. (1997). *Regions of risk: A geographical introduction to disasters*. Essex: Longman.

Hewitt, K., Burton, I. (1971). *The hazardousness of a place: A regional ecology of damaging events*. Toronto: University of Toronto.

Heyes, A. (2000). Implementing environmental regulation: Enforcement and compliance. *Journal of Regulatory Economics* 17 (2), 107–29.

Hill, E. W., Wial, H., Wolman, H. (2008). *Exploring regional economic resilience*. Berkeley: Institute of Regional Economic Development, University of California. www.econstor.eu/handle/10419/59420 [last accessed: May 22, 2017].

Hinkel, J. (2011). Indicators of vulnerability and adaptive capacity: Towards a clarification of the science – policy interface. *Global Environmental Change* 21 (1), 198–208.

Hobfoll, S. E., Schröder, K. E. E., Wells, M., Malek, M. (2002). Communal versus individualistic construction of sense of mastery in facing life challenges. *Journal of Social and Clinical Psychology* 21 (4), 362–69.

Hochrainer, S. (2006). *Macroeconomic risk management against natural disasters: Analysis focussed on governments in developing countries*. Wiesbaden: Deutscher Universitäts-Verlag.

Hochrainer, S. (2009). *Assessing the macroeconomic impacts of natural disasters – are there any?* Washington: The World Bank.

Hodgson, G. M. (1988). *Economics and institution. A manifesto for a modern institutional economics*. Cambridge, Polity Press.
Hodgson, G. M. (2000). What is the essence of institutional economics. *Journal of Economic Issues* 34, 317–29.
Hodgson, G. M. (2003). The hidden persuaders: Institutions and individuals in economic theory. *Cambridge Journal of Economics* 27 (2), 159–75.
Hodgson, G. M. (2006). What are institutions? *Journal of Economic Issues* 40 (1), 1–25.
Hodgson, G. M. (2007). Institutions and individuals: Interaction and evolution. *Organization Studies* 28 (1), 95–116.
Hodgson, G. M., Jiang, S. (2007). The economics of corruption and the corruption of economics: An institutionalist perspective. *Journal of Economic Issues* 41 (4), 1043–61.
Holling, C. S. (1973). Resilience and stability of ecological systems. *Annual Review of Ecology and Systematics* 4, 1–23.
Holling, C. S. (1978). *Adaptive environmental assessment and management*. Chichester: John Wiley & Sons Ltd.
Holling, C. S. (1986). The resilience of terrestrial ecosystems: Local surprise and global change. In W. C. Clark, R. E. Munn (eds.), *Sustainable development of the biosphere*. Cambridge: Cambridge University Press, pp: 292–317.
Holling, C. S. (1996). Engineering resilience versus ecological resilience. In P. Schulze (ed.), *Engineering within ecological constraints*. Washington: National Academy, pp: 31–44.
Holling, C. S. (2001). Understanding the complexity of economic, ecological, and social system. *Ecosystems* 4 (5), 390–405.
Holling, C. S., Gunderson, L. H., Peterson, G. D. (2002). Sustainability and panarchies. In L. H. Gunderson, C. S. Holling (eds.), *Panarchy: Understanding transformations in human and natural systems*. Washington, Covelo, London: Island press, pp: 63–102.
Holub, M., Fuchs, S. (2008). Benefits of local structural protection to mitigate torrent-related hazards. In C. A. Brebbia, E. Beriatos (eds.), *Risk analysis VI: Simulation and hazard mitigation. Southampton*. Boston: WIT Press, pp: 401–11.
Holub, M., Fuchs, S. (2009). Mitigating mountain hazards in Austria – Legislation, risk transfer, and awareness building. *Natural Hazards and Earth System Sciences* 9 (2), 523–37.
Holub, M., Suda, J., Fuchs, S. (2012). Mountain hazards: Reducing vulnerability by adapted building design. *Environmental Earth Sciences* 66 (7), 1853–70.
Höppner, C., Whittle, R., Bründl, M., Buchecker, M. (2012). Linking social capacities and risk communication in Europe: A gap between theory and practice? *Natural Hazards* 64 (2), 1753–78.
Horridge, M., Madden, J., Wittwer, G. (2005). The impact of the 2002–2003 drought on Australia. *Journal of Policy Modelling* 27 (3), 285–308.
Hossain, K., Quaik, S., Ismail, N., Rafatullah, M., Ali, I., Hatta, Z. A., Avasn, M., Pant, G., Rameeja, S. (2015). Climate change-perceived impacts on agriculture, vulnerability and response strategies for improving adaptation practice in developing countries (South asian region). *International Journal of Agricultural Research* 11 (1), 1–12.
HUD – US Department of Housing and Urban Development Department (2015). *Sandy damage estimates based on FEMA IA registrant inspection data*. https://catalog.data.gov/dataset/sandy-damage-estimates-based-on-fema-ia-registrant-inspection-data [last accessed: October 28, 2016].

HUD – US Department of Housing and Urban Development Department (2017). *National disaster resilience competition*. HUD Exchange. www.hudexchange.info/programs/cdbg-dr/resilient-recovery/ [last accessed: July 03, 2017].
Hufschmidt, G. (2011). A comparative analysis of several vulnerability concepts. *Natural Hazards* 58 (2), 621–43.
Husby, T. G., Mechler, R., Jongman, B. (2016). What if Dutch investors started worrying about flood risk? Implications for disaster risk reduction. *Regional Environmental Change* 16 (2), 565–74.
Hutter, G. (2007). Strategic planning for long-term flood risk management – some suggestions for learning how to make strategy at regional and local level. *International Planning Studies* 12 (3), 273–89.
Hutter, G. (2016). Collaborative governance and rare floods in urban regions – dealing with uncertainty and surprise. *Environmental Science & Policy* 55, 302–8.
Hutter, G., Kuhlicke, C. (2013). Resilience, talk and action – Exploring the meanings of resilience in the context of planning and institutions. *Planning Practice & Research* 28 (3), 294–306.
Hutter, G., Leibenath, M., Mattissek, A. (2014). Governing through resilience? Exploring flood protection in Dresden, Germany. *Social Sciences* 3 (2), 272–87.
Hutter, G., Schinke, R. (2016). Learning from and for rare floods in Dresden – how public officials interpret damage simulation results at the building type level. *E3S Web of Conferences* 7 (16005), 1–8.
Huxham, Ch., Vangen, S. E. (2005). *Managing to collaborate: The theory and practice of collaborative advantage*. Milton Park/New York: Routledge.
ICC – Insurance Compensation Consortium (2014). *Insurance compensation consortium*. www.consorseguros.es/web/ [last accessed: February 02, 2014].
ICPR – International Commission for the Protection of the Rhine (2001). *Rhine-Atlas*. Koblenz: International Rhine Commission for the Protection of the Rhine. www.rheinatlas.de [last accessed January 10, 2017].
IDGEC – International Dimensions of Global Environmental Change (1999). *Institutional dimensions of global environmental change*. Bonn: IHDP Report.
IFRC – International Federation of Red Cross and Red Crescent Societies (2012). *Law and regulation for the reduction of risk from natural hazards in Austria: A national law desk study*. Geneva: IFRC.
International Institute for Sustainable Development (IISD). (2017). Global Platform on DRR. *Bulletin* 141 (11), 26 May 2017. http://enb.iisd.org/isdr/globalplatform/2017/html/enbplus141num11e.html [last accessed: July 03, 2017].
IPCC – Intergovernmental Panel on Climate Change (2012). *Managing the risks of extreme events and disasters to advance climate change adaptation. A special report of Working Groups I and II of the intergovernmental panel on climate change*. Cambridge: Cambridge University Press.
IPCC – Intergovernmental Panel on Climate Change (2014a). *IPCC fifth assessment report (AR5) (No. WGII). Intergovernmental panel on climate change*. Cambridge: Cambridge University Press.
IPCC – Intergovernmental Panel on Climate Change (2014b). Annex II: Glossary. In Core Writing Team, R. K. Pachauri, L. A. Meyer (eds.), *Climate change 2014: Synthesis report. contribution of working groups I, II and III to the fifth assessment report of the intergovernmental panel on climate change*. Cambridge: Cambridge University Press, pp. 117–30.

ISTAT – Istituto nazionale di statistica (2011). *14 Censimento generale della popolazione e delle abitazioni.* http://dawinci.istat.it/MD/dawinciMD.jsp [last accessed: May 22, 2017].

Janssen, M. A., Schoon, M. L., Ke, K. W., and Borner, K. (2006). Scholarly networks on resilience, vulnerability and adaptation within the human dimensions of global environmental change. *Global Environmental Change*, 16, 240–52.

Jessop, B. (2001). Institutional re(turns) and the strategic – relational approach. *Environment and Planning A* 33, 1213–35.

Jessop, B. (2011). Metagovernance. In M. Bevir (ed.), *The SAGE handbook of governance*. London: Sage, pp: 106–123.

Joas, H., Knöbl, W. (2013). *Sozialtheorie. Zwanzig einführende Vorlesungen.* Frankfurt a. M.: Suhrkamp.

Johnson, C. (2011). *Creating and enabling environment for reducing disaster risk: Recent experience of regulatory frameworks for land, planning and building in low and middle-income countries.* Geneva: UNISDR.

Johnson, D. P., Stanforth, A., Lulla, V., Luber, G. (2012). Developing an applied extreme heat vulnerability index utilizing socioeconomic and environmental data. *Applied Geography* 35 (1–2), 23–31.

Jones, E. C., Faas, A. J., Murphy, A. D., Tobin, G. A., Whiteford, L. M., McCarty, C. (2013). Cross-cultural and site-based influences on demographic, well-being, and social network predictors of risk perception in hazard and disaster settings in Ecuador and Mexico. *Human Nature* 24 (1), 5–32.

Jones, E. C., Murphy, A. D., Faas, A. J., Tobin, G. A., McCarty, C., Whiteford, L. M. (2015). Postdisaster reciprocity and the development of inequality in personal networks. *Economic Anthropology* 2 (2), 385–404.

Jones, E. C., Tobin, G. A., McCarty, C., Whiteford, L., Murphy, A., Faas, A. J., Yepes, H. (2014). Articulation of personal network structure with gendered well-being in disaster and relocation settings. In L. W. Roeder (ed.), *Issues of Gender and Sexual Orientation in Humanitarian Emergencies.* Dordrecht: Springer, pp: 19–31.

Jongman, B., Koks, E. E., Husby, T. G., Ward, P. J. (2014). Increasing flood exposure in the Netherlands: Implications for risk financing. *Natural Hazards and Earth System Sciences* 14 (5), 1245–55.

Jongman, B., Kreibich, H., Apel, H., Barredo, J. I., Bates, P. D., Feyen, L., Gericke, A., Neal, J., Aerts, J. C. J. H., Ward, P. J. (2012). Comparative flood damage model assessment: Towards a European approach. *Natural Hazards and Earth System Sciences* 12 (12), 3733–52.

Jongman, B., Winsemius, H. C., Aerts, J. C. J. H., Coughlan de Perez, E., van Aalst, M. K., Kron, W., Ward, P. J. (2015). Declining vulnerability to river floods and the global benefits of adaptation. *Proceedings of the National Academy of Sciences of the United States of America* 112 (8), E2271–80.

Jonkman, S. N., Bockarjova, M., Kok, M., Bernardini, P. (2008). Integrated hydrodynamic and economic modelling of flood damage in the Netherlands. *Ecological Economics* 66 (1), 77–90.

Joseph, J. (2013). Resilience as embedded neoliberalism: A governmentality approach. *Resilience: International Policies, Practices and Discourses* 1 (1), 38–52.

Jun, K. S., Chung, E. S., Sung, J. Y., Lee, K. S. (2011). Development of spatial water resources vulnerability index considering climate change impacts. *Science of The Total Environment* 409 (24), 5228–42.

Kahn, M. E. (2005). The death toll from natural disasters: The role of income, gegraphy and institutions. *Review of Economics and Statistics* 87 (2), 271–84.

Kajitani, Y., Tatano, H. (2009). Estimation of lifeline resilience factors based on surveys of Japanese industries. *The Professional Journal of the Earthquake Engineering Research Institute* 25 (4), 755–76.

Kaniasty, K. (2012). Predicting social psychological well-being following trauma: The role of postdisaster social support. *Psychological Trauma: Theory, Research, Practice, and Policy* 4 (1), 22–33.

Kappes, M., Papathoma-Köhle, M., Keiler, M. (2012). Assessing physical vulnerability for multi-hazards using an indicator-based methodology. *Applied Geography* 32 (2), 577–90.

Kapucu, N., Hawkins, C. V., Rivera, F. I. (eds.) (2013). *Disaster resiliency: Interdisciplinary perspectives*. London: Routledge.

Kapucu, N., Ozerdem, A. (2011). *Managing emergencies and crises*. Burlington: Jones and Barlett.

Karagiorgos, K., Heiser, M., Thaler, T., Hübl, J., Fuchs, S. (2016). Micro-sized enterprises: Vulnerability to flash floods. *Natural Hazards* 84 (2), 1091–107.

Kasperson, R. E., Kasperson, J. X., Dow, K. (2001). Vulnerability, equity, and global environmental change. In J. X. Kasperson, R. E. Kasperson (eds.), *Global environmental risk*. Sterling: Earthscan. pp: 247–72.

Kasperson, J. X., Kasperson, R. E., Turner, B. L. (eds.) (1995). *Regions at risk: comparisons of threatened environments*. Tokyo: United Nations University.

Kasperson, R. E., Renn, O., Slovic, P., Brown, H. S., Emel, J., Goble, R., Kasperson, J. X., Ratick, S. (1988). The social amplification of risk: A conceptual framework. *Risk Analysis* 8 (2), 177–87.

Kates, R. W., Travis, W. R., Wilbanks, T. J. (2012). Transformational adaptation when incremental adaptations to climate change are insufficient. *Proceedings of the National Academy of Sciences of the United States of America* 109 (19), 7156–61.

Kaufmann, D., Kraay, A., Mastruzzi, M. (2010). *The worldwide governance indicators: Methodology and analytical issues*. Washington: The World Bank.

Kayes, D. C. (2015). *Organizational resilience. How learning sustains organizations in crisis, disaster, and breakdown*. Oxford: Oxford University Press.

Keck, M., Sakdadolrak, P. (2013). What is social resilience? Lessons learned and ways forward. *Erdkunde* 67 (1), 5–19.

Keiler, M., Zischg, A., Fuchs, S., Hama, M., Stötter, J. (2005). Avalanche related damage potential – changes of persons and mobile values since the mid-twentieth century, case study Galtür. *Natural Hazards and Earth System Sciences* 5 (1), 49–58.

Kelly, P. M., Adger, W. N. (2000). Theory and practise in assessing vulnerability to climate change and facilitating adaptation. *Climatic Change* 47 (4), 325–52.

Kelman, I., Mather, T. A. (2008). Living with volcanoes: The sustainable livelihoods approach for volcano-related opportunities. *Journal of Volcanology and Geothermal Research* 172 (3), 189–98.

Kelman, I., Spence, R. (2004). An overview of flood actions on buildings. *Engineering Geology* 73 (3–4), 297–309.

Kelman, I., Gaillard, J. C., Lewis, J., Mercer, J. (2016). Learning from the history of disaster vulnerability and resilience research and practice for climate change. *Natural Hazards* 82, S129–S143.

Khasalamwa, S. (2009). Is "build back better" a response to vulnerability? Analysis of post-tsunami humanitarian interventions in Sri Lanka. *Norwegian Journal of Geography* 63 (1), 73–88.

Kiedrzyńska, E., Kiedrzyński, M., Zalewski, M. (2014). Sustainable floodplain management for flood prevention and water quality improvement. *Natural Hazards* 76 (2), 955–77.

Kimhi, S. (2014). Levels of resilience: Associations among individual, community, and national resilience. *Journal of Health Psychology* 21 (2), 164–70.

Kind, J. M. (2014). Economically efficient flood protection standards for the Netherlands. *Journal of Flood Risk Management* 7 (2), 103–17.

King, D. (2001). Uses and limitations of socioeconomic indicators of community vulnerability to natural hazards: Data and disasters in Northern Australia. *Natural Hazards* 24 (2), 147–56.

King, D., MacGregor, C. (2000). Using social indicators to measure community vulnerability to natural hazards. *Australian Journal of Emergency Management* 15 (3), 52–7.

Klein, R. J., Nicholls, R. J., Thomalla, F. (2003). Resilience to natural hazards: How useful is this concept? *Global Environmental Change Part B: Environmental Hazards* 5 (1), 35–45.

Kleindorfer, P. R., Kunreuther, H. (1999). The complementary roles of mitigation and insurance in managing catastrophic risks. *Risk Analysis* 19 (4), 727–38.

Klinke, A., Renn, O. (2002). A new approach to risk evaluation and management: Risk-based, precaution-based, and discourse-based strategies. *Risk Analysis* 22 (6), 1071–94.

Klitgaard, R. (2010). *Addressing corruption in Haiti*. Vienna: UNDOC.

Kobayashi, T., Medina, R., Cova, T. J. (2011). Visualizing diurnal population change in urban areas for emergency management. *Professional Geographer* 63 (1), 113–30.

Koks, E. E., Jongman, B., Husby, T. G., Botzen, W. J. W. (2015). Combining hazard, exposure and social vulnerability to provide lessons for flood risk management. *Environmental Science & Policy* 47, 42–52.

Koontz, T. M. (2003). *An introduction to the institutional analysis and development (IAD) framework for forest management research*. Columbus: The Ohio State University.

Korf, B., Habullah, S., Hollenbach, P., Klem, B. (2010). The gift of disaster: The commodification of good intentions in post-tsunami Sri-Lanka. *Disasters* 34 (S1), s60–s77.

Kousky, C. (2010). Learning from extreme events: Risk perceptions after the flood. *Land Economics* 86 (3), 395–422.

Kousky, C., Kunreuther, H. (2013). *Addressing affordability in the national flood insurance program*. Washington: Resources for the Future and the Wharton Risk Management and Decision Processes Center.

Kreibich, H., Piroth, K., Seifert, I., Maiwald, H., Kunert, U., Schwarz, J., Merz, B., Thieken, A. H. (2009). Is flow velocity a significant parameter in flood damage modelling? *Natural Hazards and Earth System Sciences* 9 (5), 1679–92.

Kreibich, H., Seifert, I., Merz, B., Thieken, A. H. (2010). Development of FLEMOcs – A new model for the estimation of flood losses in the commercial sector. *Hydrological Sciences Journal* 55 (8), 1302–14.

Kreibich, H., Thieken, A. H. (2008). Assessment of damage caused by high groundwater inundation. *Water Resources Research* 44 (9), W09409.

Kreibich, H., Thieken, A. H., Petrow, T., Müller, M., Merz, B. (2005). Flood loss reduction of private households due to building precautionary measures – lessons learned from the Elbe flood in August 2002. *Natural Hazards and Earth System Sciences* 5 (1), 117–26.
Kreibich, H., van den Berg, J. C. J. M., Bouwer, L. M., Bubeck, P., Ciavola, P., Green, C., Hallegatte, S., Logar, I., Meyer, V., Schwarze, R., Thieken, A. H. (2014). Costing natural hazards. *Nature Climate Change* 4 (5), 303–6.
Kroll, C. A., Landis, J. D., Shen, Q., Stryker, S. (1991). *Economic impacts of the Loma Prieta earthquake: A focus on small business. Studies on the Loma Prieta Earthquake*. Berkely: University of California, Transportation Centre.
Krzysztofowicz, R., Davis, D. R. (1983). Category-unit loss functions for flood forecast-response system evaluation. *Water Resources Research* 19 (6), 1476–80.
Kuhlicke, C. (2013). Resilience: A capacity and a myth: Findings from an in-depth case study in disaster management research. *Natural Hazards* 67 (1), 61–76.
Kuhlicke, C., Scolobig, A., Tapsell, S., Steinführer, A., De Marchi, B. (2011). Contextualizing social vulnerability: Findings from case studies across Europe. *Natural Hazards* 58 (2), 789–810.
Kunreuther, H. (1996). Mitigating disaster losses through insurance. *Journal of Risk and Uncertainty* 12 (2–3), 171–87.
Kunreuther, H., Pauly, M. (2006). Rules rather than discretion: Lessons from Hurricane Katrina. *Journal of Risk Uncertainty* 33 (1), 101–16.
Kuttner, R. (1999). *Everything for sale*. Chicago: The University of Chicago Press.
Lammond, J., Proverbs, D., Hammond, F. (2010). The impact of flooding on the price of residential property: A transactional analysis of the UK market. *Housing Studies* 25 (3), 335–56.
Larkin, S., Fox-Lent, C., Eisenberg, D. A., Trump, B. D., Wallace, S., Chadderton, C., Linkov, I. (2015). Benchmarking agency and organizational practices in resilience decision making. *Environment Systems & Decisions*, 35 (2), 185–95.
Larson, L., Pasencia, D. (2001). No adverse impact: New direction in floodplain management policy. *Natural Hazards Review* 2 (4), 167–81.
Latora, V., Marchiori, M. (2001). Efficient behavior of small-world networks. *Physical Review Letters* 87 (19), 198701.
Lei, Y., Wang, J., Yue, Y., Zhou, H., Yin, W. 2014. Rethinking the relationships of vulnerability, resilience, and adaptation from a disaster risk perspective. *Natural Hazards*, 70, 609–27.
Leiter, A. M., Pruckner, G. J. (2009). Proportionality of willingness to pay to small changes in risk: The impact of attitudinal factors in scope tests. *Environmental and Resource Economics* 42 (2), 169–86.
Lenzen, M. (2011). Aggregation versus disaggregation in input-output analysis of the environment. *Economic Systems Research* 23 (1), 73–89.
Leone, F., Asté, J.-P., Leroi, E. (1996). L'évaluation de la vulnérabilité aux mouvements du terrain: Pour une meilleure quantification du risque. *Revue de Géographie Alpine* 84 (1), 35–46.
Leone, F., Asté, J.-P., Velásquez, E. (1995). Contribution des constats d'endommagement au développement d'une méthodologie d'évaluation de la vulnérabilité appliquée aux phénomènes de mouvements de terrain. *Bulletin de l'Association de Géographes* 1995 (4), 350–71.

Lewis, J. (2011). The hidden perpetrator of under-development and vulnerability to natural hazards and disasters. *JAMBA: Journal of Disaster Risk Studies* 3 (2), 464–75.

Linnerooth-Bayer, J., Mechler, R., Hochrainer, S. (2011). Insurance against losses from natural disasters in developing countries. Evidence, gaps and the way forward. *Journal of Integrated Disaster Risk Management* 1 (1), 59–81.

Little, R. G. (2002). Controlling cascading failure: Understanding the vulnerabilities of interconnected infrastructures. *Journal of Urban Technology* 9 (1), 109–23.

Little, R. G. (2004). A socio-technical systems approach to understanding and enhancing the reliability of interdependent infrastructure systems. *International Journal of Emergency Management* 2 (1-2), 98–110.

Liverman, D. M. (1990). Vulnerability to global environmental change. In R. E. Kasperson, K. Dow, D. Golding, J. X. Kasperson (eds.), *Understanding global environmental change: The contributions of risk analysis and management*. Worcester: Clark University, pp: 27–44.

Lo, A. Y., Spash, C. (2013). Deliberative monetary valuation: In search of a democratic and value plural approach to environmental policy. *Journal of Economic Surveys* 27 (4), 768–89.

Lobo, J., Bettencourt, L. M., Strumsky, D., West, G. B. (2013). Urban scaling and the production function for cities. *PloS One* 8 (3), e58407.

Loch, A., Adamson, D., Mallawaarachchi, T. (2014). Role of hydrology and economics in water management policy under increasing uncertainty. *Journal of Hydrology* 518 (Part A), 5–16.

Loch, A., Bjornlund, H., Wheeler, S., Connor, J. (2012). Allocation trade in Australia: A qualitative understanding or irrigator motives and behaviour. *The Australian Journal of Agricultural and Resource Economics* 56 (1), 42–60.

Lonsdale, K. G., Downing, T. E., Nicholls, R. J., Parker, D., Vafeidis, A. T., Dawson, R., Hall, J. (2008). Plausible responses to the threat of rapid sea-level rise in the Thames Estuary. *Climatic Change* 91 (1), 145–69.

Lorenz, D. F. (2013). The diversity of resilience. Contributions from social science perspective. *Natural Hazards* 67 (1), 7–24.

Lorenz, D. F., Dittmer, C. (2016). Resilience in catastrophes, disasters and emergencies. Socio-scientific perspectives. In A. Maurer (ed.), *New perspectives on resilience in socio-economic spheres*. Heidelberg: Springer, pp: 25–59.

Lotka, A. J. (1924). *Elements of physical biology*. Baltimore: Williams & Wilkins.

Loucks, D., Stedinger, J. (2007). Thoughts on the economics of floodplain development in the U.S. In O. Vasiliev, P. van Gelder, E. Plate, M. Bolgov (eds.), *Extreme hydrological events: New concepts for security*. Dordrecht: Springer, pp. 3–19.

Lüchters, G., Menkhoff, L. (1996). Human development as statistical artifact. *World Development* 24 (8), 1385–92.

Lucini, B. (2014). Multicultural approaches to disaster and cultural resilience. How to consider them to improve disaster management and prevention: The Italian case of two earthquakes. *Procedia Economics and Finance* 18, 151–6.

Luechinger, S., Raschky, P. A. (2009). Valuing flood disasters using the life satisfaction approach. *Journal of Public Economics* 93 (3–4), 620–3.

Luino, F., Cirio, C. G., Biddoccu, M., Agangi, A., Giuletto, W., Godone, F., Nigrelli, G. (2009). Application of a model to the evaluation of flood damage. *GeoInformatica* 13 (3), 339–53.

Luisetti, T., Turner, R. K., Bateman, I. J., Morse-Jones, S., Adams, C., Fonseca, L. (2011). Coastal and marine ecosystem services valuation for policy and management: Managed realignment case studies in England. *Ocean and Coastal Management* 54 (3), 212–24.

Lukes, S. (2005). *Power – a radical review*. New York: Macmillan.

Macamo, E. (2003). Nach der Katastrophe ist die Katastrophe. Die 2000er Überschwemmung in der dörflichen Wahrnehmung in Mosambik. In L. Clausen, E. M. Geenen, E. Macamo (eds.), *Entsetzliche soziale Prozesse. Theorie und Empirie der Katastrophen*. Münster: LIT, pp: 167–84.

MacKinnon, D., Derickson, K. D. (2013). From resilience to resourcefulness: A critique of resilience policy and activism. *Progress in Human Geography* 37 (2), 253–70.

Magis, K. (2010). Community resilience: An indicator of social sustainability. *Society and Natural Resources* 23 (5), 401–16.

Mallory-Hill, S., Preiser, W. F., Watson, C. G. (2012). *Enhancing building performance*. Chichester: John Wiley & Sons Ltd.

Mancini, A., Salvati, L., Sateriano, A., Mancino, G., Ferrara, A. (2012). Conceptualizing and measuring the "economy" dimension in the evaluation of socio-ecological resilience: A brief commentary. *International Journal of Latest Trends in Finance and Economic Sciences* 2 (3), 190–6.

Manuel, P., Rapaport, E., Keefe, J., Krawchenko, T. (2015). Coastal climate change and aging communities in Atlantic Canada: A methodological overview of community asset and social vulnerability mapping. *The Canadian Geographer* 59 (4), 433–46.

Manyena, S. B. (2006). The concept of resilience revisited. *Disasters*, 30 (4), 433–50.

Manyena, S. B. (2014). Disaster resilience: A question of "multiple faces" and "multiple spaces"? *International Journal of Disaster Risk Reduction* 8, 1–9.

Manyena, S. B., O'Brien, G., O'Keefe, P., Rose, J. (2011). Disaster resilience: A bounce back or bounce forward ability? *Local Environment* 16 (5), 417–24.

March, J. G. (1991). Exploration and exploitation in organizational learning. *Organization Science* 2 (1), 71–87.

Markard, J., Raven, R., Truffer, B. (2012). Sustainability transitions: An emerging field of research and its prospects. *Research Policy* 41, 955–67.

Marsh, G. P. (1869). *Man and nature or physical geography as modified by human action*. New York: Charles Scribner and Co.

Marshall, G. R. (2013). Transaction costs, collective action and adaptation in managing complex social-ecological systems. *Ecological Economics* 88, 185–94.

Mason, M. S., Haynes, K., Walker, G. (2013). Cyclone tracy and the road to improving wind-resistant design. In S. Boulter, J. Palutikof, D.-J. Karoly, D. Guitart (eds.), *Natural disasters and adaptation to climate change*. Cambridge: Cambridge University Press, pp: 87–94.

Masten, A. S. (2014). Global perspectives on resilience in children and youth. *Child Development* 85 (1), 6–20.

Masten, A., Obradović, J. (2008). Disaster preparation and recovery: Lessons from research on resilience in human development. *Ecology and Society* 13 (1), 9.

Matyas, D. and Pelling, M. (2012). *Disaster vulnerability and resilience: Theory, modelling and prospective*. London: Foresight, Government Office for Science.

Matthewman, S. (2015). *Disasters, risks and revelation. Making sense of our times*. Basingstoke: Palgrave Macmillan.

Mavrouli, O., Corominas, J. (2010). Vulnerability of simple reinforced concrete buildings to damage by rockfalls. *Landslides* 7 (2), 169–80.

Mazzorana, B., Simoni, S., Scherer, C., Gems, B., Fuchs, S., Keiler, M. (2014). A physical approach on flood risk vulnerability of buildings. *Hydrology and Earth System Sciences* 18 (9), 3817–36.

McEntire, D. A., Myers, A. (2004). Preparing communities for disasters: Issues and processes for government readiness. *Disaster Prevention and Management* 13 (2), 140–52.

Mechler, R., Bouwer, L. M. (2014). Understanding trends and projections of disaster losses and climate change: Is vulnerability the missing link? *Climatic Change* 133 (1), 23–35.

Melecky, C., Raddatz, C. (2011). *How do governments respond after catastrophes? Natural-disaster shocks and the fiscal stance*. Washington: The World Bank.

Menoni, S., Molinari, D., Parker, D., Ballio, F., Tapsell, S. (2012). Assessing multifaceted vulnerability and resilience in order to design risk-mitigation strategies. *Natural Hazards* 64 (3), 2057–82.

Merz, B., Kreibich, H., Schwarze, R., Thieken, A. H. (2010). Review article – assessment of economic flood damage. *Natural Hazards and Earth System Sciences* 10 (8), 1697–724.

Messer, N. M. (2003). *The role of local institutions and their interaction in disaster risk mitigation: A literature review*. Rome: FAO.

Messias, D. K. H., Barrington, C., Lacy, E. (2012). Latino social network dynamics and the Hurricane Katrina disaster. *Disasters* 36 (1), 101–21.

Messner, F., Meyer, V. (2005). *Flood damage, vulnerability and risk perception—challenges for flood damage research*. Leipzig: UFZ-Diskussionspapiere No. 13/2005.

Messner, F., Penning-Rowsell, E., Green, C., Meyer, V., Tunstall, S., van der Veen, A. (2007). *Evaluating flood damages: Guidance and recommendations on principles and methods*. https://repository.tudelft.nl/islandora/object/uuid:5602db10-274c-40da-953f-34475ded1755?collection=research [last accessed: May 22, 2017].

Methmann, C., Oels, A. (2014). Vulnerability. In C. Death (ed.), *Critical environmental politics*. Abingdon, Oxon: Routledge, pp: 277–286.

Meyer, J. W., Jepperson, R. L. (2000). The "actors" of modern society: The cultural construction of social agency. *Sociological Theory* 18 (1), 100–20.

Meyer, V., Becker, N., Markantonis, V., Schwarze, R., van den Bergh, J. C. J. M., Bouwer, L. M., Bubeck, P., Ciavola, P., Genovese, E., Green, C., Hallegatte, S., Kreibich, H., Lequeux, Q., Logar, I., Papyrakis, E., Pfurtscheller, C., Poussin, J., Przyluski, V., Thieken, A. H., Viavattene, C. (2013). Review article: Assessing the costs of natural hazards – state of the art and knowledge gaps. *Natural Hazards and Earth System Sciences* 13 (5), 1351–73.

Michel-Kerjan, E. O., Kousky, C. (2010). Come rain or shine: Evidence on flood insurance purchases in Florida. *Journal of Risk and Insurance* 77 (2), 369–97.

Mileti, D. (1999). *Disasters by design: A reassessment of natural hazards in the United States*. Washington: Joseph Henry Press.

Milgram, S. (1967). The small world problem. *Psychology Today* 2 (1), 60–7.

Milgram, S. (1970). The experience of living in cities. *Science* 167 (3924), 1461–8.

Miller, F., Osbahr, H., Boyd, E., Thomalla, F., Bharwani, S., Ziervogel, G., Walker, B., Birkmann, J., Van der Leeuw, S., Rockström, J., Hinkel, J.,

Downing, T., Folke, C., and Nelson, D. (2010). Resilience and vulnerability: Complementary or conflicting concepts? *Ecology and Society*, 15 (3), 11. [online] www.ecologyandsociety.org/vol15/iss3/art11/.
Mitchell, J. K., Devine, N., Jagger, K. (1989). A contextual model of natural hazard. *Geographical Review* 79 (4), 391–409.
Molinari, D., Menoni, S., Aronica, G. T., Ballio, F., Berni, N., Pandolfo, C., Stelluti, M., Minucci, G. (2014). Ex-post damage assessment: an Italian experience. *Natural Hazards and Earth System Sciences* 14 (4), 901–16.
Montz, B. E. (1992). The effects of flooding on residential property values in three New Zealand communities. *Disasters* 16 (4), 283–93.
Montz, B. E., Cross, J., Cutter, S. L. (2005). Hazards. In G. L. Gaile, C. J. Wilmott (eds.), *Geography in America at the dawn of the 21st century*. Oxford: Oxford University Press, pp: 479–91.
Moodysson, J., Jonsson, O. (2007). Knowledge collaboration and proximity: The spatial organization of biotech innovation projects. *European Urban and Regional Studies* 14 (2), 115–31.
Mudelsee, M., Börngen, M., Tetzlaff, G., Grünewald, U. (2003). No upward trends in the occurrence of extreme floods in central Europe. *Nature* 425 (6954), 166–9.
Munich, Re (2011). *NATHAN, World Map of Natural Hazards, version 2011*. München: Münchener Rückversicherungs-Gesellschaft.
Munich, Re (2017). *Topics GEO: Natural catastrophes 2016. Analyses, assessments, positions*. München: Münchener Rückversicherungs-Gesellschaft.
MURL – Ministry for Environment, Regional Planning and Agriculture of North-Rhine Westphalia (2000). *Potentielle Hochwasserschäden am Rhein in Nordrhein-Westfalen*. Düsseldorf: Landesregierung Nordrhein-Westfalen.
Mysiak, J., Testella, F., Bonaiuto, M., Carrus, G., De Dominicis, S., Ganucci Cancellieri, U., Firus, K., Grifoni, P. (2013). Flood risk management in Italy: Challenges and opportunities for the implementation of the EU Floods Directive (2007/60/EC). *Natural Hazards and Earth System Sciences* 13 (11), 2883–90.
NAS – National Academies of Science (2012). *Disaster resilience: A national imperative*. Washington: The National Academics. www.nap.edu/catalog/13457/disaster-resilience-a-national-imperative.
National Institute of Standards and Technology (NIST) (2016). *Community resilience planning guide for buildings and infrastructure systems*, Volume 1. Washington: US Department of Commerce, NIST Special Publication 1190–1. www.nist.gov/sites/default/files/community-resilience-planning-guide-volume-1_0.pdf [last accessed: July 03, 2017].
National Research Council. (2012). *Disaster resilience: A national imperative*. Washington: US National Academies of Science.
National Research Council. (2014). *Reducing coastal risks on the East and Gulf Coasts*. Washington: The National Academies Press.
Nelson, D. R., Adger, W. N., Brown, K. (2007). Adaptation to environmental change: Contributions of a resilience framework. *The Annual Review of Environment and Resources* 32, 395–419.
Nicholas, J., Holt, G. D., Proverbs, D. (2001). Towards standardising the assessment of flood damage properties in the UK. *Structural Survey* 19 (4), 163–72.
Nicholls, R. J., Hanson, S., Herweijer, C., Patmore, N., Hallegatte, S., Corfee-Morlot, J., Chateau, J., Muir-Wood, R. (2008). *Ranking port cities with high*

*exposure and vulnerability to climate extremes: Exposure estimates*. Paris: OECD Publishing.

Nilson, C., Berggren, K. (2000). Alterations of riparian ecosystems caused by river regulation. *BioScience* 50 (9), 783–92.

Nixon, R. (2014). *Slow violence and the environmentalism of the poor*. Boston: Havard University Press.

Norberg, J., Cumming, G. S. (2013). *Complexity theory for a sustainable future*. New York: Columbia University Press.

Norris, F. H., Friedman, M. J., Watson, P. J. (2002). 60,000 disaster victims speak: Part II. Summary and implications of the disaster mental health research. *Psychiatry: Interpersonal and Biological Processes* 65 (3), 240–60.

Norris, F. H., Stevens, S. P. (2007). Community resilience and the principles of mass trauma intervention. *Psychiatry: Interpersonal and Biological Processes* 70 (4), 320–8.

Norris, F. H., Stevens, S. P., Pfefferbaum, B., Wyche, K. F., Pfefferbaum, R. L. (2008). Community resilience as a metaphor, theory, set of capabilities, and strategy for disaster readiness. *American Journal of Community Psychology* 41 (1–2), 127–50.

North, D. (1990). *Institutions, institutional change and economic performance*. Cambridge: Cambridge University Press.

Noy, I. (2009). The macroeconomic consequences of disasters. *Journal of Development Economics* 88 (2), 221–31.

NRC – National Research Council (1999). *The impacts of natural disasters: A framework for loss estimation*. Washington: National Academy Press.

NRC – National Research Council (2012). *Disaster resilience – a national imperative. imperative*. Washington: The National Academies Press.

NRC – National Research Council (2014). *Reducing coastal risks on the East and Gulf Coasts*. Washington: The National Academies Press.

NRE – Department of Natural Resources and Environment (2000). *Rapid appraisal method (RAM) for floodplain management*. Melbourne: Report prepared by Read Sturgess and Associates.

NR&M (Department of Natural Resources and Mines, Queensland Government) (2002): *Guidance on the Assessment of Tangible Flood Damages, Report*. Queensland, Australia.

O'Brien, K. L., Leichenko, R., Kelkar, U., Venema, H., Aandahl, G., Tompkins, H., Javed, A., Bhadwal, S., Barg, S., Nygaard, L., West, J. (2004). Mapping vulnerability to multiple stressors: Climate change and globalization in India. *Global Environmental Change* 14, 303–13.

Oberndorfer, S., Fuchs, S., Rickenmann, D., Andrecs, P. (2007). *Vulnerabilitätsanalyse und monetäre Schadensbewertung von Wildbachereignissen in Österreich*. Vienna: Bundesforschungs- und Ausbildungszentrum für Wald, Naturgefahren und Landschaft (BfW).

O'Brien, K., Eriksen, S., Nygaard, L. P., Schjolden, A. (2007). Why different interpretations of vulnerability matter in climate change discourses. *Climate Policy*, 7 (1), 73–88.

Obrist, B., Pfeiffer, C., Henley, R. (2010). Multi-layered social resilience: A new approach in mitigation research. *Progress in Development Studies* 10 (4), 283–93.

OECD – Organization for Economic Co-operation and Development (2008). *Handbook on constructing composite indicators: Methodology and user's guide*. Paris: Organization for Economic Co-operation and Development.

OECD – Organization for Economic Co-operation and Development (2014). *Guidelines for resilience systems analysis: How to analyse risk and build a roadmap to resilience*. OECD Publishing. Available at www.oecd.org/dac/Resilience%20Systems%20Analysis%20FINAL.pdf [last accessed: July 03, 2017].

OECD – Organization for Economic Co-operation and Development (2015). *The economic consequences of climate change*. Paris: Organisation for Economic Co-operation and Development.

OECD – Organization for Economic Co-operation and Development (2016). *Strengthening economic resilience: Insights from the post-1970 record of severe recessions and financial crises*. Paris: Organisation for Economic Co-operation and Development.

Ofei-Mensah, A., Bennett, J. (2013). Transaction costs of alternative greenhouse gas policies in the Australian transport energy sector. *Ecological Economics* 88, 214–21.

O'Keefe, P., Westgate, K., Wisner, B. (1976). Taking the naturalness out of natural disasters. *Nature* 260, 566–7.

Okuyama, Y. (2009). *Critical review of methodologies on disaster impact estimation. Background paper for assessment on the economics of disaster risk reduction. The Global Facility for Disaster Research and Recovery (GFDRR)*. Genf: GFDRR.

Okuyama, Y. (2014). Disaster and economic structural change: Case study on the 1995 Kobe Earthquake. *Economic Systems Research* 26 (1), 98–117.

Okuyama, Y., Hewing, G. J. D., Sonis, M. (2004). Measuring economic impacts of disasters: Interregional input-output analysis using sequential interindustry model. In Y. Okuyama, S. E. Chang (eds.), *Modeling spatial and economic impacts of disasters*. Berlin, Heidelberg: Springer, pp: 77–102.

Oliver-Smith, A. (1991). Successes and failures in post-disaster resettlement. *Disasters* 15 (1), 12–23.

Ostadtaghizadeh, A., Ardalan, A., Paton, D., Jabbari, H., Khankeh, H. R. (2015). Community disaster resilience: A systematic review on assessment models and tools. *PLOS Currents Disasters*, Apr 8. Edition 1. doi: 10.1371/currents.dis.f224ef8efbdfcf1d508dd0de4d8210ed.

Oumeraci, H. (1994). Review and analysis of vertical breakwater failures — lessons learned. *Coastal Engineering* 22 (1–2), 3–29.

Özyurt, G., Ergin, A. (2010). Improving coastal vulnerability assessments to sea-level rise: A new indicator-based methodology for decision makers. *Journal of Coastal Research* 26 (2), 265–73.

Paavola, J. (2008). Governing atmospheric sinks: The architecture of entitlements in the global commons. *International Journal of the Commons* 2, 313–36.

Papathoma-Köhle, M. (2016). Vulnerability curves vs. vulnerability indicators: Application of an indicator-based methodology for debris-flow hazards. *Natural Hazards and Earth System Sciences* 16 (8), 1771–90.

Papathoma-Köhle, M., Kappes, M., Keiler, M., Glade, T. (2011). Physical vulnerability assessment for alpine hazards: State of the art and future needs. *Natural Hazards* 58 (2), 645–80.

Papathoma-Köhle, M., Zischg, A., Fuchs, S., Glade, T., Keiler, M. (2015). Loss estimation for landslides in mountain areas – an integrated toolbox for vulnerability assessment and damage documentation. *Environmental Modelling & Software* 63, 156–69.

Papathoma-Köhle, M., Gems, B., Sturm, M., Fuchs, S. (2017). Matrices, curves and indicators: A review of approaches to assess physical vulnerability to debris flows. *Earth-Science Reviews* 171, 272–88.

Pappenberger, F., Cloke, H. L., Parker, D. J., Wetterhall, F., Richardson, D. S., Thielen, J. (2015). The monetary benefit of early flood warnings in Europe. *Environmental Science & Policy* 51, 278–91.

Park, J., Cho, J., Rose, A. (2011). Modeling a major source of economic resilience to disasters: Recapturing lost production. *Natural Hazards* 58 (1), 163–82.

Paton, D., Anderson, E., Becker, J., Petersen, J. (2015). Developing a comprehensive model of hazard preparedness: Lessons from the Christchurch earthquake. *International Journal of Disaster Risk Reduction* 14, 37–45.

Paton, D., Johnston, D. (2006). *Disaster resilience: An integrated approach*. Charles C. Thomas Publisher: Springfield, Illinois.

Pattiaratchi, C. (1997). *Design studies and performance monitoring of an artificial surfing reef: Cable Station, Western Australia*. Sydney: Proceedings of the 1st International Surfing Reef Symposium.

Paul, B. K. (2011). *Environmental hazards and disasters–context, perspectives and management*. Chichester: John Wiley & Sons Ltd.

Pauw, K., Thurlow, J., Bachu, M., Van Seventer, D. E. (2011). The economic costs of extreme weather events: A hydrometeorological CGE analysis for Malawi. *Environment and Development Economics* 16 (2), 177–98.

Pearce, L. (2003). Disaster management and community planning: How to achieve sustainable hazard mitigation. *Natural Hazards* 28 (2–3), 211–28.

Pearce, D., Moran, D. (1994). *The economic value of biodiversity*. London: Earthscan.

Peduzzi, P., Dao, H., Herold, C., Mouton, F. (2009). Assessing global exposure and vulnerability towards natural hazards: The Disaster Risk Index. *Natural Hazards and Earth System Sciences* 9 (4), 1149–59.

Peek, L. (2008). Children and disasters: Understanding vulnerability, developing capacities, and promoting resilience—an introduction. *Children Youth and Environments* 18 (1), 1–29.

Pelling, M. (ed.). (2003). *Natural disasters and development in a globalizing world*. London: Routledge.

Pelling, M. (2003b). *The vulnerability of cities. Natural disasters and social resilience*. London: Earthscan.

Pelling, M. (2011). *Adaptation to climate change: From resilience to transformation*. London: Routledge.

Pelling, M. (2013). Review of global risk index projects: Conclusions for sub-national and local approaches. In J. Birkmann (ed.) *Measuring vulnerability to natural hazards: Towards disaster resilient societies*, 2nd Edition, Tokyo: United Nations University Press, pp. 167–96.

Penning-Rowsell, E. C., Chatterton, J. (1977). *The benefits of flood alleviation: A manual of assessment techniques*. Aldershot: Gower Technical Press.

Penning-Rowsell, E. C., Johnson, C., Tunstall, S., Tapsell, S., Morris, J., Chatterton, J., Green, C. (2005). *The benefits of flood and coastal risk management: A handbook of assessment techniques*. London: Middlesex University Press.

Penning-Rowsell, E. C., Priest, S., Johnson, C. (2014). The evolution of UK flood insurance: Incremental change over siy decades. *International Journal of Water Resources Development* 30 (4), 694–713.

Pérez-Blanco, C. D., Delacámara, G., Gómez, C. M. (2015). Water charging and water saving in agriculture. Insights from a Revealed Preference Model in a Mediterranean basin. *Environmental Modelling & Software* 69, 90–100.

Pérez-Blanco, C. D., Delacámara, G., Gómez, C. M. (2016). Revealing the willingness to pay for income insurance in agriculture. *Journal of Risk Research* 19 (7), 873–93.

Pérez-Blanco, C. D., Gómez, C. M. (2014). Insuring water: A practical risk management option in water-scarce and drought-prone regions? *Water Policy* 16 (2), 244–63.

Perrow, C. (1999). *Normal accidents: Living with high risk technologies*. Princeton: Princeton University Press.

Perrow, C. (2007). *The next catastrophe: Reducing our vulnerabilities to natural, industrial, and terrorist disasters*. Princeton: Princeton University Press.

Petrolia, D. R., Landry, C. E., Coble, K. H. (2013). Risk preferences, risk perceptions, and flood insurance. *Land Economics* 89 (2), 227–45.

Pfurtscheller, C. (2014). Regional economic impacts of natural hazards – the case of the 2005 Alpine flood event in Tyrol (Austria). *Natural Hazards and Earth System Sciences* 14 (2), 359–78.

Phillips, B. D., Hearn Morrow, B. (eds.) (2008). *Women and disasters: From theory to practice*. Bloomington: Xlibris Corporation.

Pimm, S. (1984). The complexity and stability of ecosystems. *Nature* 307, 321–26.

Pistrika, A. K., Jonkman, S. N. (2009). Damage to residential buildings due to flooding of New Orleans after hurricane Katrina. *Natural Hazards* 54 (2), 413–34.

Pistrika, A. K., Jonkman, S. N. (2009). https://link.springer.com/article/10.1007/s11069-009-9476-y.

Plate, E. J. (2002). Flood risk and flood management. *Journal of Hydrology* 267 (1–2), 2–11.

Pomonis, A. (2002). The Mount Parnitha (Athens) earthquake of September 7, 1999: A disaster management perspective. *Natural Hazards* 27 (1), 171–99.

Poole, M. S., Van de Ven, A. H. (eds.) (2004). *Handbook of organizational change and innovation*. Oxford: Oxford University Press.

Pope, J. (1997). Responding to coastal erosion and flooding damages. *Journal of Coastal Research* 13 (3), 704–10.

Pope, J., Dean, J. L. (1986). *Development of design criteria for segmented breakwaters. Proceedings, 20th International Conference Coastal Engineering*. New York: ASCE, pp: 2144–58.

Prettenthaler, F., Amrusch, P., Habsburg-Lothringen, C. (2010). Estimation of an absolute flood damage curve based on an Austrian case study under a dam breach scenario. *Natural Hazards and Earth System Sciences* 10 (4), 881–94.

Priest, S. (2014). *Review of international flood insurance and recovery mechanisms: Implications for New Zealand and the resilience of older people*. London: Flood Hazard Research Centre, Middlesex University.

Prince, S. H. (1920). *Catastrophe and social change: Based upon a sociological study if the Halifax Disaster*. London: King and Son.

Provan, K. G., Kenis, P. (2008). Modes of network governance: Structure, management, and effectiveness. *Journal of Public Administration Research and Theory* 18 (2), 229–52.

Pugh, J., 2014. Resilience, complexity and post-liberalism. *Area* 46 (3), 313–19.

Queensland Government (1988). *Queensland parliamentary debates* [Hansard] (Legislative Assembly, Parliamentary Debate No. Thursday, August 25, 1988). Brisbane: Queensland Government.

Queensland Government (2016). *Natural hazards, risk and resilience. Guidance for considering natural hazards, risk and resilience when designating land for community infrastructure. Technical manual. State Planning Policy – state interest guidance*. Queensland, Australia: Department of Infrastructure, Local Government and Planning. www.dilgp.qld.gov.au/resources/guideline/spp/spp-technical-manual-considering-natural-hazards-when-designating-land.pdf [last accessed: May 01, 2017].

Queensland Government (2017). *2017 Brisbane river catchment flood study map*. www.qld.gov.au/environment/water/catchments/brisbane-river-studies/ [last accessed: March 29, 2017].

Queensland Water Commission (2010). *South East Queensland water strategy*. Brisbane: Queensland WaterCommission.

Rabin, M., Kőszegi, B. (2007). Mistakes in choice-based welfare analysis. *American Economic Review* 97 (2), 477–81.

Ramachandran, V., Walz, J. (2012). *Haiti: Where has all the money gone?* Washington: Centre for Gloval Development.

Raschky, P. A. (2008). Institutions and the losses from natural disasters. *Natural Hazards and Earth System Sciences* 8 (4), 627–34.

Raschky, P. A., Schwarze, R., Schwindt, M., Zahn, F. (2013). Uncertainty of governmental relief and the crowding out of flood insurance. *Environmental and Resource Economics* 54 (2), 179–200.

Raschky, P. A., Schwindt, M. (2009). *Aid, natural disasters and the Samaritan's dilemma*. Washington: The World Bank.

Raschky, P. A., Weck-Hannemann, H. (2007). *Charity hazard – a real hazard to natural disaster insurance?* Innsbruck: University of Innsbruck.

Reed, D. A., Kapur, K. C., Christie, R. D. (2009). Methodology for assessing the resilience of networked infrastructure. *IEEE Systems Journal* 3 (2), 174–80.

Reid, C. E., Mann, J. K., Alfasso, R., English, P. B., King, G. C., Lincoln, R. A., Margolis, H. G., Rubado, D. J., Sabato, J. E., West, N. L., Woods, B., Navarro, K. M., Balmes, J. R. (2012). Evaluation of a heat vulnerability index on abnormally hot days: An environmental public health tracking study. *Environmental Health Perspectives* 120 (5), 715–20.

Reissman, D. B., Spencer, S., Tanielian, T. L., Stein, B. D. (2005). Integrating behavioral aspects into community preparedness and response systems. *Journal of Aggression, Maltreatment & Trauma* 10 (3–4), 707–20.

Renn, O. (2008a). Concepts of risk: An interdisciplinary review – part 1: Disciplinary risk concepts. *Gaia* 17 (1), 50–66.

Renn, O. (2008b). Concepts of risk: An interdisciplinary review – part 2: Integrative approaches. *Gaia* 17 (2), 196–204.

Renn, O., Klinke, A., van Asselt, M. (2011). Coping with complexity, uncertainty and ambiguity in risk governance: A synthesis. *Ambio* 40 (2), 231–46.

Republic of Austria (1966). *Katastrophenfondgesetz. BGBl 207/1966*. Vienna: Bundesgesetzblatt.

Republic of Austria (1975). *Forstgesetz. BGBl. 440/1975*. Vienna: Bundesgesetzblatt.

Republic of Austria (1996). *Katastrophenfondsgesetz BGBl. 201/1996*. Vienna: Bundesgesetzblatt.

Republic of Austria (2014). *Wasserrechtgesetz 1959. BGBl. I 54/2014*. Vienna: Bundesgesetzblatt.

Resilience (2009). *National infrastructure advisory council*. Washington: Homeland Security.

Rivera, F. I., Kapucu, N. (2015). *Disaster vulnerability, hazards and resilience: Perspectives from Florida*. New York: Springer.

Roberts, N. J., Nadim, F., Kalsnes, B. (2009). Quantification of vulnerability to natural hazards. *Georisk* 3 (3), 164–73.
Robinson, G. M., Carson, D. A. (2016). Resilient communities: Transitions, pathways and resourcefulness. *The Geographical Journal* 182 (2), 114–22.
Rose, A. (1995). Input-output economics and computable general equilibrium model. *Structural Change and Economic Dynamics* 6 (3), 295–304.
Rose, A. (2004). Economic principles, issues and research priorities in hazard loss estimation. In Y. Okuyama, S. E. Chang (eds.), *Modeling spatial and economic impacts of disasters*. Berlin, Heidelberg: Springer, pp: 13–36.
Rose, A. (2007). Economic resilience to natural and man-made disasters: Multidisciplinary origins and contexts. *Environmental Hazards* 7 (4), 383–92.
Rose, A. (2009). *Economic resilience to disasters*. Los Angeles: Community and Regional Resilience Institute.
Rose, A., Benavides, J., Chang, S. E., Szczesniak, P., Lim, D. (1997). The regional economic impact of an earthquake: Direct and indirect effects of electricity lifeline disruptions. *Journal of Regional Science* 37 (3), 437–58.
Rose, A., Krausmann, E. (2013). An economic framework for the development of a resilience index for business recovery. *International Journal of Disaster Risk Reduction* 5, 73–83.
Rose, A., Liao, S. Y. (2005). Modelling regional economic resilience to disasters: A computable general equilibrium analysis of water service disruption. *Journal of Regional Science* 45 (1), 75–112.
Rose, A., Oladosu, G., Liao, S. (2007). Business interruption impacts of a terrorist attack on the electric power system of Los Angeles: Customer resilience to a total blackout. *Risk Analysis* 27 (3), 513–31.
Rufat, S. (2013). Spectroscopy of urban vulnerability. *Annals of the Association of American Geographers* 103 (3), 505–25.
Rufat, S., Tate, E., Burton, C. G., Maroof, A. S. (2015). Social vulnerability to floods: Review of case studies and implications for measurement. *International Journal of Disaster Risk Reduction, 14, Part* 4, 470–86.
Russell, C. (1970). Losses from natural hazards. *Land Economics* 43, 383–93.
Ryan, A. M., Spash, C. L. (2011). Is WTP an attitudinal measure? Empirical analysis of the psychological explanation for contingent values. *Journal of Economic Psychology* 32 (5), 674–87.
Rygel, L., O`Sullivan, D., Yarnal, B. (2006). A method for constructing a social vulnerability index. *Mitigation and Adaptation Strategies for Global Change* 11 (3), 741–64.
Saisana, M., Tarantola, S. (2002). *State-of-the-art report on current methodologies and practices for composite indicator development*. Ispra, JRC – Joint Research Centre. European Commission.
Samuelson, P. (1938). A note on the pure theory of consumers' behaviour. *Economica* 5 (17), 61–71.
Sarewitz, D., Pielke, R., Keykhah, M. (2003). Vulnerability and risk: Some thoughts from a political and policy perspective. *Risk Analysis* 23 (4), 805–810.
Saunders, W. S. A., Becker, J. S. (2015). A discussion of resilience and sustainability: Land use planning recovery from the Canterbury earthquake sequence, New Zealand. *International Journal of Disaster Risk Reduction* 14, 73–81.
Scawthorn, C., Flores, P., Blais, N., Seligson, H., Tate, E., Chang, S., Mifflin, E., Thomas, W., Murphy, J., Jones, C., Lawrence, M. (2006a). HAZUS-MH flood loss estimation methodology. I: Overview and flood hazard characterization. *Natural Hazards Review* 7 (2), 60–71.

Scawthorn, C., Flores, P., Blais, N., Seligson, H., Tate, E., Chang, S., Mifflin, E., Thomas, W., Murphy, J., Jones, C., Lawrence, M. (2006b). HAZUS-MH flood loss estimation methodology. II. damage and loss assessment. *Natural Hazards Review* 7 (2), 72–81.

Schindler, S., Sebesvari, Z., Damm, C., Euller, K., Mauerhofer, V., Schneidergruber, A., Biró, M., Essl, F., Kanka, R., Lauwaars, S. G., Schulz-Zunkel, C., Sluis, T. van der, Kropik, M., Gasso, V., Krug, A., Pusch, M. T., Zulka, K. P., Lazowski, W., Hainz-Renetzeder, C., Henle, K., Wrbka, T. (2014). Multifunctionality of floodplain landscapes: Relating management options to ecosystem services. *Landscape Ecology* 29 (2), 229–44.

Schmidtlein, M. C., Deutsch, R. C., Piegorsch, W. W., Cutter, S. L. (2008). A sensitivity analysis of the social vulnerability index. *Risk Analysis* 28 (4), 1099–114.

Schmidtlein, M. C., Shafer, J. M., Berry, M., Cutter, S. L. (2011). Modeled earthquake losses and social vulnerability in Charleston, South Carolina. *Applied Geography* 31 (1), 269–81.

Schmidt-Thomè, P. (2006). *Natural and technological hazards and risks affecting the spatial development of european regions.* Helsinki: Geological Survey of Finland.

Schneiderbauer, S., Ehrlich, D. (2006). *Social levels and hazard (in)dependence in determining vulnerability. Measuring vulnerability to natural hazards: Towards disaster resilient societies. J. Birkmann.* Tokyo: United Nations University Press, pp: 78–102.

Schuster, R., Fleming, R. (1986). Economic losses and fatalities due to landslides. *Bulletin of the Association of Engineering Geologists* 23 (1), 11–28.

Schwarz, J., Maiwald, H. (2007). Prognose der Bauwerksschädigung unter Hochwassereinwirkung. *Bautechnik* 84 (7), 450–64.

Schwarz, J., Maiwald, H. (2008). Damage and loss prediction model based on the vulnerability of building types. 4th International Symposium on Flood Defence, Toronto, Canada. http://gfzpublic.gfz-potsdam.de/pubman/item/escidoc:6057:6/component/escidoc:6058/74_Schwarz.pdf [last accessed: January 10, 2017].

Scott, W. R. (2008). *Institutions and organizations.* Los Angeles: Sage Publication.

Scott, W. R. (2014). *Institutions and organizations. Ideas, interests, and identities.* Thousand Oaks: Sage Publication.

Scussolini, P., Aerts, J. C. J. H., Jongman, B., Bouwer, L. M., Winsemius, H. C., de Moel, H., Ward, P. J. (2016). FLOPROS: An evolving global database of flood protection standards. *Natural Hazards and Earth System Sciences* 16 (5), 1049–61.

Sen, A. (1981). *Poverty and famines: An essay on entitlement and deprivation.* Oxford: Clarendon Press.

Seqwater. (2016). *Weekly water security and consumption update* (No. February 12, 2016). Brisbane: Seqwater.

Seqwater. (2017). 2017 *Dam levels.* www.seqwater.com.au/water-supply/dam-levels [last accessed: April 26, 2017].

Shabman, L. A., Damianos, D. I. (1976). Flood hazard effects on residential property values. *Journal of Water Resources and Planning* WR1, 151–62.

Shakra, V. A. (2016). *Measuring the feasibility of elevating residential structures in flood vulnerable areas at the parcel level as a hazard mitigation approach: The case of 100-year inundated areas in Galveston County, Texas.* Unpublished Master Thesis, Galveston: Texas A&M University.

Sharifi, A., Yamagata, Y. (2016). On the suitability of assessment tools for guiding communities towards disaster resilience. *International Journal of Disaster Risk Reduction*, 18, 115–24.

Shaw, D., Scully, J., Hart, T. (2014). The paradox of social resilience: How cognitive strategies and coping mechanisms attenuate and accentuate resilience. *Global Environmental Change* 25, 194–203.

Shaw, R., Goda, K. (2004). From disaster to sustainable civil society: The Kobe experience. *Disasters* 28 (1), 16–40.

Sherrieb, K., Norris, F. H. and Galea, S. (2010). Measuring capacities for community resilience. *Social Indicators Research*, 99, 227–47.

Shinozuka, M., Rose, A., Eguchi, R. (eds.) (1998). *Engineering and socioeconomic impacts of earthquakes: An analysis of electricity lifeline disruptions in the new Madrid area*. Buffalo: Multidisciplinary Centre for Earthquake Engineering Research.

Siebeneck, L. (2016). Examining social, physical, and environmental dimensions of tornado vulnerability in Texas. *Journal of Emergency Management* 14 (2), 139–51.

Siebeneck, L., Arlikatti, S., Andrew, S. A. (2015). Using provincial baseline indicators to model geographic variations of disaster resilience in Thailand. *Natural Hazards* 79 (2), 955–75.

Simarmata, H. A. (2015). How informal institutions manage flood-risk at community level: An empirical study of Kampung Muara baru, Jakarta. In J. Velasquez, A. Panda (eds.), *The "state of DRR at the local level": A 2015 report on the pattern of disaster risk reduction actions at local level*. Geneva: UNISDR.

Skantz, T., Strickland, T. (1987). House prices and a flood event: An empirical investigation of market efficiency. *Journal of Real Estate Research* 2 (2), 75–83.

Skoufias, E. (2003). Economic crises and natural disasters: Coping strategies and policy implications. *World Development* 31 (7), 1087–102.

Smit, B., Wandel, J. (2006). Adaptation, adaptive capacity and vulnerability. *Global Environmental Change*, 16 (3), 282–92.

Smith, D. I. (1981). Actual and potential flood damage: A case study for urban Lismore, NSW, Australia. *Applied Geography* 1, 31–9.

Smith, D. I. (1994). Flood damage estimation – a review of urban stage-damage curves and loss functions. *Water SA* 20 (3), 231–8.

Smith, K. (2013). *Environmental hazards. Assessing risk and reducing disaster*. Abingdon: Routledge.

Smith, K., Petley, D. N. (2009). *Environmental hazards – assessing risk and reducing disaster*. London: Routledge.

Smith, K., Tobin, G. A. (1979). *Human adjustment to the flood hazard*. London: Longman.

Smith, K., Ward, R. (1998). *Floods: Physical processes and human impacts*. Chichester: John Wiley & Sons Ltd.

Spiekermann, R., Kienberger, S., Norton, J., Briones, F., Weichselgartner, J. (2015). The disaster-knowledge matrix – reframing and evaluating the knowledge challenges in disaster risk reduction. *International Journal of Disaster Risk Reduction* 13, 96–108.

SSHAC – Senior Seismic Hazard Analysis Committee (1997). *Recommendations for probabilistic seismic hazard analysis: Guidance on uncertainty and use of experts*. Washington: U.S. Nuclear Regulatory Commission.

Standard & Poor's (2015a). *Storm alert: Natural disasters can damage sovereign creditworthiness*. New York: Standard & Poor's.

Standard & Poor's (2015b). *The heat is on: How climate change can impact sovereign ratings*. New York: Standard & Poor's.

Stein, J., Moreno, P., Conrad, D., Ellis, S. (2000). *Troubled waters: Congress, the corps of engineers, and wasteful water projects*. Washington: Taxpayers for common sense and National Wildlife Federation.

Steinführer, A., Kuhlicke, C. (2007). *Social vulnerability and the 2002 flood. Country Report Germany (Mulde River)*. Leipzig: UFZ Centre for Environmental Research.

Stiefelmeyer, H., Hlatky, T. (2008). HORA – An Austrian platform for natural hazards as a new way in risk communication. In M. Mikoš, J. Hübl, G. Koboltschnig (eds.), *International symposion interpraevent*. Klagenfurt: Internationale Forschungsgesellschaft Interpraevent, pp: 229–36.

Stocker, T. F., Qin, D., Plattner, G.-K., Tignor, M., Allen, S. K., Boschung, J., Nauels, A., Xia, Y., Bex, V., Midgley, P. M. (eds.) (2013) *Climate change 2013: The physical science basis. Contribution of Working Group I to the fifth assessment report of the intergovernmental panel on climate change*. Cambridge: Cambridge University Press.

Strübing, J. (2013). *Qualitative Sozialforschung. Eine komprimierte Einführung für Studierende*. München: Oldenbourg Verlag.

Stumpp, E. M. (2013). New in town? On resilience and "resilient cities". *Cities* 32, 164–6.

Sugarman, S. D. (2006). *Roles of government in compensating disaster victims*. Berkeley: The Berkeley Electronic Press.

Surminski, S. (2009). *How can the insurance industry promote climate change adaptation? A case study from the UK*. London: Chartered Insurance Institute.

Surminski, S., Aerts, J. C. J. H., Botzen, W. J. W., Hudson, P., Mysiak, J., Pérez-Blanco, C. D. (2015). Reflections on the current debate on how to link flood insurance and disaster risk reduction in the European Union. *Natural Hazards* 79 (3), 1451–79.

Surminski, S., Oramas-Dorta, D. (2014). Flood insurance schemes and climate adaptation in developing countries. *International Journal of Disaster Risk Reduction* 7, 154–64.

Sutcliffe, K. M., Vogus, T. J. (2003). Organizing for resilience. In K. Cameron, J. E. Dutton, R. E. Quinn (eds.), *Positive organizational scholarship*. San Francisco: Berrett-Koehler, pp: 94–110.

Swiss, Re (2016). *Natural catastrophes and man-made disasters in 2015*. Zurich: Swiss Reinsurance Company.

Tapsell, S., McCarthy, S., Faulkner, H., Alexander, M. (2010). *Social vulnerability to natural hazards*. London: Flood Hazard Research Centre, Middlesex University.

Tapsell, S., Penning-Rowsell, E. C., Tunstall, S. M., Wilson, T. L. (2002). Vulnerability to flooding: Health and social dimensions. *Philosophical Transactions of the Royal Society A – Mathematical, Physical and Engineering Sciences* 360 (1796), 1511–25.

Tate, E. (2012). Social vulnerability indices: A comparative assessment using uncertainty and sensitivity analysis. *Natural Hazards* 63 (2), 325–47.

Tate, E. (2013). Uncertainty analysis for a social vulnerability index. *Annals of the Association of American Geographers* 103 (3), 526–43.

Telford, J. (2007). The international humanitarian system and the 2004 Indian Ocean earthquake and tsunamis. *Disasters* 31 (1), 1–28.

Thaler, T. (2015). *Rescaling in flood risk governance – new spatial and institutional arrangements and structures*. Unpublished dissertation, Middlesex University.

Thaler, T., Boteler, B., Dworak, T., Holen, S., Lago, M. (2014). Investigating the use of environmental benefits in the policy decision process: A qualitative study focusing on the EU water policy. *Journal of Environmental Planning and Management* 57 (10), 1515–30.

Thaler, T., Priest, S. J., Fuchs, S. (2016). Evolving inter-regional co-operation in flood risk management: Distances and types of partnership approaches in Austria. *Regional Environmental Change* 16 (3), 841–53.

Thieken, A. H., Müller, M., Kreibich, H., Merz, B. (2005). Flood damage and influencing factors: New insights from the August 2002 flood in Germany. *Water Resource Research* 41 (12), W12430.

Thieken, A. H., Olschewski, A., Kreibich, H., Kobsch, S., Merz, B. (2008). Development and evaluation of FLEMOps – a new flood loss estimation model for the private sector. In D. Proverbs, C. A. Brebbia, E. C. Penning-Rowsell (eds.), *Flood recovery, innovation and response*. London: WIT Press, pp: 315–24.

Thywissen, K. (2006). *Components of risk. A comparative glossary*. Bonn: United Nations University Press.

Tierney, K. (1997). Business impacts of the Northridge Earthquake. *Journal of Contingencies and Crisis Management* 5 (2), 87–97.

Tierney, K. (2014). *The social roots of risk. Producing disasters, promoting resilience*. Stanford: Stanford University Press.

Tierney, K. (2015). Resilience and the neoliberal project: Discourses, critiques, practices – and Katrina. *American Behavioral Scientist* 59 (10), 1327–42.

Timmerman, P. (1981). *Vulnerability, resilience and the collapse of society: A review of models and possible climatic applications*. Toronto: University of Toronto.

Tobin, G. A. (1999). Sustainability and community resilience: The holy grail of hazards planning?. *Global Environmental Change Part B: Environmental Hazards* 1 (1), 13–25.

Tobin, G. A. (2015). Risk mitigation: We are all going to die. In U. Fra Paleo (ed.), *Risk governance. The articulation of hazard, politics and ecology*. Dordrecht: Springer, pp: 495–7.

Tobin, G. A., Montz, B. E. (1988). Catastrophic flooding and the response of the real estate market. *The Social Science Journal* 25 (2), 167–77.

Tobin, G. A., Montz, B. E. (1995). The flood hazard and dynamics of the urban residential land market. *Water Resource Bulletin* 30 (4), 673–85.

Tobin, G. A., Montz, B. E. (1997). *Natural hazards: Explanation and integration*. New York: The Guilford Press.

Tobin, G. A., Montz, B. E. (2015). *Evolving approaches to understanding natural hazards*. Newcastle: Cambridge Scholars Publishing.

Tobin, G. A., Whiteford, L. M., Jones, E. C., Murphy, A. D., Garren, S. J., Padros, C. V. (2011). The role of individual well-being in risk perception and evacuation for chronic vs. acute natural hazards in Mexico. *Applied Geography* 31 (2), 700–11.

Tobin, G. A., Whiteford, L. M., Murphy, A. D., Jones, E. C., McCarty, C. (2014). Modeling social networks and community resilience in chronic disasters: Case studies from volcanic areas in Ecuador and Mexico. In P. Gasparini, M. Daetano, D. Asprone (eds.), *Resilience and sustainability in relation to natural disasters: A challenge for future cities*. Dordrecht: Springer, pp: 13–44.

Tompkins, E. L., Penning-Rowsell, E. C., Parker, D., Platt, S., Priest, S., So, E., Spence, R. (2012). *Institutions and disaster outcomes: Successes, weaknesses and significant research needs*. www.gov.uk/government/publications/disaster-outcomes-role-of-institutions [last accessed: May 01, 2017].

Torres, H. R. (2017). *How lessons from a past disaster can influence resilience and climate adaptation in Broward County, Florida*. Unpublished dissertation, Tampa: University of South Florida.

Totschnig, R., Fuchs, S. (2013). Mountain torrents: Quantifying vulnerability and assessing uncertainties. *Engineering Geology* 155, 31–44.

Totschnig, R., Sedlacek, W., Fuchs, S. (2011). A quantitative vulnerability function for fluvial sediment transport. *Natural Hazards* 58 (2), 681–703.

Transparency International (2010). *What is transparency international?* Retrieved 24 November 2016, from www.transparency.org/about_us [last accessed: May 01, 2017].

Tsertsvadze, A., Clar, C., Court, R., Clarke, A., Mistry, H., Sutcliffe, P. (2014). Cost-effectiveness of manual therapy for the management of musculoskeletal conditions: A systematic review and narrative synthesis of evidence from randomized controlled trials. *Journal of Manipulative Physiological Therapeutics* 37 (6), 343–62.

Tuccillo, J. V., Buttenfield, B. P. (2016). Model-based clustering of social vulnerability to urban extreme heat events. In J. A. Miller, D. O'Sullivan, N. Wiegand (eds.), *Geographic information science: 9th international conference, GIScience 2016, Montreal, QC, Canada, September 27–30, 2016*. Cham: Springer, pp: 114–29.

Turner II, B. L. (2010). Vulnerability and resilience: Coalescing or paralleling approaches for sustainability science? *Global Environmental Change* 20 (4), 570–6.

Turner II, B. L., Kasperson, R. E., Matson, P. A., McCarthy, J. J., Corell, R. W., Christensen, L., Eckley, N., Kasperson, J. X., Luers, A., Martello, M. L., Polsky, C., Pulsipher, A., Schiller, A. (2003). A framework for vulnerability analysis in sustainability science. *Proceedings, National Academy of Sciences of the United States of America* 100 (14), 8074–9.

Turner, R. K. (2007). Limits to CBA in UK and European environmental policy: Retrospects and future prospects. *Environmental and Resource Economics* 37 (1), 253–69.

Turnpenny, J., Nilsson, M., Russel, D., Jordan, A., Hertin, J., Nykvist, B. (2008). Why is integrating policy assessment so hard? A comparative analysis of the institutional capacities and constraints. *Journal of Environmental Planning and Management* 51 (6), 759–75.

Twigg, J. (2015). *Disaster risk reduction. Good practice review*. London: Overseas Development Institute.

UK Government, Office for Science. (2012). *Foresight reducing risks of future disasters for decision makers, final project report*. London: The Government Office for Science. www.gov.uk/government/uploads/system/uploads/attachment_data/file/286476/12-1289-reducing-risks-of-future-disasters-report.pdf [last accessed: July 03, 2017].

UK Government. (2014). Resilience in society: Infrastructure, communities and businesses. www.gov.uk/guidance/resilience-in-society-infrastructure-communities-and-businesses [last accessed: July 03, 2017].

UNDRO (1980). *Natural disasters and vulnerability analysis*. Geneva: Office of the United Nations Disaster Relief Co-ordinator.

UNDRO (1984). *Disaster prevention and mitigation – a compendium of current knowledge. Preparedness aspects*. Volume 11. New York: Office of the United Nations Disaster Relief Co-ordinator.

Unger, D. G. and Powell, D. R. (1980). Supporting families under stress: The role of social networks. *Family Relations* 29 (4), 566–74.

UNISDR – United Nations International Strategy for Disaster Reduction. (1994). *Yokohama strategy and plan of action for a safer world: Guidelines for natural disaster prevention, preparedness and mitigation*. Geneva: UNISDR.

UNISDR – United Nations International Strategy for Disaster Reduction. (2004). *Living with Risk: A Global Review of Disaster Reduction Initiatives Volume 1 and 2*. New York and Geneva: United Nations.

UNISDR – United Nations International Strategy for Disaster Reduction. (2005). *Hyogo framework for action 2005-2015. Building the resilience of nations and communities to disasters*. Geneva: UNISDR.

UNISDR – United Nations International Strategy for Disaster Reduction. (2009). *Terminology on disaster risk reduction*. Geneva: United Nations.

UNISDR – United Nations International Strategy for Disaster Reduction. (2011). *Hyogo framework for action 2005–2015 mid-term review. United Nations Office for DisasterRisk Reduction*. Geneva: UNISDR.

UNISDR – United Nations International Strategy for Disaster Reduction. (2012a). *Number of climate-related disasters around the world* (1980–2011). www.preventionweb.net/files/20120613_ClimateDisaster1980-2011.pdf [last accessed: February 1, 2014].

UNISDR – United Nations International Strategy for Disaster Reduction. (2012b). *Terminology on DisasterRisk Reduction. United Nations Office for DisasterRisk Reduction*. Geneva: UNISDR.

UNISDR – United Nations International Strategy for Disaster Reduction. (2014). *Progress and challenges in DisasterRisk Reduction*. Geneva: UNISDR.

UNISDR – United Nations International Strategy for Disaster Reduction. (2015a). *Sendai framework for DisasterRisk Reduction*. Geneva: UNISDR.

UNISDR – United Nations International Strategy for Disaster Reduction. (2015b). *Global assessment report on disaster risk reduction 2015 (GAR15). United Nations Office for DisasterRisk Reduction*. Geneva: UNISDR.

UNISDR – United Nations International Strategy for Disaster Reduction. (2017). Terminology, www.unisdr.org/we/inform/terminology#letter-m [last accessed: July 03, 2017].

United Nations – Department of Economic and Social Affairs. (2014). *World urbanization prospects: The 2014 revision*. https://esa.un.org/unpd/wup/ [last accessed: March 02, 2017].

United Nations General Assembly. (1989). *International decade for natural disaster reduction. United Nations General Assembly Resolution 236 session 44 of 22 December 1989*. New York: United Nations General Assembly.

United Nations. (2014). *Zero draft of the post-2015 framework for Disaster Risk Reduction. United Nations, Sendai (Japan)*. New York: United Nations.

UNU Wider – United Nations University World Institute for Development Economics Research. (2008). *Economic vulnerability and resilience*. Helsinki: UNU-WIDER.

Url, T., Sinabell, F. (2008). Flood risk exposure in Austria–options for bearing risk efficiently. *Schnollers Jahrbuch: Journal of Applied Social Sciences Studies/ Zeitschrift für Wirtschafts- und Sozialwissenschaften* 128 (4), 593–614.

US Economic Development Administration (2016). *Comprehensive economic development strategy (CEDS) content guidelines: Recommendations for creating an impactful CEDS*. Washington: US Economic Development Administration.

Van de Ven, A. H. (2007). *Engaged scholarship. A guide for organizational and social research*. Oxford: Oxford University Press.

Van de Ven, A. H., Polley, D., Garud, R., Venkataraman, S. (1999). *The innovation journey*. Oxford: Oxford University Press.

van den Honert, R. C., McAneney, J. (2011). The 2011 Brisbane floods: Causes, impacts and implications. *Water* 3 (4), 1149–73.

van der Vink, G. E. (2007). Democracy, GDP and natural disasters. *Geotimes: Earth, Energy and Environment News*. www.geotimes.org/oct07/article.html?id=feature_democracy.html [last accessed: January 10, 2017].

van Deth, J. W., Zmerli, S. (2010). Introduction: Civicness, equality, and democracy—A "Dark Side" of social capital? *American Behavioral Scientist* 53 (5), 631–9.

Varduhn, V., Mundani, R. P., Rank, E. (2015). Multi-resolution models: Recent progress in coupling 3D geometry to environmental numerical simulation. In M. Breunig, M. Al-Doori, E. Butwilowski, P. V. Kuper, J. Benner, K. H. Haefele (eds.), *3D geoinformation science*: Cham: Springer International Publishing, pp: 55–69.

Varian, H. R. (1992). *Microeconomic analysis*. New York: WW Norton and Company.

Varian, H. R. (2012). Revealed preference and its applications. *The Economic Journal* 122 (560), 332–8.

Varnes, D. (1984). *Landslide hazard zonation: A review of principles and practice*. Paris: UNESCO.

Vatn, A. (2009). An institutional analysis of methods for environmental appraisal. *Ecological Economics* 68, 2207–15.

Vatn, A., Bromley, D. W. (1994). Choices without prices without apologies. *Journal of Environmental Economics and Management* 26 (2), 129–48.

Veblen, T. B. (1899). *The theory of the leisure class: An economic study in the evolution of institutions*. New York, Macmillian.

Veblen, T. B. (1914). *The instinct of workmanship and the state of the industrial arts*. New York, Macmillian.

Veblen, T. B. (1919). *The place of science in modern civilisation and other essays*. New York, Huebsch.

Villagrán de León, J. C. (2006). *Vulnerability. A conceptual and methodological review*. Bonn: United Nations University Press.

Vink, K., Takeuchi, K. (2013). International comparison of measures taken for vulnerable people in disaster risk management laws. *International Journal of Disaster Risk Reduction* 4, 63–70.

Visser, H., Petersen, A. C., Ligtvoet, W. (2014). On the relation between weather-related disaster impacts, vulnerability and climate change. *Climate Change* 125 (3), 461–77.

Visualtheasurus (2017). *Thinkmap Visualtheasurus*. www.visualthesaurus.com/ [last accessed: February 23, 2017].

Vogel, C., Moser, S. C., Kasperson, R. E., Dabelko, G. D. (2007). Linking vulnerability, adaptation, and resilience science to practice: Pathways, players, and partnerships. *Global Environmental Change*, 17 (3–4), 349–64.

Vogt-Schilb, A., Hallegatte, S. (2014). Marginal abatement cost curves and the optimal timing of mitigation measures. *Energy Policy* 66, 645–53.

von Ciriacy-Wantrup, S. (1971). The economics of environmental policy. *Land Economics* 47 (1), 36–45.

Von Peter, G., Von Dahlen, S., Saxena, S. C. (2012). *Unmitigated disasters? New evidence on the macroeconomic cost of natural catastrophes.* www.bis.org/publ/work394.htm [last accessed: May 23, 2017].

Voss, M. (2008). The vulnerable can't speak. An integrative vulnerability approach to disaster and climate change research. *Behemoth* 1 (3), 39–56.

Voss, M. (2009). Vulnerabilität. In C. Hammerl, T. Kolnberger (eds.), *Naturkatastrophen: Rezeption – Bewältigung – Verarbeitung*. Vienna: Studien-Verlag, pp: 103–21.

Voss, M., Funk, L. (2015). Participative vulnerability and resilience assessment and the example of the Tao People (Taiwan). In F. Krüger, G. Bankoff, T. Cannon, B. Orlowski, L. Schipper (eds.), *Cultures and disasters. Understanding cultural framings in disaster risk reduction*. London: Routledge, pp: 255–76.

Voss, M., Wagner, K. (2010). Learning from (small) disasters. *Natural Hazards* 55 (3), 657–69.

Vulturius, G. (2013). Policy change and policy learning for enhanced flood policies at local, subnational and EU level. In E. C. H. Keskitalo (ed.), *Climate change and flood risk management. Adaptation and extreme events at the local level.* Cheltenham/Northampton: Edward Elgar, pp: 121–49.

Walker, B., Holling, C. S., Carpenter, S. R., Kinzig, A. (2004). Resilience, adaptability and transformability in social-ecological systems. *Ecology and Society* 9 (2), 5.

Wang, B., Ruel, M., Couture, R., vanDine, D., Bobrowsky, P., Blais-Stevens, A. (2013). Current status of landslide guidelines around the world. In C. Margottini, P. Canuti, K. Sassa (eds.), *Landslide Science and Practice*, Vol. 7. Berlin: Springer, pp: 271–7.

Ward, P. J., Jongman, B., Sperna Weiland, F. C., Bouwman, A., van Beek, R., Bierkens, M. F. P., Ligtvoet, W., Winsemius, H. C. (2013). Assessing flood risk at the global scale: Model setup, results, and sensitivity. *Environmental Research Letters* 8 (4), 044019.

Warner, K., Ranger, N., Surminski, S., Arnold, M., Linnerooth-Bayer, J., Michel-Kerjan, E., Kovacs, P., Herweijer, C. (2009). *Adaptation to climate change: Linking disaster risk reduction and insurance*. Geneva: UNISDR.

Watts, M. J., Bohle, H. G. (1993). The space of vulnerability: The causal structure of hunger and famine. *Progress in Human Geography* 17 (1), 43–67.

Watts, D. J., Strogatz, S. H. (1998). Collective dynamics of "small-world" networks. *Nature* 393 (6684), 440–2.

WEF – World Economic Forum (2016). *Global risks 2016*. Davos: World Economic Forum.

Weichselgartner, J. (2001a). Disaster mitigation: The concept of vulnerability revisited. *Disaster Prevention and Management* 10 (2), 85–94.

Weichselgartner, J. (2001b). *Naturgefahren als soziale Konstruktion*. Unpublished dissertation, Bonn: University of Bonn.

Weichselgartner, J., Kelman, I. (2015). Geographies of resilience challenges and opportunities of a descriptive concept. *Progress in Human Geography* 39 (3), 249–67.

Weick, K. E. (1993). The collapse of sensemaking in organizations. The Mann Gulch Disaster. *Administration Science Quarterly* 38 (4), 628–52.

Weick, K. E. (1995). *Sensemaking in organizations*. London: Sage.

Weick, K. E. (2001). *Making sense of the organization*. Oxford/Malden: Blackwell.

Weick, K. E., Sutcliffe, K. M. (2015). *Managing the unexpected. Sustained performance in a complex world*. Chichester: John Wiley & Sons Ltd.

Weick, K. E., Sutcliffe, K. M., Obstfeld, D. (1999). Organizing for high reliability: Processes of collective mindfulness. *Research in Organizational Behavior* 21, 81–124.

Weick, K. E., Sutcliffe, K. M., Obstfeld, D. (2005). Organizing and the process of sensemaking. *Organization Science* 16 (4), 409–21.

Weil, F., Lee, M. R., Shihadeh, E. S. (2012). The burdens of social capital: How socially-involved people dealt with stress after Hurricane Katrina. *Social Science Research* 41 (1), 110–19.

Wein, A., Rose, A. (2011). Economic resilience lessons from the ShakeOut Earthquake scenario. *Earthquake Spectra* 27 (2), 559–73.

Welsh, M. (2014). Resilience and responsibility: Governing uncertainty in a complex world. *The Geographical Journal* 180 (1), 15–26.

Werner, E. (1971). *The children of Kauai. A longitudinal study from the prenatal period to age ten*. Honolulu: University of Hawaii Press.

Werner, E. (1993). Resilience and recovery: Findings from the Kauai longitudinal study. *Development and Psychopathology* 5 (4), 503–15.

Werner, E., Smith, R. (1982). *Vulnerable but invincible: A study of resilient children*. New York: McGraw-Hill.

Westley, F., Carpenter, S. R., Brock, W. A., Holling, C. S., Gunderson, L. H. (2002). Why systems of people and nature are not just social and ecological systems. In L. H. Gunderson, C. S. Holling (eds.), *Panarchy. Understanding transformations in human and natural systems*. Washington: Island Press, pp: 103–20.

White, G. F. (1936). Limit of economic justification for flood protection. *Journal of Land and Public Utility Economics* 12, 133–48.

White, G. F. (1945). *Human adjustment to floods. A geographical approach to the flood problem in the United States*. Chicago: The University of Chicago Working Paper No. 29.

White, G. F. (1973). Natural hazards research. In R. J. Chorley (ed.), *Directions in geography*. London: Methuen, pp: 193–216.

White, G. F. (1978). Natural hazards and the third world—a reply. *Human Ecology* 6 (2), 229–31.

White, G. F., Kates, R. W., Burton, I. (2001). Knowing better and losing even more: The use of knowledge in hazards management. *Global Environmental Change Part B: Environmental Hazards* 3 (3), 81–92.

Whiteford, L. M., Tobin, G. A. (2009). If the pyroclastic flow doesn't kill you, the recovery will: Cascading impacts of Mt. Tungurahua's eruptions in rural Ecuador. http://scholarcommons.usf.edu/geo_facpub/55 [last accessed: May 22, 2017].

WHO – World Health Organisation (2014). *Making choices in health: WHO guide to cost-effectiveness analysis*. Geneva: World Health Organization.

Wildavsky, A. (1988). *Searching for safety*. New Brunswick: Bowling Green State University.

Williamson, O. E. (2000). The new institutional economics: Taking stock, looking ahead. *Journal of Economic Literature* 38 (3), 595–613.

Wilson, G. (2010). Multifunctional "quality" and rural community resilience. *Transactions of the Institute of British Geographers* 35 (3), 364–81.

Wilson, N. J. (2014). The politics of adaptation: Subsistence livelihoods and vulnerability to climate change in the Koyukon Athabascan village of Ruby, Alaska. *Human Ecology* 42 (1), 87–101.

Winsemius, H. C., Aerts, J. C. J. H., van Beek, L. P. H., Bierkens, M. F. P., Bouwman, A., Jongman, B., Kwadijk, J. C. J., Ligtvoet, W., Lucas, P. L., van Vuuren, D. P., Ward, P. J. (2016). Global drivers of future river flood risk. *Nature Climate Change* 6, 381–5.

Winter, S. C., May, P. J. (2001). Motivation for compliance with environmental regulations. *Journal of Policy Analysis and Management* 20 (4), 675–98.

Winter, M. G., Smith, J. T., Fotopoulou, S., Pitilakis, K., Mavrouli, O., Corominas, J., Argyroudis, S. (2014). Recommendations for the quantitative analysis of landslide risk. *Bulletin of Engineering Geology and the Environment* 73 (2), 291–305.

Wirth, P., Hutter, G., Schanze, J. (2010). Flood risk management and regional governance – The case of Weisseritz Regio (Germany). In T. Kluvánková-Oravská (ed.), *From government to governance? New governance for water and biodiversity in an enlarged Europe*. Pargue: Alfa Nakladatelství, pp: 128–41.

Wisner, B. (1998). Marginality and vulnerability: Why the homeless of Tokyo don't 'count' in disaster preparations. *Applied Geography* 18 (1), 25–33.

Wisner, B. (2016). Vulnerability as concept, model, metric, and tool. *Natural Hazard Science: Oxford Research Encyclopedias*. doi: 10.1093/acrefore/9780199389407.013.25 http://naturalhazardscience.oxfordre.com/view/10.1093/acrefore/9780199389407.001.0001/acrefore-9780199389407-e-25 [last accessed: March 02. 2017].

Wisner, B., Blaikie, P., Cannon, T. and Davis, I. (2004). *At risk. Natural hazards, people's vulnerability and disasters*. 2nd edition. London: Routledge.

Wisner, B., Gaillard, J. C., Kelman, I. (eds.) (2012). *Handbook of hazards and disaster risk reduction and management*. Abingdon, Oxon: Routledge.

Wisner, B., Luce, H. R. (1993). Disaster vulnerability: Scale, power and daily life. *GeoJournal* 30 (2), 127–40.

Wolch, J. R., Byrne, J., Newell, J. P. (2014). Urban green space, public health, and environmental justice: The challenge of making cities "just green enough". *Landscape and Urban Planning* 125, 234–44.

Wymann von Dach, S., Bachmann, F., Alcántara-Ayala, I., Fuchs, S., Keiler, M., Mishra, A., Sötz, E. (eds.) (2017). *Safer lives and livelihoods in mountains: Making the Sendai Framework for Disaster Risk Reduction work for sustainable mountain development*. Bern: Centre for Development and Environment (CDE), University of Bern, with Bern Open Publishing (BOP).

Yin, R. K. (2014). *Case study research. Design and methods*. Thousand Oaks: Sage.

Young, O. R. (1989). *International cooperation: Building regimes for natural resources and the environment*. Ithaca: Cornell University Press.

Young, O. R. (2010). Institutional dynamics: Resilience, vulnerability and adaptation in environmental and resource regimes. *Global Environmental Change* 20 (3), 378–85.

Young, O. R., Berkhout, F., Gallopin, G. C., Janssen, M. A., Ostrom, E., van der Leeuw, S. (2006). The globalization of socio-ecological systems. An agenda for scientific research. *Global Environmental Change* 16 (3), 304–16.

Young, P. H. (1996). The economicy of convention. *The Journal of Economic Perspective* 10 (2), 105–22.

Zander, M. (2010). Armes Kind – starkes Kind? *Die Chance der Resilienz*. Wiesbaden: VS – Verlag für Sozialwissenschaften.

Zautra, A. J., 2009. Resilience: One part recovery, two parts sustainability. *Journal of Personality* 77 (6), 1935–43.

Zebrowski, C. R. (2013). The nature of resilience. *Resilience: International Policies, Practices and Discourses* 1 (3), 59–173.

Zeller, C. (2004). North Atlantic innovative relations of Swiss pharmaceuticals and the importance of regional biotech arenas. *Economic Geography* 80 (1), 83–111.

Zijderveld, A. C. (2000). *The institutional imperative – The interface of institutions and networks*. Amsterdam: Amsterdam University Press.

Zimmermann, K. (2010). Der veränderte Stellenwert von Wissen in der Planung. Ein Beitrag zu einem wissensbasierten Verständnis von Planung. *Raumforschung und Raumordnung* 68 (2), 115–25.

Zimmermann, M., Keiler, M. (2015). International frameworks for disaster risk reduction: Useful guidance for sustainable mountain development? *Mountain Research and Development* 35 (2), 195–202.

# Index